Polarized Sources, Targets and Polarimetry

Proceedings of the 13th International Workshop

Polarized Sources, Targets and Polarimetry

Proceedings of the 13th International Workshop

Ferrara, Italy, 7 – 11 September 2009

edited by

G Ciullo
INFN – Ferrara & University of Ferrara, Italy

M Contalbrigo
INFN – Ferrara, Italy

P Lenisa
INFN – Ferrara & University of Ferrara, Italy

World Scientific

NEW JERSEY • LONDON • SINGAPORE • BEIJING • SHANGHAI • HONG KONG • TAIPEI • CHENNAI

Published by

World Scientific Publishing Co. Pte. Ltd.

5 Toh Tuck Link, Singapore 596224

USA office: 27 Warren Street, Suite 401-402, Hackensack, NJ 07601

UK office: 57 Shelton Street, Covent Garden, London WC2H 9HE

British Library Cataloguing-in-Publication Data
A catalogue record for this book is available from the British Library.

Fresco on the front cover:
Part of the ceiling frescoes (1503–1506), by Benvenuto Tisi said 'Garofalo', located in the Costabili Palace, said of 'Ludovico il Moro (the Moor)', housing the National Archaeological Museum, devoted to the etruscan town of Spina (end of VI — half III century B.C.) — Ferrara. On concession of the Office for the Goods and the Cultural Activities of the Italian Republic.

The background of the cover:
Bugnato of the exterior wall — Palazzo dei Diamanti-Ferrara (build from 1493, designed by Biagio Rossetti), courtesy of Comune di Ferrara.

POLARIZED SOURCES, TARGETS AND POLARIMETRY
Proceedings of the 13th International Workshop

ISBN-13 978-981-4324-91-5
ISBN-10 981-4324-91-4

Printed in Singapore by Mainland Press Pte Ltd.

CONTENTS

Fig. 1. Conference photo.

Fig. 2. Poster of the "XIII International Workshop on Polarized Sources, Targets & Polarimetry", September 07-11, 2009 - Ferrara (Italy).

PREFACE

The XIII International Workshop on Polarized Sources - PST 2009, Targets & Polarimetry, was held in Ferrara on September 07th-11th, 2009.

The Workshop, sponsored by the International Spin Physics Committee, has been hosted for more than 20 years, regularly circulating between the USA, Europe and Japan. Although dedicated meetings exist for some of the fields covered, the aim of Workshop was to give a general, but comprehensive view of the technology involved in the polarization experiments and applications and to encourage discussions and exchange of information between the different fields. The schedule included a round-table discussion on the intensity limitations of various polarized atomic sources and possible paths for future progress. Visibility has been given to new ideas, proposals and medical applications.

The topics covered in the Workshop included:

- Polarized Electron Sources.
- Polarized Proton and Deuterium Sources.
- Polarized Internal Targets.
- Polarized ^{3}He Ion Sources and Targets.
- Polarimetry (e, p, d) at Low and High Energy.
- Polarized antiprotons.
- Polarized Solid Targets.

The site for the workshop has been chosen to be the Hall of the Camera di Commercio facing the XV century Castello Estense, in the heart of Ferrara. Over 80 physicists took part to the Workshop, 50 of them made presentations.

The sources of finance for the workshop, besides the International Spin Physics Committee, were the University of Ferrara, the Istituto Nazionale di Fisica Nucleare (INFN) and the Virtual Institute for spin and QCD.

G. Ciullo (Ferrara)
M. Contalbrigo (Ferrara)
P. Lenisa (Ferrara, chair)
March, 2010

ORGANISING COMMITTEES

Local Organising Committee

Giuseppe Ciullo	Università di Ferrara and INFN
Marco Contalbrigo	INFN of Ferrara
Paola Ferretti Dalpiaz	Università di Ferrara and INFN
Paolo Lenisa (chair)	Università di Ferrara and INFN
Erhard Steffens	University of Erlangen

Program Committee

P. Lenisa	(Ferrara, chair)
P.F. Dalpiaz	(Ferrara)
E. Steffens	(Erlangen)
Polarized electron sources	(K. Aulenbacher, Mainz)
Polarized proton and deuterium sources	(A. Belov, Moscow)
Polarized internal targets	(A. Nass, Erlangen)
Polarimetry	(G. Ciullo, Ferrara)
Polarized antiprotons	(F. Rathmann, Jülich)
Polarized solid targets	(D. Crabb, Virginia)

International Spin Physics Committee

K. Imai, Kyoto (chair)	T. Roser, Brookhaven (past-chair)
E. Steffens, Erlangen (chair-elect)	M. Anselmino, Torino
F. Bradamante, Trieste	E.D. Courant*, BNL
D.G. Crabb, Virginia	A.V. Efremov, JINR
G. Fidecaro*, CERN	H. Gao, Duke
W. Haeberli*, Wisconsin	K. Hatanaka, RCNP
A.D. Krisch, Michigan	G. Mallot, CERN
A. Masaike*, JSPS	R.G. Milner, MIT
R. Prepost, Wisconsin	C.Y. Prescott*, SLAC
F. Rathmann, Jeulich	H. Sakai, Tokyo
Yu.M. Shatunov, Novosibirsk	V. Soergel*, Heidelberg
E.J. Stephenson, Indiana	N.E. Tyurin, IHEP
W.T.H. van Oers*, Manitoba	(* honorary members)

REMARKS ON THE HISTORY OF WORKSHOPS ON "SPIN TOOLS"

E. Steffens

Physikalisches Institut, University of Erlangen–Nürnberg,
D–91058 Erlangen, Germany
E-mail: steffens@physik.uni-erlangen.de

This workshop is part of a series of workshops on techniques required for experiments to measure spin-dependent observables in the scattering of energetic particles for nuclear or particle physics experiments. The aim of this talk is to show how these workshops developed from the beginning and which impact they had on the field of spin physics in nuclear and particle physics.

Keywords: Experimental techniques; spin-dependent observables; history of spin workshops.

1. Introduction

This Workshop PST2009 on polarized ion sources, gas targets and polarimetry is part of a series of meetings which is sponsored by the International Committee for Spin Physics Symposia (ICSP). The present composition of the ICSP can be found elsewhere in these proceedings. The workshops deal with techniques required for measurements of spin-dependent quantities in scattering experiments with polarized ion beams from accelerators. They have evolved into their present form over the last 20–30 years. The aim of this talk is to show the roots of these meetings which date back to the early 1960s when the first methods to produce and accelerate polarized beams where developed. With the advent of electron or ion storage rings, polarized window-less gas targets became a viable alternative to solid polarized targets. Open gas targets of highest density are based on the storage cell proposed by W. Haeberli already in 1965.[1] Suchlike cells are fed with a beam of polarized atoms from Atomic Beam Sources (ABS), well known from polarized ion sources. That is why workshops on polarized ion sources have often been combined with the subject of polarized gas targets.

1.1. *Symposia on polarization phenomena in nuclear reactions*

The first meeting took place in 1960 in Basel, organized by Huber and Meyer.[2] The meeting reflected the intense work at several laboratories to produce beams of spin polarized nuclei. The immediate cause was the successful implementation of a polarized source for deuterium ions inside the terminal of a high-voltage generator at Basel. For the first time, nuclear reactions initiated by accelerated polarized ions from a source could be studied. The concept of an ABS had already been proposed by the Erlangen group in 1956,[3] but their attempts to produce a beam of polarized hydrogen ions were unsuccessful due to the limited vacuum technology available at that time. The Basel group went for a deuterium beam, resulting in a much lower contribution from ionization of the residual gas.[4]

The first polarization symposium was followed by a series of meetings every five years, starting with Karlsruhe 1965, Madison 1970, Zürich 1975, Santa Fe 1980, Osaka 1985 and Paris 1990. At the Paris meeting, organized by J. Arvieux, a resolution was passed to merge the polarization symposia with another series, the high energy spin physics symposia. The 8th and last polarization symposium took place in 1994 in Bloomington, in parallel with the 11th symposium on high energy spin physics.

1.2. *Symposia on high energy spin physics*

This series was triggered by the successful acceleration of polarized protons to several GeV in the ZGS, the zero gradient synchrotron at the Argonne National Laboratory.[5] The first meeting took place at Argonne in 1974.[6] The biennial follow-up meetings were Argonne 1976 and 1978, Lausanne 1980, Brookhaven 1982, Marseille 1984, Protvino 1986, Minneapolis 1988, Bonn 1990, Nagoya 1992, Bloomington 1994 (parallel to the Polarization Symposium), Amsterdam 1996 and Protvino 1998.

1.3. *Joint symposia on spin physics*

The first joint meeting took place in Osaka in 2000, organized by H. Ejiri and K. Hatanaka (RCNP Osaka). The International Committee, chaired by A.D. Krisch (Michigan), was composed of members from the high energy spin and the nuclear polarization physics communities. Due to the many ideas and techniques commonly used by both communities, they grew together in a rather short time. The following meetings were Brookhaven 2002, Trieste 2004, Kyoto 2006 and Charlottesville (VA) 2008, reflecting the latest results on particle and nuclear spin physics, and the related technologies. The next meeting, SPIN2010, will be organized by the research center Jülich (FZJ).[7]

1.4. *How is the spin physics community linked together?*

Spin symposia are uniquely dedicated to the fundamental role of spin in nuclear and particle physics. In nuclei, the force on a single nucleon moving in a potential well which is caused by the other nucleons contains a strong non-central part depending on the relative orientation of nucleon spin and orbital angular momentum, the so-called spin-orbit term. In contrast to level splitting in an atom, the splitting of the nuclear levels is very strong, giving rise to pronounced gaps in the scheme of ground-state energies as function of the nucleon number. The corresponding model, the shell model, was able to explain for the first time the magic numbers of nuclei, i.e. proton or neutron numbers with exceptional stability. Similar arguments hold for the energy of hadrons which depend on their partonic spin states. In collisions between particles their spins determine to a large extent which reactions will proceed . Scattering processes are studied both in nuclear and particle physics.

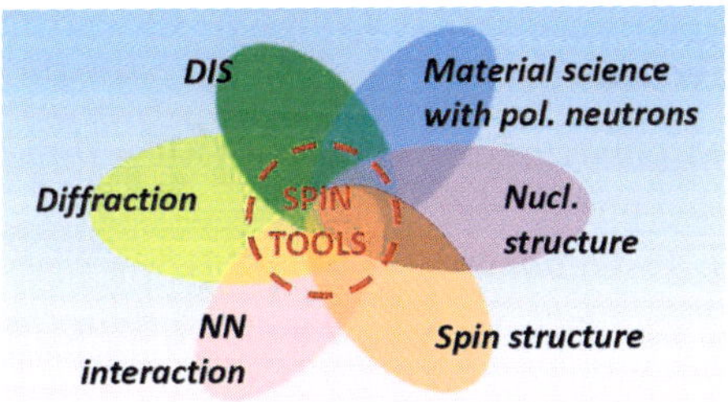

Fig. 1. A common aspect of experiments involving spin dependent observables are the polarization techniques or *spin tools*.

All these experiments have in common that they require certain tools which allow for the study of spin degrees of freedom. Suchlike tools include beams or targets of spin-polarized particles. Polarized beams of nucleons (p, n) and (light) nuclei (d,^{3}He,^{6}Li,^{7}Li,^{23}Na) have been used, as well as electrons, muons and photons. As polarized target nuclei, protons and deuterons have been used, both within a crystal or as pure hydrogen gas target. ^{3}He can be used as a polarized gas target, too. Most of the spin experiments require techniques to measure the polarization of beams and/or targets. Precise *polarimetry* is of growing importance as it determines the ultimate precision in the measurement of polarization observables. Beam polarimetry is also an integral part of acelleration of polarized beams.

The various fields of spin physics are linked by these tools which enable us to perform *spin* or *polarization* experiments as illustated in figure 1. Spin tools are discussed at workshops, and they are also reviewed at the spin physics symposia. The first meeting on spin tools with broad attendence known to me was the Saclay conference in 1966.

2. International conference on polarized targets and ion sources, Saclay 1966

The meeting[8] with nearly 170 participants was chaired by A. Abragam, author of the famous book *The Principles of Nuclear Magnetism*[9]. He also co-invented the non-adiabatic or Abragam-Winter RF-transitions[10] induced between hyperfine states of atoms in flight, enabling complete exchange of populations of hfs states. There were several highlights at this early meeting, like:

(i) Solid polarized targets.
(ii) Spin-filtering of neutrons.
(iii) Polarized ion sources.

2.1. *Solid polarized targets*

The advent of solid polarized targets had at that time enabled the measurement of analyzing powers e.g. in elastic scattering, induced by energetic unpolarized beams. The minimum beam energy was high due to the energy loss within the rather thick targets. The underlying method called Dynamical Nuclear Polarization (DNP) was invented in the early 1960s by C.D. Jeffries[11] and applied with great success at Argonne, Berkeley, Saclay, CERN, Dubna, Rutherford Lab and other places. It is based on spin-spin interaction of unpaired electrons of a crystal lattice with protons, e.g. of water within the crystal. At typical conditions of low temperature and high magnetic field, e.g. $T = 1$ K and $B = 2.5$ T, electrons in thermal equilibrium take on a very high polarization. When irradiated with microwaves at ESR conditions, the population is equalized and deviates largely from thermal equilibrium. By mutual e-p spin flips, electrons and protons acquire polarization. Due to their long T_1 relaxation times, a large number of protons, e.g. from the water of crystallation, can be polarized by a single unpaired electron or *paramagnetic center*. This situation was illustrated in the review talk by A. Abragam using a drawing which shows King Solomon and his

700 princesses (see fig. 2). The high level of polarized target technology was described by H.H. Atkinson (Rutherford Lab), followed by an overview of experimental results of remarkable quality, presented by O. Chamberlain (Berkeley).

Fig. 2. King Solomon and his 700 princesses, illustrating spin-exchange between the unpaired electron and the many protons as performing in the DNP process. The picture was taken from A. Abragam: *Polarized Targets: How?* Talk given at the Saclay meeting.[8]

2.2. *Spin-filtering of neutrons*

In his talk, F.L. Shapiro (Dubna) reviewed the work at the Joint Institute on polarizing neutrons by spin-dependent transmission through a polarized proton target, as illustrated in figure 3. The best target at that time was LMN(Nd), a $La_2Mg_3(NO_3)_{12} \cdot 24H_2O$ compound, doped with a small fraction of Nd. The polarization obtained for protons of the water of crystallization was around $P = 0.7$. The neutron polarization obtained in the resonance energy region is of the same order, i.e. very high. Suchlike beams have been utilized to measure the spin-dependence in the interaction of neutrons with nuclei. Spin-filtering of neutrons has the great advantage that there is no long-range Coulomb force which in the case of ions tends to cause a strong energy loss and beam blow-up which limits the thickness

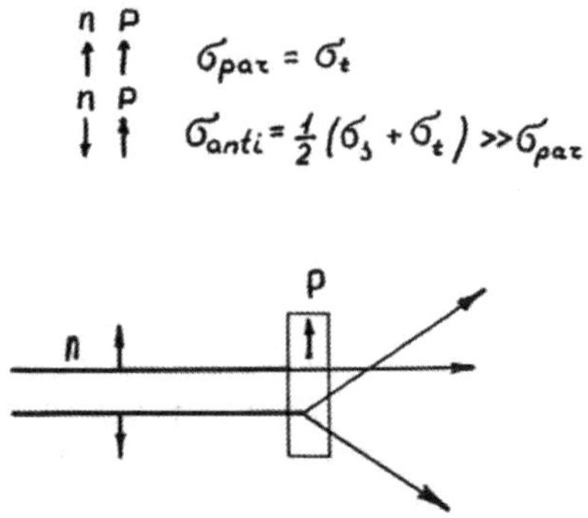

Fig. 3. Total cross section for the scattering of neutrons by protons with parallel and anti-parallel spins. Note the large difference between them resulting in nearly unattenuated transmission of the *parallel* neutrons. The picture was taken from F.L. Shapiro: *Polarized Nuclei and Neutrons*. Talk given at the Saclay meeting.[8]

of the filter target considerably. Spin-filtering of ions has been studied in a Cooler Storage Ring (CSR),[12] and this method will be discussed at the PST2009 workshop later this week.

2.3. *Polarized ion sources*

This talk by R. Beurtey (Saclay) reflects the rather advanced status of polarized ion sources in the mid 1960s. A key question was, which concept will be best for proton and deuteron sources: the Atomic Beam Source (ABS) with Abragam-Winter transitions, or the Lamb-shift source based on the properties of the metastable H(2s)-states? This conflict is illustrated in figure 4. Today we know that the intensity of Lamb-shift sources is finally limited by quenching of the metastables due to the space charge. In addition, thermal beams from ABS's can be used for storage cell targets. Apart from many technical details, Beurtey gives a strong pleading for performing experiments *polarized*, and not objecting "the complications spin introduces to the experiment and analysis" rather than acknowledging "the enormous benefit due to the existence of spin".

In his conclusion, Beurtey states that for nuclear physics at low and intermediate energies, the experimentalists will demand more and more intense polarized beams. He points to the big and at that time unresolved difficulties of accelerating such beams to very high energies. Today we know, e.g. from the HERA and RHIC projects, that high energy machines have to be designed *spin transparent* from the beginning to be successful. If

Fig. 4. Ancient discussion about the pros and cons of the Atomic Beam Source vs. Lamb-shift source. The picture was taken from R. Beurtey: *Polarized Ion Sources.* Talk given at the Saclay meeting.[8]

available, such beams turn out to be very powerful, sometimes opening up new areas in physics.

3. Early meetings relevant to spin tools

The second conference on polarized (solid) targets took place in Berkeley in 1971, chaired by O. Chamberlain. Work on spin tools was discussed at the polarization symposia e.g. at

- Karlsruhe 1965: First ideas of the storage cell and of the Colliding Beam Source (CBS) by W. Haeberli (Wisconsin)
- Madison 1970: Achromatic focusing of thermal atomic beams by means of a compressor sextupole magnet by H.F. Glavish (Stanford)
- Zürich 1975: Polarized electrons from a GaAs cathode by D.T. Pierce and F. Meier (ETH)
- Santa Fe 1980: First experimental demonstration of the CBS by the Madison group.

4. Workshops on spin tools 1981–1990

There were a number of topical workshops in the 1980s, mostly initiated by the high energy spin physics community.

- Ann Arbor 1981 and Vancouver 1983 on *High Intensity Polarized Proton Sources.*

- Abingdon 1981, Brookhaven 1982 and Bad Honnef 1984 on *Polarized Target Materials and Techniques.*
- Bodega Bay 1985 on *Polarized Antiprotons.*
- Montana 1986 on *Polarized Sources and Targets.*
- Minneapolis 1988 and Bonn 1990: several satellite workshops to SPIN1988 and SPIN1990, incl. *Polarized Gas Targets.*
- KEK (Tsukuba) 1990 on *Polarized Ion Sources and Gas Jets.*

This period saw steady improvements in polarized electron and ion sources, and in polarized solid proton and deuteron targets, culminating in the EMC polarized target which was instrumental for detecting the deficit in quark polarization of the nucleon, giving rise to the *spin crisis*. The ion sources gained tremendously in realibilty and became an integral part of accelerators. With the advent of electron and ion storage rings, there was an increasing need for internal polarized jets or open gas targets, and this subject was added to the workshop menu, starting in Montana and continued in Minneapolis 1988 and in Tsukuba and Bonn 1990. In 1988, the first results on a storage cell target for polarized hydrogen gas in an electron storage ring were presented.[13]

5. Workshops on polarized beams and targets after 1990

After 1990, these workshops took place predominantly in the uneven years, between the spin physics symposia. A selection is listed below.

- Heidelberg 1991 (gas targets only). Emphasis was on polarized gas targets for storage rings, which were operating or under design in Boston, Heidelberg, Madison, Novosibirsk and elsewhere.
- Madison 1993. The first results on the FILTEX H/D target at the TSR Heidelberg and the CE-25 ^{3}He target at the IUCF Cooler were presented.
- Cologne 1995. Successful tests of the PINTEX target at the IUCF Cooler, and of the HERMES ^{3}He target at the HERA electron ring were reported.
- Urbana 1997. About six polarized gas targets in storage were operational: at AmPS, EDDA at COSY, HERMES at DESY, two targets at IUCF: PINTEX and first results on the LDS, and the VEPP-3 H/D target in Novosibirsk.
- Erlangen 1999. This workshop was dedicated to Rudolf Fleischmann (Erlangen) ∗1903, †2002, a pioneer in spin physics (see fig. 5 and ref. 3).
- Nashville 2001. Successful operations of the TRIUMF and BNL OPPIS optically pumped polarized proton sources were reported. First operation of the PINTEX target with deuterium. Status reports on the HER-

MES H/D and BLAST targets were given. Ongoing development on laser driven source and ultracold source projects. Many other reports on solid polarized targets, electron sources and polarimetry were presented.

- Novosibirsk 2003. Highlights were the SLAC polarized electron sources and beams, and the new polarized deuterium source for the VEPP-3 target with superconducting magnets.
- Tokyo 2005. Among the topics were cryogenic targets, optical pumping methods and polarimetry. Emphasis was on polarization of radioactive beams and related methods.
- Brookhaven 2007. Here the focus was on the proton injector for the RHIC-spin complex and its outstanding performance, on polarimetry at low and high energies, and on the acceleration and storage of polarized protons. Reports of studies on a double-polarized Electron-Ion Collider (EIC) were presented.

Fig. 5. Rudolf Fleischmann (right), to whom the workshop PST99 was dedicated, and Willy Haeberli at the workshop in Erlangen 1999.

Now we are looking forward to a new workshop on Polarized Sources and Targets, and polarimetry (PST2009), and to exiting talks and discussions on spin tools!

6. Outlook

The methods for studying the spin-dependence of scattering processes between spin-polarized beams and/or targets have been developed over more than 50 years. A large variety of tools and applications have emerged:

- Atomic beams.
- Ion sources and beams.
- Electron sources and beams.

- Photon beams.
- Solid cryogenic or organic targets.
- Open gas targets for storage rings.
- Closed high pressure gas targets.

In addition, we have methods at our disposal to detect and apply spin polarization. *Polarimetry* of beams of ions, neutrons and electrons, and of recoils and decay particles has been developed to very high standards and precision. The field of *application of spin* has grown considerably, in *medicine* e.g. Magnetic Resonance Imaging (MRI) based on protons, polarized ^{3}He or ^{129}Xe gas, and other nuclei, in *material research*, e.g. muon spin rotation, or in *electronics and information technology*, which is often called *Spintronics*, e.g. the Giant Magnetic Resistance GMR) effect dicovered by A. Fert (Paris) and P. Grünberg (Jülich) which is utilized in modern hard disks for mass storage of data.

Let me finish by saying, that our workshops have an impressive record with many innovations for spin experiments. There are new initiatives needed in order to keep the field up and active! I look forward to a creative meeting in a stimulating atmosphere!

References

1. W. Haeberli, in *Proc. 2nd Int. Symp. on Polarization Phenomena of Nucleons, Karlsruhe 1965*, eds. P. Huber and H. Schopper, Experientia, Suppl. **3** (Birkhäuser Verlag, Basel 1966).
2. P. Huber and K. P. Meyer (eds.), *Proc. Int. Symp. on Polarization Phenomena of Nucleons, Basel 1960*, Helv. Phys. Acta Suppl. **VI** (Birkhäuser Verlag, Basel 1961).
3. G. Clausnitzer *et al.*, *Z. Physik* **144**, 336 (1956).
4. H. Rudin *et al.*, *Helv. Phys. Acta* **34**, 58 (1961).
5. E. F. Parker *et al.*, in *Proc. Particle Accelerator Conference, 1975.*
6. G. Thomas *et al.*, in *Proc. Symposium on High Energy Physics with Polarized Beams, Argonne 1974* (Atomic Energy Commission 1974).
7. Available at `http://www.fz-juelich.de/ikp/spin2010` .
8. *Proc. Int. Conf. on Polarized Targets and Ion Sources*, (Centre d'Etudes Nucleaire de Saclay, Saclay 1966).
9. A. Abragam, *The Principles of Nuclear Magnetism.* (Oxford University press, London, 1961).
10. A. Abragam and J.M. Winter, *Phys. Rev. Lett.* **10**, 374 (1958).
11. C. D. Jeffries, *Dynamical Nuclear Orientation* (Interscience, New York 1963).
12. F. Rathmann *et al.*, *Phys. Rev. Lett.* **71**, 1379 (1993).
13. S. I. Mishnev *et al.*, in *Proc. 8th Int. Symp. on High Energy Spin Physics, Minneapolis 1988*, ed. K. J. Heller, AIP Conf. Proc. **187**, 1286 (AIP, New York, 1989).

POLARIZED PROTON BEAMS IN RHIC

A. Zelenski

Brookhaven National Laboratory, Upton, NY
E-mail: aulenbac@kph.uni-mainz.de

The polarized beam for RHIC is produced in the optically-pumped polarized H^- ion source and then accelerated in LINAC to 200 MeV for strip-injection to booster and further accelerated 24.3 GeV in AGS for injection in RHIC. In 2009 run polarized protons was successfully accelerated to 250 GeV beam energy. The beam polarization of about 60 % at 100 GeV beam energy and 36-42 % at 250 GeV energy was measured with the H-jet and p-carbon CNI polarimeters. The gluon contribution to the proton spin was studied in collisions of longitudinally polarized proton beams at 100×100 GeV. At 250×250 GeV an intermediate boson W production with the longitudinally polarized beams was studied for the first time.

Keywords: Polarimetry; collider; RHIC; polarized beams.

1. Introduction

Collisions of protons at energies $\sqrt{S}$ = 200–500 GeV and large transfer momentum ($p_T > 10$ GeV/c) are described as parton collisions (quarks, gluons) and for polarized proton beams these partons are polarized too. The analyzing powers for polarized parton scattering can be directly calculated in the frame of perturbative QCD. This provides a unique opportunity for proton spin structure studies, fundamental tests of QCD predictions with possible extension to probe the physics beyond the "Standard Model".[1,2] RHIC is the first collider where the "siberian snake" technique was successfully implemented to avoid the resonance depolarization during beam acceleration in AGS and RHIC[3] (see fig. 1). A luminosity of a $60 \cdot 10^{30}$ cm^{-2} s^{-1} (peak $120 \cdot 10^{30}$) for polarized proton collisions in RHIC at $\sqrt{S} = 500$ GeV energy was produced by colliding 112 bunches in each ring at $1.4 \cdot 10^{11}$ protons/bunch intensity.

For the first time the intensity of the polarized beams produced in an Optically Pumped Polarized H^- Ion Source (OPPIS) was sufficient to charge RHIC to the maximum intensity limited by the beam-beam inter-

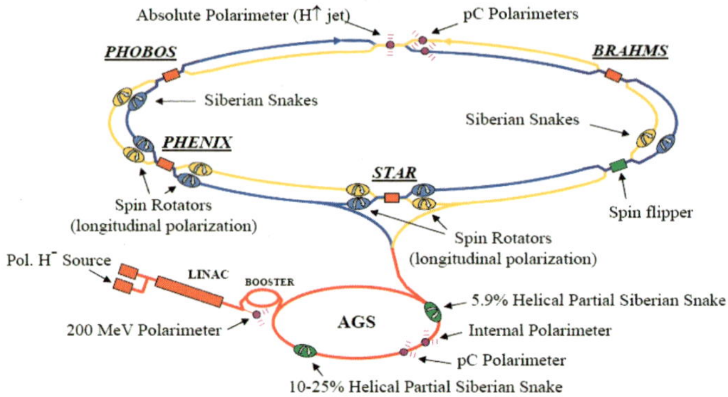

Fig. 1. Accelerator-collider complex RHIC polarization hardware layout.

action. Polarimetry is another essential component of the polarized collider facility. A complete set of polarimeters includes a Lamb-shift polarimeter at the source energy, a 200 MeV polarimeter after the LINAC, and polarimeters in AGS and RHIC based on proton-carbon scattering in the Coulomb-Nuclear Interference (CNI) region. A polarized hydrogen-jet polarimeter was used for the absolute polarization measurements in RHIC.[4]

Longitudinally polarized beams for the STAR and PHENIX experiments are produced with the spin rotators, which are tuned using "local polarimetrs" based on asymmetry in neutron production for pp collisions. The STAR and PHENIX detectors provide complementary coverage of the different polarization processes.

2. Polarized sources development at RHIC

Feasibility studies of new techniques for the production of polarized electron, H^- (proton), D^- (D^+) and $^3He^{++}$ ion beams are in progress at BNL. The OPPIS delivered beams for polarization studies in RHIC. The polarized deuteron beam will be required for the future deuteron Electron Dipole Moment (EDM) experiment, and the polarized electron (see I. Ben-Zvi talk in this Workshop) and $^3He^{++}$ ion beam is a part of the experimental program for the future eRHIC (electron-ion) collider.

The present operational polarized source (OPPIS) for RHIC is based on spin-transfer proton (or atomic hydrogen) collisions in the optically-pumped Rb vapor cell. In the BNL OPPIS, an ECR-type source produces primary proton beam of a 2.0–3.0 keV energy, which is converted to electron-spin polarized H atoms by electron pick-up in an optically pumped Rb vapor

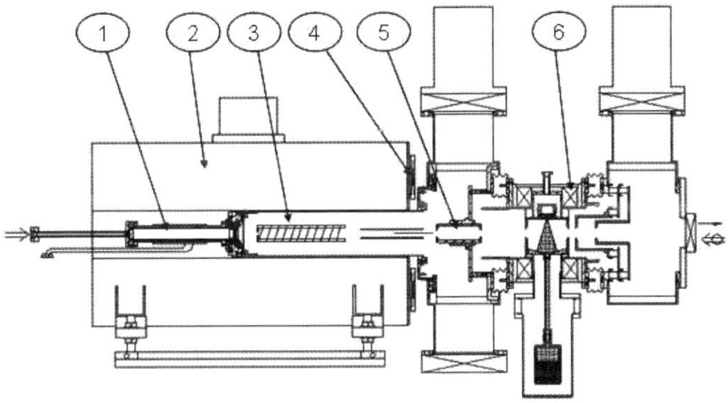

Fig. 2. Layout of the RHIC OPPIS: 1) electron-cyclotron resonance primary proton source; 2) super-conducting solenoid; 3) optically-pumped Rb cell; 4) correction coil; 5) Sona-shield; 6) Na-jet ionizer.

cell (see fig. 2). A pulsed Cr:LISAF laser of a 1 kW peak at 500 μs pulse duration is used for optical pumping. Electrostatic deflection plates downstream of the polarized alkali remove any residual H^+ or other charged species. The electron-polarized H^0 beam then passes through a magnetic field reversal region, where the polarization is transferred to the nucleus via the hyperfine interaction (Sona-transition). The nuclear polarized H atoms are then negatively ionized in a Na-jet vapor cell to form nuclear polarized H^- ions. Alternatively, the H atoms can be ionized in a He gaseous cell to form polarized protons. After ionization polarized H^- ions are accelerated from 3.0 keV to 35 keV energy by a negative 32 kV pulsed voltage applied to the ionizer cell.

The OPPIS technique is a multi-step polarization-transfer process. At each step there is some loss of polarization. Detailed studies of polarization losses in the RHIC OPPIS and the source parameters optimization resulted in the OPPIS polarization increase to 86-90 %. The significant gain of about 5 % was obtained from experimental studies and numerical simulations of the Sona-transition efficiency. The simulations were done using code, which was developed at INR, Moscow.[5]

The OPPIS produces routinely 0.5–1.0 mA (maximum 1.6 mA) H^- ion current with 400 μs pulse duration and 80-85 % polarization.[6] The beam is accelerated to 200 MeV with an RFQ and LINAC for strip-injection to the Booster. About 60 % of the OPPIS beam intensity can be accelerated to 200 MeV. The 400 μs H^- ion pulse is captured in a single booster bunch which contains about $4 \cdot 10^{11}$ polarized protons. Single bunch is accelerated in the booster to 2.5 GeV kinetic energy and then transferred to the AGS, where it is accelerated to 24.3 GeV for injection to RHIC.

The OPPIS initial longitudinal polarization was converted to the transverse direction while the beam passes two bending magnets, before injection into the RFQ. The second of these bending magnets (47.4 °) is pulsed to switch injection between polarized and unpolarized (of about 100 mA intensity) H^- ion beams. A pulsed focusing solenoid in front of the RFQ is tuned for the optimal transmission for either beam. This solenoid also rotates the polarization direction by about 360 °, but keeps it in the transverse plane. Final polarization alignment to the vertical direction was adjusted by a spin-rotator solenoid in the 750 keV beam transport line before injection to the LINAC. The drawbacks of this injection scheme were: a) unnecessary spin rotation, which may cause polarization losses, b) poor matching between the RFQ and LINAC, which causes beam loss and beam emittance degradation. The injector was upgraded for the 2009 polarized run. The RFQ was moved closer to the LINAC and the MEBT (750 keV) beam line was rebuilt to improve RFQ-LINAC beam matching. Only one bending magnet was used, thus eliminating a 180 ° spin rotation. The spin-rotator was moved from the 750 keV line to the 35 keV line to align the spin to the vertical direction before injection to the RFQ. The upgrade resulted in reduced spin precession, better optics matching between RFQ and LINAC, which reduced the emittance degradation in the LINAC (in both transverse and longitudinal phase space). This smaller emittance was propagated through the accelerator chain, and resulted in smaller emittance and higher polarization in RHIC.

The AGS cycle for polarized beam operation is 3 seconds, while OPPIS usually operates at 1 Hz repetition rate. Pulses not sent to the booster are directed to a 200 MeV p-carbon polarimeter for spin-rotator tuning and continuous polarization monitoring, by a pulsed bending magnet in the high-energy beam transport line. The inclusive p-carbon polarimeter was calibrated in 2002-03 runs by comparison with elastic p-deuteron scattering in the additional detector arms. A low deuteron average counting rate was observed in these measurements due to small duty factor. The limitation of detector peak single counting rates did not allow use of available high beam intensity. There is a plan for a polarimeter upgrade for the 2010 polarized run in AGS to improve the absolute accuracy of polarization measurements to +/-1 %.

2.1. *OPPIS upgrade with the atomic hydrogen beam injector*

The ECR proton source is operated in high magnetic field. It has low hydrogen gas consumption, which makes possible a dc OPPIS operation with intensity in excess of 1.0 mA.

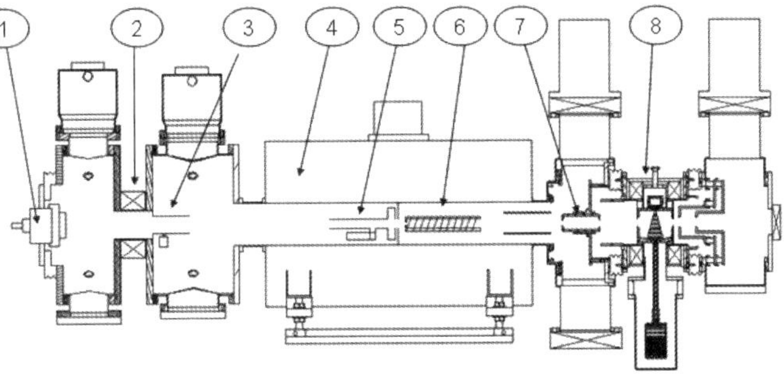

Fig. 3. Layout of the OPPIS with atomic hydrogen injector: 1) high-brightness proton source; 2) focusing solenoid; 3) pulsed hydrogen neutralization cell; 4) super conducting solenoid 30 kG; 5) Pulsed He ionizer cell; 6) optically-pumped Rb cell; 7) Sona shield; 8) sodium-jet ionizer cell.

However, the proton beam produced in the ECR source has a comparatively low emission current density and high beam divergence. This limits further current increase and gives rise to inefficient use of the available laser power for optical pumping. In fact only about 15 % of the electron-spin polarized hydrogen atoms produced in the Rb cell is within the ionizer cell acceptance.

In pulsed operation, suitable for application at high-energy accelerators and colliders, the ECR source limitations can be overcome by using instead of ECR, a high brightness proton source outside the magnetic field.[7] Following neutralization in hydrogen, the high brightness 5.0–8.0 keV atomic H^0 beam is injected into a superconducting solenoid, where both a He ionizer cell and an optically-pumped Rb cell are situated in the same 25–30 kG solenoid field, which is required to preserve the electron-spin polarization. The injected H atoms are ionized in the He cell with 80 % efficiency to form a low emittance intense proton beam, which enters the polarized Rb vapor cell (fig. 3). The protons pick up polarized electrons from the Rb atoms to become a beam of electron-spin polarized H atoms (similar to ECR based OPPIS). A negative bias of about 2.0–5.0 kV applied to the He cell decelerates the proton beam produced in the cell to the 3.0 keV beam energy optimal for the charge-exchange collisions in the Rb and sodium cells. This also would allow the energy separation of the polarized hydrogen atoms produced after lower energy proton neutralization in Rb vapor and residual hydrogen atoms of the primary beam. Residual higher energy atoms will be neutralized with lower efficiency in Rb cell (due to cross-section decrease at higher energy) and unpolarized component will be further suppressed by lower H^- ion yield at 5.0–8.0 keV atomic beam energy. The H^- ion beam acceleration (by -32 kV pulsed voltage applied to the ionizer cell) will produce polarized H^- ion beam of a 35 keV beam energy and unpolarized beam of a 40–43 keV beam energy. Further suppression of unpolarized higher energy ion beam can be done in the Low Energy Beam Transport Line (LEBT).

Atomic hydrogen beam current densities greater than 100 mA/cm^2 can be obtained at the Na jet ionizer location (about 200 cm from the source) by using a very high brightness fast atomic beam source developed at BINP, Novosibirsk. This was tested in experiments at TRIUMF, where more than 10 mA polarized H^- and 50 mA proton beam intensity was demonstrated.[6] Higher polarization is also expected with the fast atomic beam source due to: a) elimination of neutralization in residual hydrogen; b) better Sona-transition transition efficiency for the smaller $\sim$ 1.5 cm diameter beam; c) use of a higher ionizer field (up to 3.0 kG), while still keeping the beam emittance below 2.0 π mm $\cdot$ mrad, because of the smaller beam – 1.5 cm diameter. All these factors combined will further increase polarization in the pulsed OPPIS to $\sim$ 90 % and the source intensity to over 10 mA.

The RHIC polarized source upgrade for higher intensity and polarization is approved and fully funded for implementation in 2009–12. The source will provide the high intensity low emittance (high brightness) beam for the polarized RHIC luminosity upgrade and for future eRHIC facilities.

2.2. *Proposal for polarized $^3He^{++}$ source for eRHIC*

Polarized neutrons collisions can be studied with the deuteron beams. Unfortunately due to the small deuteron anomalous magnetic moment G-factor (G=-0.143), the "siberian snakes" and spin rotators will be too weak to be effective for the polarized deuteron acceleration in AGS and RHIC. Polarized beams of $^3He^{++}$ ions also contain the polarized neutron component (as a deuteron) and its G-factor: G=-4.184 is even larger than the proton G_p value, therefore the AGS and RHIC siberian snake can be operated at a lower field to preserve polarization during acceleration.[7] In this case the difficulty is the polarized $^3He^{++}$ source. The proposed polarized $^3He^{++}$ acceleration in RHIC (and also for the future RHIC upgrade to electron ion collider eRHIC) will require about $2 \cdot 10^{11}$ ions in the source pulse and about 10^{11} ions in the RHIC bunch. To deliver these ions in a 20μ s pulse duration for injection to the booster, the source peak current has to be about 1000 μ A, which is 10000 times higher than ever achieved in any $^3He^{++}$ ion sources. A new technique has been proposed for production of high intensity $^3He^{++}$ ion beams. It is based on ionization of nuclear polarized ^3He gas in the Electron Beam Ion Source (EBIS)[8] (see fig. 4). The highest ^3He nuclear polarization in excess of 80 % so far was achieved by the metastability exchange technique. In this method, ^3He gas at typically 1 torr pressure is contained in a glass bulb and a weak RF discharge is maintained in the gas. Metastable atoms in the $2\ ^3S_1$ state are produced in the discharge and may be polarized by means of optical pumping with circularly polarized (2^3S_1 - 2^3P_0) 1083 nm light. The polarization is transferred to the ground state atoms by the metastability-exchange collisions. In the cesium coated quartz cell the long (>100 h) polarization relaxation time was obtained resulting in high ground state polarization. In the proposed technique, the polarized ^3He atoms consumption for injection to an ionizer (another polarized atoms loss factor) is very small, of the order of 10^{13}–10^{14} He atoms s^{-1} and high polarization is expected.

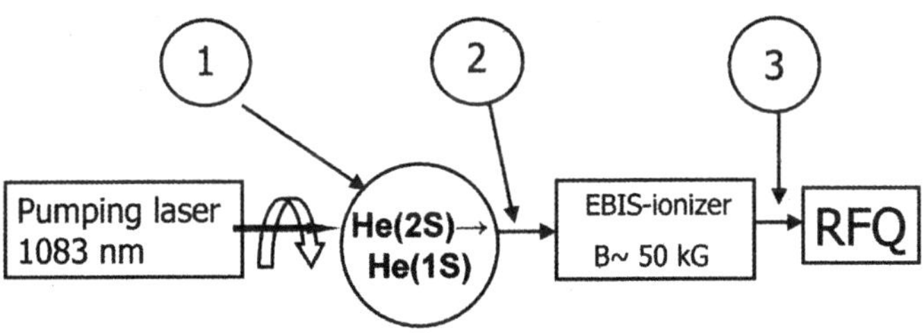

Fig. 4. Schematic diagram of the polarized $^3He^{++}$ ion source. 1: metastability-exchange polarizing cell; 2: ^3He transfer line; 3: 10^{12} polarized $^3He^{++}$ ions for injection to RFQ.

An EBIS is under construction at BNL as an alternative to the Tandem heavy ion injector for RHIC. It is proposed to use the EBIS to produce $^3He^{++}$ by ionization of the polarized ^3He gas, which is fed from the polarizing cell. The ionization in the EBIS is produced in a 50 kG magnetic field, which preserves the nuclear ^3He polarization while in the intermediate single-charged He^+ state. The ionization efficiency to the double-charged

$^3He^{++}$ will be close to 100 % and the number of ions is limited to the maximum charge, which can be confined in the EBIS. From experiments with Au^{32+} ion production, one expects about $2.5 \cdot 10^{11}$ $^3He^{++}$ ions/pulse to be produced and extracted for subsequent acceleration and injection to RHIC.

After $^3He^{++}$ acceleration to a few MeV/nucleon, He-D or He-carbon collisions can be used for polarization measurements. The Lamb-shift polarimeter at the source energy of 10–20 keV can be used in the feasibility studies (similar to the OPPIS polarimeter). In this technique $^3He^{++}$ ions are partially converted to He^+ $(2S)$ - metastable ions in the alkali vapor cell. Then the hyperfine sublevel populations can be analyzed in the spin-filter device to extract the primary $^3He^{++}$ nuclear polarization.

A study of limitation on the maximum attainable nuclear polarization in the metastability exchange technique (at the very low polarized ^3He gas consumption rate) will be required to define the maximum attainable polarization. Possible depolarization effects during polarized ^3He gas injection to existing EBIS prototype and multi-step ionization process should be also studied. The expected $^3He^{++}$ ion beam intensity is in excess of $2 \cdot 10^{11}$ ions/pulse with polarization in excess of 70 %.

3. Polarized beam acceleration in AGS and RHIC

3.1. *Polarized beams in AGS*

Two partial siberian snakes were installed in AGS to preserve polarization during beam acceleration from booster energy 2.4 GeV to 24 GeV (see T. Roser talk at PST2007[9]). With this snake configuration the polarization losses at all imperfection and vertical intrinsic resonances was eliminated except a few vertical intrinsic resonances near injection energy. As a result a polarized proton beam with $1.5 \cdot 10^{11}$ intensity/bunch and 65 % polarization was delivered for injection to RHIC in the 2006–2009 runs. A spin tracking simulations showed that the remaining polarization losses ~ 15 % (beam polarization at injection to AGS was about 80-85 %) can be caused by: the loss due to horizontal intrinsic resonances along the ramp; loss associated with snake resonances near the strongest 36+v: the loss from vertical intrinsic resonances near injection. All these losses can be reduced by the smaller beam emittance. In the 2009 run, smaller beam emittance out of the LINAC, shorter strip-injection time to the booster (see above) and better beam injection matching from the booster to AGS, resulted in smaller beam emittance in AGS and RHIC. Also a new lattice with the vertical tune in the spin tune gap near injection has been used to reduce the loss near injection energy. As a result 65–70 % polarization was consistently delivered for injection to RHIC (see fig. 5). New pulsed quadrupoles were installed in AGS to produce fast tune jump during the energy ramp in an attempt to eliminate all residual depolarization at the horizontal spin resonances.[10]

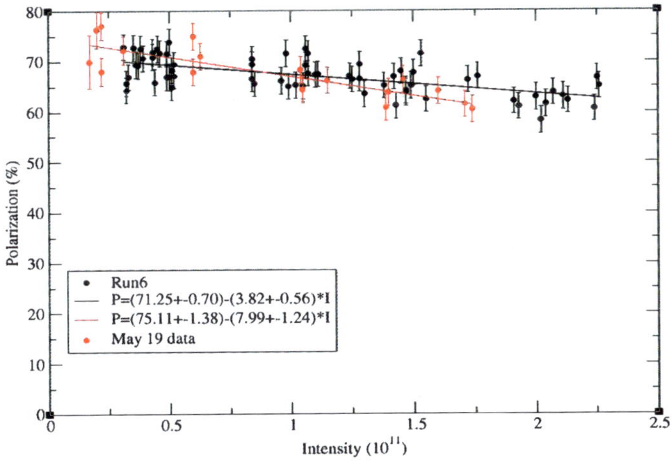

Fig. 5. Polarization vs bunch intensity in AGS.

3.2. *Polarized beams in RHIC*

The RHIC heavy ion collider is the first high-energy machine where polarized proton acceleration was included in the primary design. The RHIC "full siberian snake" (which rotates spin direction for 180 °) is a superconducting helical magnet system of about 10 m long. Two 90 ° helical spin rotators in each ring produce the longitudinal polarization for experiments in STAR and PHENIX detectors.

Up to 120 beam bunches can be accelerated and stored in each ring. The polarization direction of every RHIC bunch is determined by the spin-flip control system in the polarized ion source. Every single source pulse is accelerated and becomes the RHIC bunch of the requested polarity. By loading selected patterns of spin direction sequences in the rings (such as: +-+- in one ring and +–+ in another) the experiments have all possible spin-direction combinations for colliding bunches which greatly enhance the systematic error control.

Since 2005, RHIC has successfully accelerated polarized protons up to 100 GeV with no polarization loss by carefully controlling the betatron tunes and the vertical orbit distortion. A record polarization of 60 % was reached during the RHIC polarized proton operation in 2006 and routinely delivered in run 9. However, between 100 GeV and 250 GeV, there are three strong intrinsic spin resonances which may cause polarization loss. These resonances are more than a factor of two stronger than the strong spin resonances below 100 GeV. Since the stronger the intrinsic resonance, the stronger the derived snake resonances, the tolerance on the nearby imperfection resonance, i.e. the vertical closed orbit distortion, is tighter. The numerical simulation shows that the imperfection resonance strength should be below 0.075 to avoid depolarization at these three strong intrinsic resonances.[11] This corresponds to a closed orbit distortion of 0.3 mm RMS value. These accuracies of the orbit and tune control were achieved in the 2009 run. As a result the proton polarization transfer efficiency (from injection to store) was improved to about 80 % and on average about 42 % polarization was measured during the 250 GeV physics stores.

A maximum polarized beam luminosity of a $45 \cdot 10^{30}$ cm^{-2}s^{-1} (about $30 \cdot 10^{30}$ cm^{-2} s^{-1} averaged over 8 hours store time) was obtained with 109 bunches of $1.6 \cdot 10^{11}$ bunch intensity in the 2009 run.

4. Polarization measurements in AGS and RHIC

Proton polarization measurements in the AGS and RHIC are based on proton-carbon and proton-proton elastic scattering in the Coulomb Nuclear Interference (CNI) region.[12] This process has a large cross-section and sizable analyzing power of a few percents which is expected to have weak energy dependence in the 24–250 GeV energy range. A very thin (5 μg/cm^2, 5–10 μm wide) carbon strip target in the high intensity circulating beam produces a high collision rate and a highly efficient DAQ system acquires up to $5 \cdot 10^6$ carbon events/sec. High recoil carbon nuclei intensity from the scattering of the circulated proton beam in the thin carbon target is efficiently utilized in the silicon strip detectors and data acquisition system, which is capable of analyzing the event rate up to a few millions/second. This gives unique possibilities for the fast, practically non-destructive polarization measurements. The polarization measurements during the beam energy ramp were also implemented in AGS and RHIC, which provides an insight into the polarization losses pattern. Polarimeter operation in the scanning mode also gives the polarization profile and beam profile (beam emittances including bunch by bunch measurements).

The CNI polarimeters were upgraded for the 2009 run. Two identical target motion mechanisms and detectors assemblies were installed in new vacuum chambers in each ring. One polarimeter was used for the vertical polarization and intensity profile measurements and the other for the horizontal profile measurements (or vice versa). As a result, the systematic polarization, polarization profiles and emittance measurements were obtained for both transverse planes. This is in contrast to previous runs where measurements were limited to one plane due to long target switch times.

The absolute beam polarization at 100 GeV beam energy was measured with a polarized H-jet polarimeter.[4,13] The simultaneous measurements in p-carbon and H-jet polarimeters provide the calibration for p-Carbon analyzing power. Fast p-carbon polarimeter measures possible polarization losses during the energy ramp and possible polarization decay during the RHIC store. The 250 GeV beam polarization measurements averaged over the fill duration (~9 hours) corrected for polarization profiles and normalized by the H-jet polarimeter measurements are shown in figure 6.

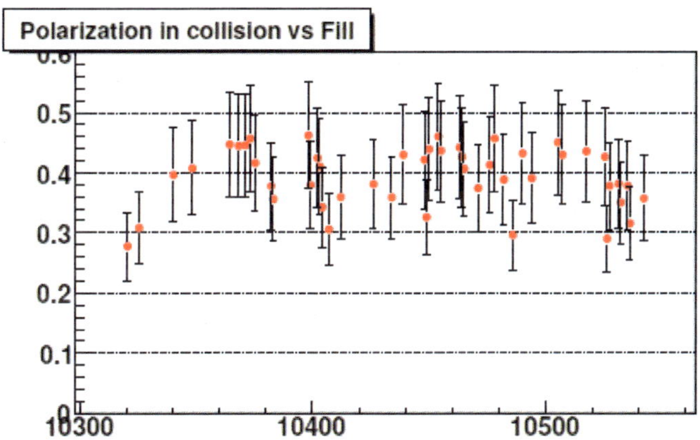

Fig. 6. Fill by fill polarization measurements in yellow ring at 250 GeV beam energy in the run 2009 (normalized by the H-jet measurements).

5. Summary

The RHIC spin program is a great beneficiary of the latest development in the polarized ion source and polarized target technology. For the first time the polarized proton beam intensity in the high-energy accelerator is not limited by the polarized source intensity. In the 2009 run polarized protons were successfully accelerated to 250 GeV beam energy. The beam polarization of 60 % at 100 GeV beam energy and 42 % at 250 GeV beam energy was measured with the polarized H-jet and p-carbon CNI polarimeters. The gluon contribution to the proton spin was studied in collisions of longitudinally polarized proton beams at $\sqrt{S}$=200 GeV. At $\sqrt{S}$=500 GeV an intermediate boson W production with the longitudinally polarized beams was studied for the first time.

References

1. G. Bunce *et al.*, Prospects for Spin Physics in RHIC, *Ann. Rev. Nucl. Part. Sci.* **50**, 525 (2000).
2. N. Saito *et al.*, in *Proc. 16th Int. Spin Physics Symposium (SPIN2004)*, 58 (World Scientific, Singapore, 2005).
3. I. Alekseev *et al.*, *Nucl. Instr. Meth. A* **499**, 392 (2003).
4. A. Zelenski *et al.*, *Nucl. Instr. Meth. A* **536**, 248 (2005).
5. E. Antishev and A. Belov, AIP Conf. Proc. **980**, 263 (2008).
6. A. Zelenski *et al.*, AIP Conf. Proc. **570**, 179 (2000).
7. A. Zelenski *et al.*, *Nucl. Instr. Meth. A* **245**, 223 (1986).
8. A. Zelenski and J. Alessi, *ICFA Beam Dynamics Newsletter* **30**, 39 (2003).
9. T. Roser, AIP Conf. Proc. **980**, 15 (2008).
10. H. Huang *et al.*, AGS Polarized Proton operation in Run 2009, BNL-81823-2009-CP (2009).
11. M. Bai *et al.*, in *Proc. PAC 2007*, 745.
12. I. Nakagava *et al.*, AIP Conf. Proc. **980**, 380 (2008).
13. H. Okada *et al.*, *Phys. Lett. B* **638**, 450 (2006).

THE COSY/JÜLICH POLARIZED H^- AND D^- ION SOURCE

O. Felden, R. Gebel* and R. Maier

Institute for Nuclear Physics at Forschungszentrum Jülich, Jülich, 52425, Germany
**E-mail: r.gebel@fz-juelich.de*
www.fz-juelich.de/ikp

The polarized ion source at the cooler synchrotron facility COSY of the research center Jülich, Germany, delivers negative polarized proton or deuteron ions for investigation of hadron structure and dynamics in the momentum range from 0.3 GeV/c to 3.8 GeV/c. The polarized negative ion source at COSY is based on the colliding beams principle, using an intense pulsed neutralized cesium beam for charge exchange with a pulsed highly polarized hydrogen or deuterium beam. The source is operated at 0.5 Hz repetition rate with 20 ms pulse length which is the maximum useful length for the stripping injection into the synchrotron. The paper summarizes the status and activities at the polarized ion source at the COoler SYnchrotron COSY/Jülich.

Keywords: Polarized ion source; synchrotron; polarized protons and deuterons.

Introduction

Since 1996 polarized H^- ions have been delivered to the cooler synchrotron COSY[1] at the IKP of the Forschungszentrum Jülich. The layout of the synchrotron facility with its subsystems is described in references 2–4. The Colliding Beams Source (CBS) itself provides polarized negatively charged protons or deuterons. The source is operated at 0.5 Hz repetition rate with 20 ms pulse length which is the maximum useful length for the stripping injection into the synchrotron. The principle of the source is to collide a pulsed polarized hydrogen or deuterium beam from a ground state atomic beam source with a pulsed neutral cesium beam[5] having a kinetic energy of about 45 keV.[6–9] In a charge exchange reaction, taking place in a solenoidal field, negatively charged hydrogen or deuteron ions are created at a potential of 4.5 to 8 kV and accelerated toward the extraction elements. Then the ions are bent magnetically by 90 °, pass a Wien-filter and enter the transporting source beam line that guides them into the cyclotron. The CBS in its actual configuration is described in more details in reference 4. Since introduction of the pulsed operation of all main components of the source, about 1 μA of polarized H^- ions are routinely delivered for charge exchange injection into COSY.[10] Polarized D^- ions, in sequences of up to fifteen different deuteron polarization states, have been provided to experiments since 2003.[11] The distribution of delivered beam species from the years 2000 to 2009 is displayed in figure 1.

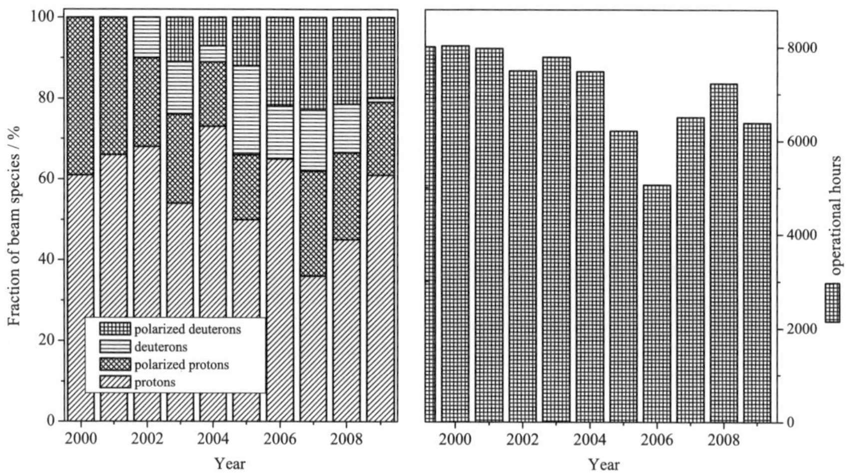

Fig. 1. The distribution of beam species at COSY since 2000.

Atomic beam studies

The pulsing concept was further developed to include the atomic beam part as well. This was also crucial to increase the uptime of the ion source. The dissociator, producing atomic hydrogen or deuterium beams, is a prime component of the source. In a pyrex-tube containing the gas, a high frequency discharge breaks the molecular bond. A stream of atomic gas leaves this tube through a nozzle cooled down to 36 K. The pulsing encompasses the dissociator RF-discharge as well as the gas supply. The latter one comprises three gases. The main gas is hydrogen, respectively deuterium, and small additions of nitrogen and oxygen are needed with respect to the working point of the dissociator. For each the exact flow and timing was investigated to obtain optimal performance. For this component 14 parameters have to be adjusted. Difficulties are amplified by the necessity to condition the components for longer time periods as surfaces and vacuum change during the tuning process. Pulsed polarized ion sources of INR, Moscow[12] and IUCF, Bloomington[13] produced pulsed polarized atomic hydrogen beams with peak intensities of 2×10^{17} H^0/s and a duration of about 2 ms. The COSY CBS RF discharge dissociator works with a low RF power of 250–350 W and long pulse duration compared to 2–4 kW of the INR or the IUCF pulsed sources. The active COSY CBS uses the magnet configuration of table 1 and is running with peak intensities of 0.75×10^{17} H^0/s in pulses with 20–100 ms downstream of the hexapole magnets. In order to

Table 1. Setup for old dissociator 0.75×10^{17} H^0/s with 2.5 mm nozzle at 36 K.

Element	Drift [mm]	Aperture [mm]	Length [mm]	Aperture [mm]	Field [T]
1.	31	7.6	11.0	9.8	1.32
2.	15	11.0	25.0	15.0	1.35
3.	15	16.2	25.0	23.0	1.35
4.	15	25.0	70.0	25.0	1.35
5.	220	30.0	40.0	30.0	1.35
6.	19	30.0	80.0	30.0	1.35
8.	66	10.0	10.0	10.0	TOF MS

study and to improve the pulsed atomic beam part, a copy of the atomic beam part of the CBS is used. This source on a test bench is equipped with diagnostic tools like a beam chopper, iris diaphragm, beam shutter, Time-Of-Flight Mass Spectrometer (TOF-MS), Quadrupole Mass Spectrometer (QMS) and Compression Tubes (CT). The density of the atomic hydrogen beam was measured with a TOF-MS. The velocities of atoms and molecules were measured with the TOF method using a chopper wheel with two slits and a QMS installed 880 mm downstream of the chopper wheel.

The dissociator with its cooled nozzle is available in a version optimized for higher repetition rate and low gas consumption. In a very fruitful collaboration with the Institute for Nuclear Research (INR, Troitsk, Russia) progress was realized. The achieved improvement is described in contributions to the annual reports of the IKP.[14–16] Proper tuning of the discharge parameters, optimization of the cooling channels and the hexapole configuration resulted in a substantial improvement. With the new dissociator and spare hexapoles from the EDDA atomic beam target a peak inten-

Table 2. Setup for new dissociator 1.1×10^{17} H^0/s with 2.5 mm nozzle at 77 K.

Element	Drift [mm]	Aperture [mm]	Length [mm]	Aperture [mm]	Field [T]
1.	35	6.0	25.0	20.0	Skimmer
2.	45	11.0	25.0	15.0	1.35
3.	15	16.2	70.0	23.0	1.35
4.	25	25.0	70.0	25.0	1.35
5.	360	30.0	80.0	30.0	1.35
6.	15	30.0	80.0	30.0	1.35
7.	15	30.0	40.0	30.0	1.35
8.	15	30.0	40.0	30.0	1.35
9.	300	10.0	10.0	10.0	TOF MS

sity of 1.1×10^{17} H^0/s has been realized at the test bench. The improved configuration of the permanent hexapoles is listed in table 2.

Polarimetry

Until now it is only possible to determine the polarization by using the in-beam p-carbon polarimeter in the transfer beam line from the cyclotron to the synchrotron based on the elastic scattering of H^- or D^- beams on a carbon target. For optimizing and tuning of the polarized beams the cyclotron had to be used and COSY operation is delayed. During preparation of a COSY experimental run all transition units necessary for providing the requested polarization sequences have to be tested. This is realized by installing each transition unit between the two hexapole groups and measuring the beam intensity as a function of the unit's parameters. Two examples of the transition unit's functionality for hydrogen and deuterium are given in figure 2. At a fixed frequency the magnetic field of the unit is scanned for intensity reduction. For hydrogen the intensity should be reduced to about 50%. For the deuteron beam one can expect a reduction by 33% to 66%, depending on the resonance frequency of the tunable resonator. Routinely, a performance close to these expectations is reached.

These results from each of the transition units give the start points for the optimization of the transition units with the low energy polarimeter behind the cyclotron, mainly to provide proper spin alignment to the cyclotron and COSY. Spin alignment is provided by a Wien-filter and solenoids in the injection beam line towards the cyclotron. Sample spectra from the p-carbon polarimeter for deuterons and protons are provided in figure 3. H^- or D^- ions are scattered from a 0.5 mm carbon wire and detected in NaJ scintillators at angles with a high figure-of-merit and acceptable signal to background ratios.

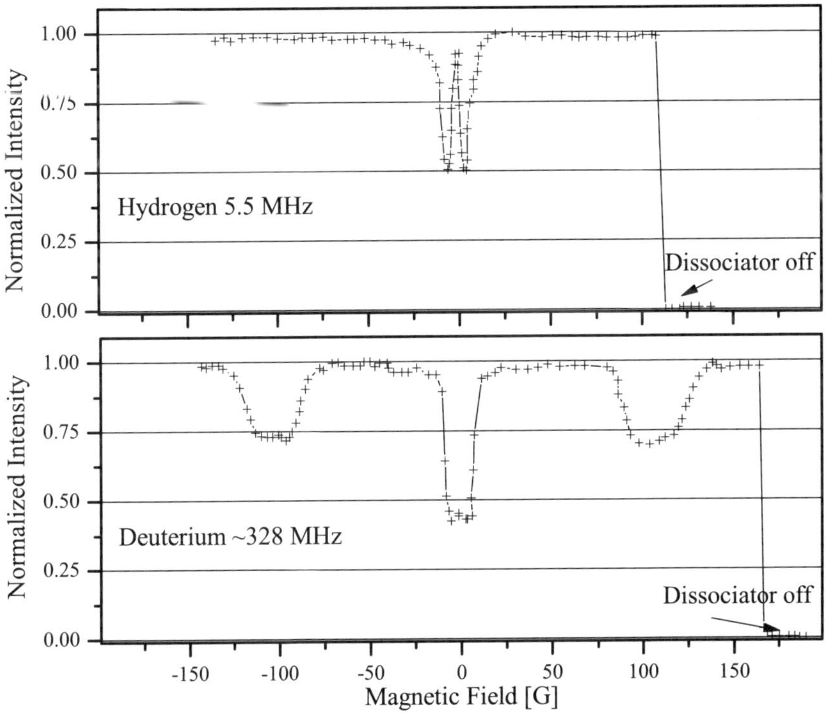

Fig. 2. Examples for the verification of the transition units performance for H and D beams at the CBS.

The determination of the beam's polarization is achieved by counting the events in the elastic channel of the p-carbon reaction. For tuning purposes single channel analyzers and computer controlled counters are used. Pulse height analysis and fit routines are used for setup of the amplifiers and analyzers and precision measurements. A statistical error of about 1% is achieved after 200 to 300 injection cycles. Since the beginning of 2009 the data acquisition is enabled to collect data during long COSY cycles. The beam from the ion source can be transmitted to the p-carbon polarimeter and is dumped behind the polarimeter. With a maximum repetition rate of 0.5 Hz the asymmetries are determined. This monitoring feature enables the judgment of the stability of the system up to the polarimeter. Figure 4 shows an example for polarized deuterons. The data is collected over a period of about 11 days during an experimental run. Besides the efforts before and during an experiment with polarized beams it is obvious that there is a need to improve tuning and preparation, especially for

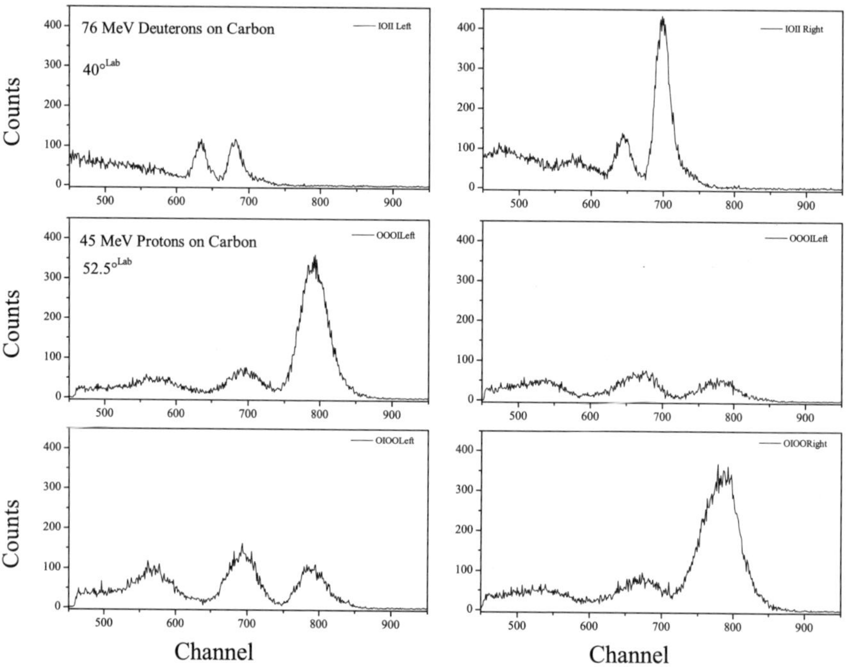

Fig. 3. Low energy polarimeter spectra for protons and deuterons.

polarized deuteron beams. Therefore a Lamb-Shift Polarimeter (LSP) has been constructed.

The LSP is connected to the injection beam line of the cyclotron. The new setup consists of a gaseous stripping target in a solenoidal field, an electrostatic deflector, a cesium charge exchange cell, a spin filter and a photon detector. In order to match the beam to the acceptance of the setup focusing and steering devices have been installed.

The LSP is a copy of the operating ANKE polarimeter.[17] Recently new 135 ° deflectors have been constructed and installed successfully. These deflectors are needed to get the beam through the existing transfer beam line without large losses, which made it impossible to reach sufficient intensities for double charge exchange in the LSP setup. H^- and D^- beams from the COSY CBS have been transported to the LSP and first spectra have been measured. The excellent resolution for deuterons is depicted in figure 5. The commissioning will be continued after finishing the scheduled COSY runs with polarized beams.

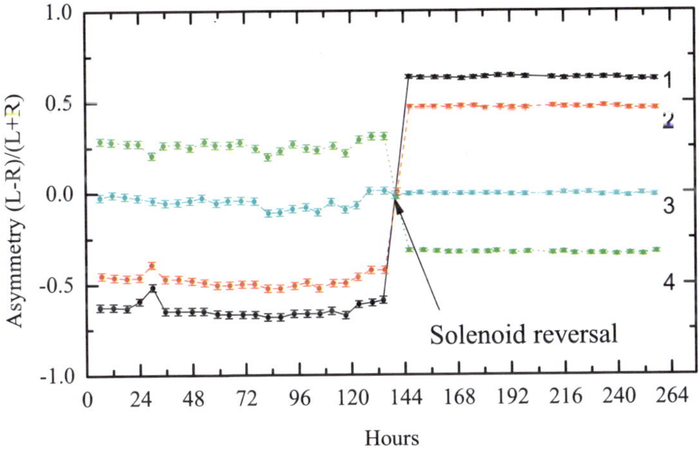

Fig. 4. Monitoring of deuteron beam asymmetries behind the cyclotron with the p-carbon polarimeter. The averages of the left-right asymmetries are plotted for 11 days. Four different states are displayed.

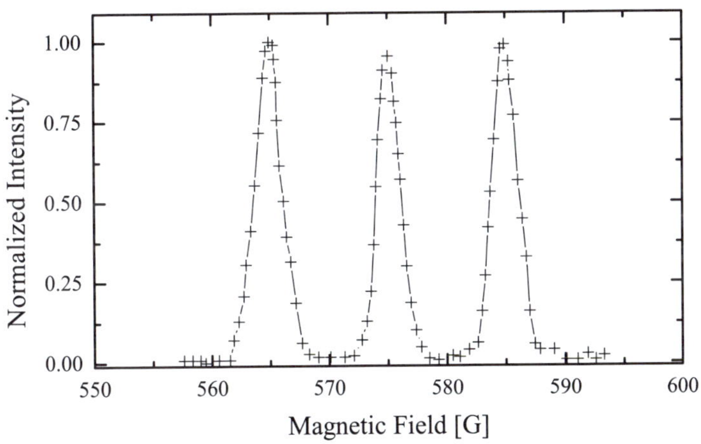

Fig. 5. The resolution of the Lamb-shift polarimeter for deuterons. The current from the multiplier is recorded as a function of the main field of the spin filter magnet.

Acknowledgments

The authors are grateful to L. Barion, A.S. Belov, R. Engels, G. d'Orsaneo and M. Westig for their help and advice, and thank the injector staff members R. Brings, A. Kieven and N. Rotert, as well as H. Hadamek and his staff of the IKP workshop, for their effective support of our endeavors in keeping the ion sources and cyclotron operating reliably and to raising their performance.

References

1. R. Maier, *Nucl. Instr. Meth. A* **390**, 1 (1997).
2. W. Bräutigam *et al.*, in *Proc. Int. Conf. on Cyclotrons and their Applications, Caen, France, 1998*, eds. E. Baron and M. Lieuvin, 654 (IOP publishing, Bristol, 1999).
3. W. Bräutigam *et al.*, in *Proc. 16th Int. Conf. on Cyclotrons and their Applications, 2001, East Lansing*, AIP Conf. Proc. **600**, 123 (AIP, New York, 2001).
4. O. Felden *et al.*, in *Proc. of the Workshop PSTP 2008*, AIP Conf. Proc. **980**, 231 (AIP, New York, 2008).
5. M. Eggert *et al.*, *Nucl. Instr. Meth. A* **453**, 514 (2000).
6. W. Haeberli, *Nucl. Instr. Meth.* **62**, 355 (1968).
7. P. D. Eversheim *et al.*, in *Proc. of PST 1995, Köln*, 224.
8. P. D. Eversheim *et al.*, in *Proc. of SPIN 1996, Amsterdam.*
9. R. Weidmann *et al.*, *Rev. Sci. Instr.*, **67**, 1357 (1996).
10. O. Felden *et al.*, in *Proc. 11th EPAC, Edinburgh, 2006*, 1705.
11. O. Felden *et al.*, *Nucl. Instr. Meth. A* **536**, 278 (2005).
12. A.S. Belov *et al.*, *Nucl. Instr. Meth. A* **255**, 442 (1987).
13. V.P. Derenchuk and A. S. Belov, in *Proc. PAC 2001, Chicago, USA*, 2093.
14. A.S. Belov *et al.*, in *IKP Annual Report 2004, Jülich* 4168 ISSN 0944-2952, 91 (2005).
15. A.S. Belov *et al.*, in *IKP Annual Report 2005; Jülich* 4212 ISSN 0944-2952, 129 (2006).
16. A.S. Belov *et al.*, in *IKP Annual Report 2006; Jülich* 4234 ISSN 0944-2952, 131 (2007).
17. R. Engels *et al.*, *Rev. Sci. Instr.*, **74**, 4607 (2003).

THE NEW SOURCE OF POLARIZED IONS FOR THE JINR ACCELERATOR COMPLEX

V. V. Fimushkin*, A. D. Kovalenko, L. V. Kutuzova, Yu. A. Plis,
Yu. V. Prokofichev and V. P. Vadeev
Joint Institute for Nuclear Research,
Dubna, Russia
**E-mail: fimushkin@jinr.ru*
www.jinr.ru

A. S. Belov
Institute for Nuclear Research of Russian Academy of Sciences, Moscow, Russia
E-mail: belov@inr.ru

The project assumes the design and construction of a universal high-intensity source of polarized deuterons (protons) using a charge-exchange plasma ionizer.[1] The output $\uparrow D^+$ ($\uparrow H^+$) current of the source is expected to be at a level of 10 mA. The polarization will be up to 90 % of the maximal vector (± 1) for $\uparrow D^+$ ($\uparrow H^+$) and tensor (+1,−2) for $\uparrow D^+$ polarization. Realization of the project is carried out in close cooperation with INR of RAS (Moscow). The equipment available from the CIPIOS ion source (IUCF, Bloomington, USA) is partially used for the Dubna device. The new source at the JINR NUCLOTRON accelerator facility will make it possible to increase the polarized deuteron beam intensity up to the level of 10^{10} d/pulse. Previous test runs on acceleration of polarized deuterons at the NUCLOTRON up to about 1 GeV/u and slow extraction of the beam to the beam transfer lines, have shown the absence of depolarization resonances. The first dangerous resonance is predicted at the beam energy of 5.6 GeV/u. The source could be transformed into a source of polarized negative ions if necessary.

Keywords: Polarized atomic beams; polarized ion sources; deuterons.

1. Project motivation

Studies of the structure of light nuclei, including the deuteron, and features of strong interactions using beams of polarized deuterons accelerated at the synchrophasotron - weak-focusing 10 GeV proton synchrotron have been carried out at the Laboratory of High Energies (LHE, JINR) since the middle of the 1980s.[2,3] The unique 4.5 GeV/c per nucleon polarized deuteron beams with the intensity of up to 5×10^9 d/pulse attracted many collaborators from different countries. A lot of new experimental data, which could not be described with the existing popular theoretical models, were obtained at that time. Experiments with polarized deuteron beams resulted in the discovery of some effects, which had clearly demonstrated the necessity to revise the standard model of the nuclei as a set of nucleons, in the kinematical region where the distances between the nucleons were less than their sizes. Since 2003 these studies have been continued at the NUCLOTRON - a strong focusing superconducting 6 A·GeV heavy ion synchrotron that was put into operation in 1993.[4] The accelerator can also provide the proton beam as well at maximum magnetic rigidity of about 50 T·m. The basic problem of the new machine in comparison with the synchrophasotron is one-turn injection. The NUCLOTRON injection time is limited to about 8.36 μs whereas it was about 200 μs for the old accelerator. That's why the construction of a new high-pulsed current polarized ion source is considered as a very important high priority task. The new flagship JINR project in the domain of high energy nuclear physics, NICA (NUCLOTRON-based Ion Collider fAcility), aimed at the study of phase transitions in strongly interacting nuclear matter at the highest possible baryon density, was started in 2006.[5] Such conditions are obtained in heavy ion collisions in the energy range of $\sqrt{s_{NN}} \sim$ 4–11 GeV. The NICA program consists of several subprojects. The first one is the project NUCLOTRON-M, where the new polarized ion source is included. The realization of the project was started in 2008 and is supposed to be completed in 2011. Physics with polarized light ion beams is considered to be an important part of the NICA collider program. It is supposed to realize collisions of different polarized particles (p,d) with different orientations of their spin and in both head-on and merging collisions. The expected luminosity is planned at the level of (10^{30}–10^{31}) $cm^{-2}{\cdot}s^{-1}$. Some proposals for the NICA research program are collected in the "NICA White Book".[6]

The source of polarized deuterons used up to now (0.4 mA D^+ cryogenic source POLARIS[2,3]) cannot provide some of the key parameters of the beams necessary for the NUCLOTRON/NICA facility. That is why the JINR scientific leaders were impressed by the performances of the Cooler Injector Polarized IOn Source (CIPIOS) developed at the Indiana University Cyclotron Facility (IUCF, USA) in cooperation with the Institute for Nuclear Research, Russian Academy of Sciences (INR, Moscow) in 1999 and expressed the interest to use this source at the JINR Accelerator Complex.

The upgraded CIPIOS will provide high-quality polarized deuteron and proton beams at the NUCLOTRON-M. Installation of the new polarized source at the accelerator complex will soon allow us to continue the studies with polarized beams at a much higher level of investigations. The goal of the first stage is to accelerate the 10 mA polarized D^+ beam to 5 GeV/u, i.e. to increase the beam intensity up to 10^{10} d/pulse.

2. Source of polarized ions

The project realization includes the following stages:

- development of the universal high-intensity source of polarized deuterons (protons) using a charge-exchange plasma ionizer with the output current up to 10 mA of $\uparrow D^+(\uparrow H^+)$,
- complete tests of the source,
- modification of the linac pre-accelerator high voltage platform and power station,
- adaptation of the existing remote control system (console of LU-20) of the polarized ion source to operate under the high voltage,
- mounting of the source and equipment at the pre-accelerator platform, linac LU-20 runs with polarized beams and polarization measurements at the linac output.

The Source of Polarized Ions (SPI) consists of an atomic beam section that uses sextupole magnets for focusing, and radio-frequency transition units to polarize the atoms before they are focused into the ionizer. SPI uses a set of permanent sextupoles ($B = 1.4$ T), a conventional electromagnet sextupole ($B = 0.9$ T) and radio-frequency units for nuclear polarization of the atomic beam. This allows one to get nuclear ± 1 vector for $\uparrow D^+(\uparrow H^+)$ and $+1, -2$ for $\uparrow D^+$ tensor polarization. The cryocooler is used for cooling the atomic beam. The resonant charge-exchange ionizer[7,8] produces pulses of positive ion plasma inside the solenoid. The atomic beam pulse is focused through the extraction system into the solenoid where atoms are ionized by a highly efficient charge-exchange reaction.

Nearly resonant charge-exchange reactions to produce polarized protons and deuterons are as follows:

$$H^0 \uparrow + D^+ \Rightarrow H^+ \uparrow + D^0$$

$$D^0 \uparrow + H^+ \Rightarrow D^+ \uparrow + H^0$$

$$\sigma \sim 5 \times 10^{-15}\ \text{cm}^2.$$

Spin orientation of $\uparrow D^+(\uparrow H^+)$ ions at the output of SPI is vertical. The ion beam is formed with a 25 kV extraction potential in a 100 μs wide pulse at a rate of 0.14 Hz. At the moment the specificity of the NUCLOTRON is that one-turn injection is used for it and this machine allows one to accelerate only positive ions.

The important fact is as follows: depolarization resonances are absent in the total energy range of the NUCLOTRON but only for the deuteron beam. Therefore as the first step of the offered project it is expedient to use a source of positive polarized deuterium ions. It is known that INR of RAS (Moscow) has developed a source of polarized protons with a charge-exchange plasma ionizer[9] and the polarized atom storage in the ionization volume.[10,11] The intensity and polarization of the beam from the INR source are as high as 6 mA, 85 %[9] without polarized atom storage and 11 mA, 80 %[11] with the storage. The ionizer with the storage of polarized atoms allows one not only to increase intensity of the polarized ion beam but also to reduce emittance of the polarized beam and reduce considerably H_2^+ current, which is difficult to be separated from polarized deuterons due to the similar mass of the ions. The number of the polarized ions of the source at the intensity of the 10 mA beam and the 8 μs pulse duration is 5×10^{11}, which meets the requirements of the given project. The normalized emittance of the source beam is 1.2 π mm mrad,[12] which is much smaller than the acceptance of the LINAC.

Taking into account the above, we assume to convert the charge-exchange ionizer of CIPIOS into the ionizer using a storage of polarized deuterium atoms and production of polarized deuterons by resonance recharging in the hydrogen plasma. Within the frame of the project we suppose to develop and fabricate the missing parts for the future source, which are mainly elements of the Atomic Beam Source (ABS), see figure 1:

- a vacuum chamber of the ABS, dissociator,
- a channel of the atomic deuterium beam cooling,
- a pulse valve of molecular deuterium injection into the dissociator bulb,
- a high-frequency pulse generator with the pulse power up to 5 kW operating in the 50 MHz self-excitation circuit,
- a modulator of the high-frequency generator with a maximum voltage up to 4.5 kV a pulse current up to 2 A,
- a power supply of the pulse gas valve.

To optimize the atomic beam intensity, it is also necessary to do the following: to measure the atomic beam density in the pulse mode using the

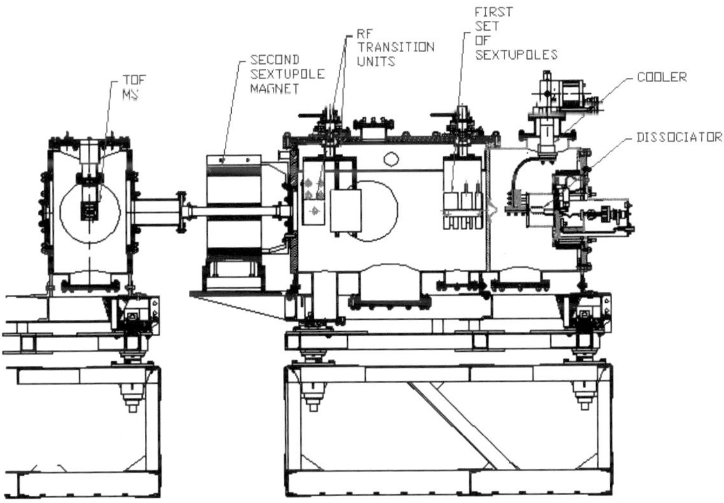

Fig. 1. Atomic beam source general view. A pulsed dissociator (INR-type), nozzle cooling to 70 K, set of permanent magnet sextupoles and one electromagnet sextupole (CIPIOS), WF and SF RF transitions units. Expected intensity of polarized deuteron beam is 1.5×10^{17} $\sec^{-1}$ (3 ms pulse), polarized hydrogen beam - 2×10^{17} $\sec^{-1}$.

time-of-flight mass-spectrometer; to know the atomic beam velocity distribution; to compute the optimum location of the skimmer and nozzle on the basis of the measurements.

The RF-transition units will be checked and tuned with a sextupole electromagnet as an analyzing device. The purpose is to get the atomic D beam with the pulse density of 2.5×10^{10} atoms/cm^3 at the distance of 150 cm from the cooling channel outlet and the most probable velocity of 1.5×10^5 cm/s.

The design and manufacture of ABS parts, optimization of the intensity of the atomic beams, and functional test of the cells of the nuclear polarization of deuterium (hydrogen) atoms will be performed on the agreement with INR. The beginning of tests of ABS is planned at the end of 2010.

The work to be carried out at LHEP JINR, includes the following:

- assembly and tests of the charge-exchange plasma ionizer, including the storage cell in the region of ionization and transportation of hydrogen plasma with the flow of unpolarized protons up to 100 mA through the storage cell,
- optimization of the ion-optical system up to 25 keV and transportation of the high-current proton beam,

• long-term tests with the storage cell in the ionizer.

It is necessary to develop electronic control system components for primary analysis and data acquisition, and fiber optic connection with the computer.

3. Status of the project realization

Intensive work on preparation of the source for tests is carried out at INR. Special attention during the tests will be focused on the problems of the atomic beam formation process and study of the RF-transitions efficiency under nuclear polarization of the atomic deuterium beam. Operation of the ionizer with a storage cell at room temperature is planned at JINR for the fall of 2010. Technical specification for necessary reconstruction of the for-injector hall, will have also been prepared by this time.

References

1. V. V. Fimushkin *et al.*, *Eur. Phys. J., Special Topics* **162**, 275 (2008).
2. N. G. Anischenko *et al.*, *Journ. de Phys., Colloquia C2*, **46**, C2-703 (1985).
3. V. P. Ershov *et al.*, in *Proc. Int. Workshop on Polarized Beams and Polarized Gas Target Cologne, 1995*, 193 (World Scientific, Singapore 1996).
4. A. D. Kovalenko, Status of the Nuclotron, in *Proc. EPAC'94*, 194 (World Scientific, Singapore 1995).
5. A. Sissakian *et al.*, The Project NICA/MPD at JINR: Search for the Mixed Phase of Strongly Interacting Matter at Nuclotron-based Ion Collider Facility, in *Int. Conf. on Lepton-Photon Interactions, LP' 07, August 2007, Daegu, Korea.*
6. The NICA Wbook `http://theor.jinr.ru/twiki-gi/view/NICA/WebHome` .
7. V. P. Derenchuk and A. S. Belov, in *Proc. 2001 Particle Accelerator Conference Chicago*, 2093.
8. A. S. Belov *et al.*, *Nucl. Instr. Meth. A* **333**, 256 (1993).
9. A. S. Belov *et al.*, *Rev. Sci. Instr.* **77**, 03A522 (2006).
10. A. S. Belov *et al.*, *Nucl. Instr. Meth. A* **255**, 442 (1987).
11. A. S. Belov *et al.*, in *7th Int. Workshop on Polarized Gas Targets and Polarized Beams, Urbana, 1997*, eds. Roy J. Holt and M. A. Miller, AIP Conf. Proc. **421**, 362 (AIP, New York, 1998).
12. A. S. Belov *et al.*, in *13th Int. Symposium on High Energy Spin Physics* 1998, eds. N.E. Tyurin *et al.*, 622 (World Scientific, Singapore, 1999).

RESONANCE EFFECTS IN NUCLEAR DICHROISM — AN INEXPENSIVE SOURCE OF TENSOR-POLARIZED DEUTERONS

H. Seyfarth*, R. Engels, F. Rathmann and H. Ströher

Institut für Kernphysik, Jülich Center for Hadron Physics, Forschungszentrum Jülich, Leo–Brandt–Str. 1, D-52425 Jülich, Germany
**E-mail: H.Seyfarth@fz-juelich.de*

V. Baryshevsky and A. Rouba

Research Institute for Nuclear Problems, Bobruiskaya Str. 11, 220050 Minsk, Belarus

C. Düweke, R. Emmerich and A. Imig

Institut für Kernphysik, Universität zu Köln, Zülpicher Str. 77, D-50937 Köln, Germany

K. Grigoryev and M. Mikirtychiants

Institut für Kernphysik, Jülich Center for Hadron Physics, Forschungszentrum Jülich, Leo–Brandt–Str. 1, D-52425 Jülich, Germany and Petersburg Nuclear Physics Institute, 188300 Gatchina, Russia

A. Vasilyev

Petersburg Nuclear Physics Institute, 188300 Gatchina, Russia

The effect of nuclear spin dichroism, predicted by theoretical studies as the appearance of tensor polarization in initially unpolarized beams behind unpolarized or spinless targets, has been studied by measurements with the use of 9.5 to 18.7 MeV unpolarized deuteron beams from the Köln tandem accelerator and graphite targets of areal densities from 36 to 188 mg/cm^2. Distinct deviations from the predicted weak effect were observed with a maximum vale of $p_{zz} = -(0.28 \pm 0.03)$, measured behind a 129 mg/cm^2 target at 14.8 MeV initial beam energy. The present results will allow one to produce tensor-polarized deuteron beams with p_{zz} about -0.30 or +0.25 from initially unpolarized beams by graphite targets of appropriate thickness.

Keywords: Deuteron beam; Tensor polarization.

1. Introduction

Nuclear spin dichroism[1] on the analogy to optical dichroism results in the appearance of tensor polarization in an initially unpolarized beam of particles with nuclear spin $S \geq 1$ behind unpolarized or $S = 0$ targets. The transmitted beam is confined to a narrow forward cone around the direction of the primary beam direction, which defines the quantization axis. The beam behind the target is described by the component p_{zz} of the tensor polarization ($p_{xx} + p_{yy} = -p_{zz}$). In a beam of unpolarized deuterons ($S = 1$), the m=+1, m=0, and m=-1 substates are equally populated. The interaction with the $S = 0$ target nuclei is described by total elastic and inelastic cross sections $\sigma_{\pm 1}(E)$ for the deuterons in the m=+1 and m=-1 substates and $\sigma_0(E)$ for those in the m=0 substate. Any difference between them results in a non-zero p_{zz}. For a thin target of atomic density ρ and thickness d_t under the condition $\rho d_t \cdot [\sigma_{\pm 1} - \sigma_0] \ll 1$

$$p_{zz}(\rho d_t) = -\frac{2\rho d_t}{3} \cdot [\sigma_{\pm 1}(E) - \sigma_0(E)]. \tag{1}$$

Calculations for 5 to 20 MeV deuterons and carbon targets yields positive values of $\sigma_{\pm 1} - \sigma_0$ for energies below ~13 MeV, reaching 5 b around 8 MeV, and negative values above, reaching -5 b at 20 MeV. The change of sign occurs due to interference terms in the nuclear proton(neutron)-carbon and the Coulomb proton-carbon interaction. For a 100 mg/cm^2 carbon target ($\rho d_t = 5 \cdot 10^{21}$ atoms/cm^2) the average values correspond to $p_{zz} \approx -0.01$ below and $\approx +0.01$ above 13 MeV.[2] Cross-section differences in the order of 10^{-2} b, as predicted for relativistic energies,[3] were confirmed by a recent measurement with 5.5 GeV/c deuterons interacting with carbon targets.[4]

2. Experimental setup

The measurements were performed with unpolarized deuteron beams from the HVEC FN Van-de-Graaff tandem accelerator of the Institut für Kernphysik of Universität zu Köln. The polarizing effect by seven carbon targets was studied with the use of a polarimeter based on the reaction $\vec{d}+{}^3\mathrm{He}\rightarrow \mathrm{p}+{}^4\mathrm{He}$.[5] For each target a set of primary beam energies E_{in} with steps of 0.1 MeV was utilized chosen such that the deuteron energies in the ^{3}He gas cell of the polarimeter, E_{cell}, were between 5 and 8 MeV (see table 1), where the tensor analyzing powers of the polarimeter reaction are large.[5]

The 3x3 cm^2 target foils were cut from graphite sheets produced by rolling out expanded graphite.[6] The nominal 99% carbon purity was con-

Table 1. Target material M and thickness d_t, target label, run (I: 2003, II: 2006), ranges of the primary deuteron-beam energies (E_{in}), of the mean deuteron energies behind the targets (E_{out}), and of the mean deuteron energies in the ^{3}He gas cell of the polarimeter (E_{cell}).

M	d_t [mg/cm^2]	Label	Run	E_{in} [MeV]	E_{out} [MeV]	E_{cell} [MeV]
Au	5.0 ± 0.3	Au5	II	6.20 ... 7.90	6.02 ... 7.74	5.56 ... 7.36
C	35.90 ± 0.19	C36	II	9.50 ... 10.50	6.49 ... 7.79	6.06 ... 7.41
C	57.69 ± 0.32	C58	I	10.80 ... 12.20	6.18 ... 8.20	5.73 ... 7.83
C	93.59 ± 0.37	C94	II	13.00 ... 14.00	6.17 ... 7.84	5.72 ... 7.46
C	129.49 ± 0.42	C129	II	14.80 ... 15.90	5.88 ... 7.93	5.41 ... 7.55
C	152.63 ± 0.75	C153	I	16.20 ... 16.70	6.38 ... 7.37	5.94 ... 6.97
C	165.39 ± 0.46	C165	II	16.70 ... 17.50	6.16 ... 7.77	5.70 ... 7.39
C	187.93 ± 0.74	C188	I	17.50 ... 18.70	5.60 ... 8.16	5.11 ... 7.78

firmed by own analyses.[7] Transmission X-ray diagrams[7] show that the graphite foils contain comparable parts with layers oriented parallel to the surface, with disoriented layers, and of amorphous material. Except C36 and C58 (see tab. 1), the targets consisted of stacks of at least two foils of available thickness (0.2, 0.3, and 0.5 mm). The measurement with the thin gold foil was performed to obtain reference data with a target producing no polarization, but simulating the multiple Coulomb scattering by the carbon targets.

The beam diameter at the targets was 1.5 mm. By three diaphragms D_1, D_2, and D_3 of 2.0, 2.5, and 3.0 mm diameter and positioned 132, 187, and 251 mm, respectively, behind the targets, the emission of deuterons into the polarimeter cell was confined to polar angles $\leq 0.5^\circ$ (see fig. 1). The distance from D_3 to the entrance window of the polarimeter cell, a 6.5 μm Havar foil, was 48 mm. The 100 μm tantalum exit window and a 300 μm tantalum foil on a sliding ladder, added during tuning of the primary beam, were sufficiently thick to stop the deuterons behind the cell. The ^{3}He cell and the diaphragms were used to monitor the beam currents. The primary beam for all targets was tuned to deliver ~7 nA to the polarimeter cell.

The polarimeter is equipped with four side detectors to measure protons emitted under polar angles of $\theta = (24.5 \pm 2.9)^\circ$ and azimuthal angles $\varphi = 0^\circ, 90^\circ, 180^\circ$, and 270° (labeled L (left), U (up), R (right), and D (down), respectively, as seen in beam direction). A forward detector (labeled F) measures the protons emitted under $\theta \leq 0^\circ$. The detector acceptances are defined by pairs of tantalum apertures.

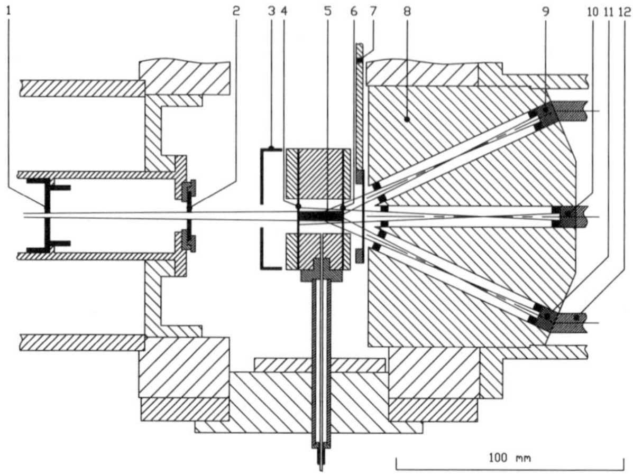

Fig. 1. Top view along the horizontal mid plane of the polarimeter with the beam coming from the left (1,2: diaphragms D_2 and D_3, 3: electron-backbending electrode, 4: Havar entrance window, 5: polarimeter-gas cell, filled by ^{3}He of 3 bar, with the gas-feeding tube from below, 6: tantalum exit window, 7: tantalum foil on a sliding ladder, 8: aluminum shielding block, 9-11: left (9), forward (10), and right (11) NaJ(Tl) detector, 12: one of the light guides to the Philips XP1911 secondary-electron multipliers).

3. Measurements and results

All measured proton spectra in a consistent way were fitted by three Gaussians in the peak region, to account for the asymmetric peak shape, together with an exponential and a modified error function to fit the background. For each of the detectors the count number in the peak is

$$N_{\rm i}(E_{\rm cell}) = \int j_{\rm cell}(t)dt \cdot \rho_{\rm He} \cdot l_{\rm i} \cdot \Omega_{\rm i} \cdot \epsilon_{\rm i} \cdot \sigma_0(E_{\rm cell}, \theta_{\rm i}) \times [1 + \frac{1}{2} \cdot p_{zz}(E_{\rm cell}) \cdot A_{zz}(E_{\rm cell}, \theta_{\rm i})]. \quad (2)$$

The two first terms give the total current to the polarimeter cell during a run and the cell-gas density. The $l_{\rm i}$ are the widths of the reaction volume along the axes of the diaphragms. $\Omega_{\rm i}$ and $\epsilon_{\rm i}$ denote the detector acceptances and detection efficiencies. $\sigma_0(E_{\rm cell}, \theta_{\rm i})$ and $A_{\rm zz}(E_{\rm cell}, \theta_{\rm i})$ are the unpolarized differential cross sections[10] of the polarimeter reaction with proton emission under 0 and 24.5 ° and the tensor analyzing power[5,8,9] at 0 ° and that[5,9,10] at 24.5 °. The deuteron energy in the polarimeter reaction, $E_{\rm cell}$, resulting from the primary $E_{\rm in}$ by the energy losses in the targets, the Havar window, and 13 mm within the ^{3}He of 3 bar, was calculated with the use of the Bethe-Bloch energy-loss formula.

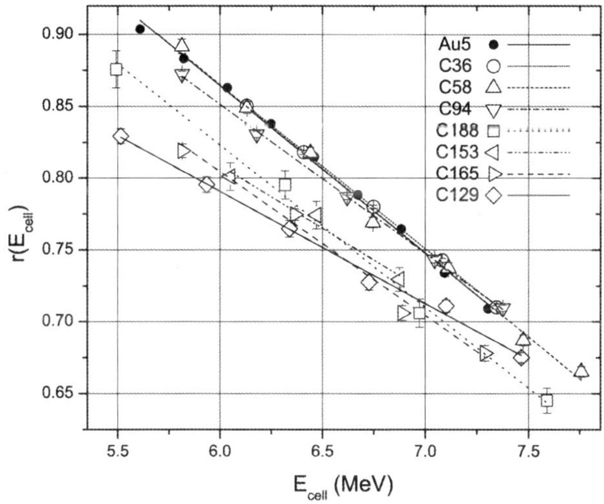

Fig. 2. Count-number ratios $r(E_{\text{cell}})$ as defined in Eqn. 3 as function of the mean deuteron energy in the polarimeter cell, E_{cell}, and the linear fit lines. Data points were combined to ease readability.

To compensate the dependence of the detector count rates on the focusing of the beam, observed as fluctuations of ~5%, the proton-peak ratios

$$r(E_{\text{cell}}) = \frac{N_{\text{L}}(E_{\text{cell}}) + N_{\text{U}}(E_{\text{cell}}) + N_R(E_{\text{cell}}) + N_D(E_{\text{cell}})}{N_F(E_{\text{cell}})} \tag{3}$$

were used in the data analysis. The ratios, measured with the gold target and the seven carbon targets are collected in figure 2. Linear fit functions, as expected for the gold target according to the unpolarized cross sections, were applied to fit the carbon data sets, too. The parameters of the fit functions are found in the table 2. Because all the measured ratios for a target have comparable errors, the uncertainties of the fits in a data point are given by the sample standard deviation s.[11] Especially the deviation from the gold data, observed with the C129 target, indicates a significant p_{zz} in the beam behind the target.

With the use of equation 2 and the fit functions, seven sets of p_{zz} are derived from the double ratios

$$\frac{r_{\text{fit}}^{Cx}(E_{\text{cell}})}{r_{\text{fit}}^{\text{Au5}}(E_{\text{cell}})} = \frac{1 + \frac{1}{2}p_{zz}^{Cx}(E_{\text{cell}})A_{zz}(E_{\text{cell}}, 24.5^\circ)}{1 + \frac{1}{2}p_{zz}^{Cx}(E_{\text{cell}})A_{zz}(E_{\text{cell}}, 0^\circ)}, \tag{4}$$

yielding

$$p_{zz}^{Cx}(E_{\text{cell}}) = \frac{2 \cdot [r_{\text{fit}}^{\text{Au5}}(E_{\text{cell}}) - r_{\text{fit}}^{Cx}(E_{\text{cell}})]}{r_{\text{fit}}^{Cx}(E_{\text{cell}})A_{zz}(E_{\text{cell}}, 0^\circ) - r_{\text{fit}}^{\text{Au5}}(E_{\text{cell}})A_{zz}(E_{\text{cell}}, 24.5^\circ)}. \tag{5}$$

Table 2. Parameters of the linear fit functions $r_{\rm fit}(E_{\rm cell}) = r(0) + k \cdot E_{\rm cell}$[MeV] to the measured proton-peak ratios (Fig. 2) and the sample standard deviations s.

target	r(0)±δr(0)	k±δk [MeV^{-1}]	s
Au5	1.561±0.010	-0.1162±0.0015	0.0049
C36	1.553±0.011	-0.1146±0.0016	0.0029
C58	1.570±0.017	-0.1173±0.0024	0.0088
C94	1.467±0.010	-0.1027±0.0016	0.0042
C129	1.262±0.011	-0.0784±0.0017	0.0057
C153	1.347±0.049	-0.090±0.008	0.0097
C165	1.408±0.023	-0.1005±0.0036	0.0082
C188	1.508±0.024	-0.1141±0.0035	0.014

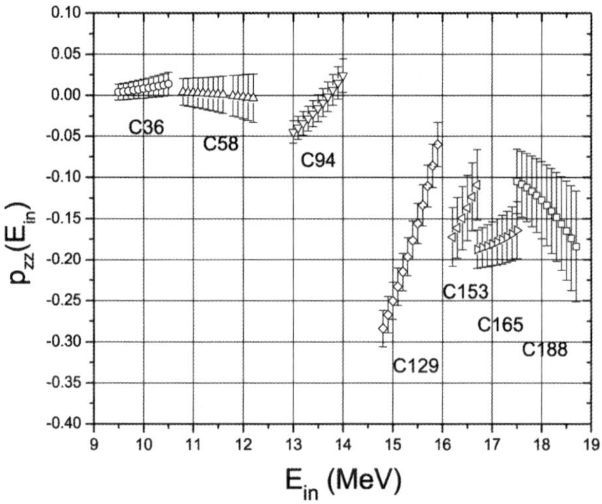

Fig. 3. The sets of tensor polarizations $p_{zz} \pm \Delta p_{zz}$ produced by the seven carbon targets as function of the primary deuteron-beam energies.

Here $r_{\rm fit}^{\rm Au5}(E_{\rm cell})$ serves as reference function with p_{zz}=0. For all targets, the sets of $E_{\rm in}$ were chosen such that those of $E_{\rm out}$ and $E_{\rm cell}$ were similar (conf. tab. 1. The differences in p_{zz}, measured with one of the targets and the next-heavier at similar $E_{\rm cell}$ (fig. 2), have to be attributed to the additional energy range during deceleration in the heavier target. Therefore, the figure 3 shows the seven sets of p_{zz} as function of $E_{\rm in}$. The error bars reflect the uncertainties of the target thicknesses given in table 1. Their upper and lower limits result in changes of the calculated values of E_{cell} and, via the energy dependence of the terms in equation 5, yield modified values of p_{zz}.

The present values of p_{zz} around E_{in}=15 MeV are appreciably larger than ≈ 0.01, expected from calculations.[2] The most distinct difference in p_{zz} is found between deceleration from E_{in}=14.8 MeV in the C129 target and from 14.0 MeV in the C94 target. It can be explained by a cross-section difference $\sigma_{\pm1}(E) - \sigma_0(E) > 0$ for deuterons decelerated in the target from 14.8 to 14.0 MeV. A simple Gaussian distribution was taken to describe the energy dependence of $\sigma_{\pm1}(E) - \sigma_0(E)$. The decrease of p_{zz} above E_{in}=14.8 MeV can be described by such a distribution with $\sigma_{\pm1}(E) - \sigma_0(E) < 0$ for deceleration from 15.8 to 14.8 MeV. These two cross section distributions, centered at 14.4 and 15.4 MeV, and additional ones, centered at 9.8 and 13.8 MeV ($\sigma_{\pm1}(E) - \sigma_0(E) < 0$) and 10.8 and 12.5 MeV ($\sigma_{\pm1}(E) - \sigma_0(E) > 0$), allow one to describe the $p_{zz}(E_{\text{in}})$ of figure 3 for $E_{in} \leq 15.8$ MeV. The excitation energies in the compound system d+^{12}C,

$$E^*(^{14}\text{N}) = \frac{m_{^{12}\text{C}}}{m_{\text{d}} + m_{^{12}\text{C}}} \cdot E_{\text{lab}} + m_{\text{d}} + m_{^{12}\text{C}} - m_{^{14}\text{N}}, \tag{6}$$

which correspond to these six deuteron laboratory energies, are in surprising agreement with the energies of known states above 19 MeV in ^{14}N.[12] These states lie in the range of the ^{14}N giant resonance, spread around the peak at 23 MeV.[13]

4. An application of the present results

The present data indicate that it is possible to produce tensor-polarized deuterons from an initially unpolarized beam by graphite targets of appropriate thickness. The cross-section differences, described by Gaussian distributions, enable one to predict the expected p_{zz} for primary beam energies of 14.8 and 15.8 MeV, as shown in figure 4. An unpolarized primary deuteron beam of E_{in}=14.8 MeV is predicted to leave a carbon target with p_{zz} around -0.30 and E_{out} depending on the target thickness. For E_{in}=15.8 MeV, a thin carbon target of 20 mg/cm^2 would yield p_{zz}=+0.21 at E_{out}=14.8 MeV. In a thicker target, the positive polarization is compensated by the negative polarization produced during deceleration below 14.8 MeV. If the deceleration from 14.8 to 14.0 MeV is caused by an intermediate, sandwiched material without polarizing effect, also beams of lower energy and even slightly increased positive p_{zz} can be produced.

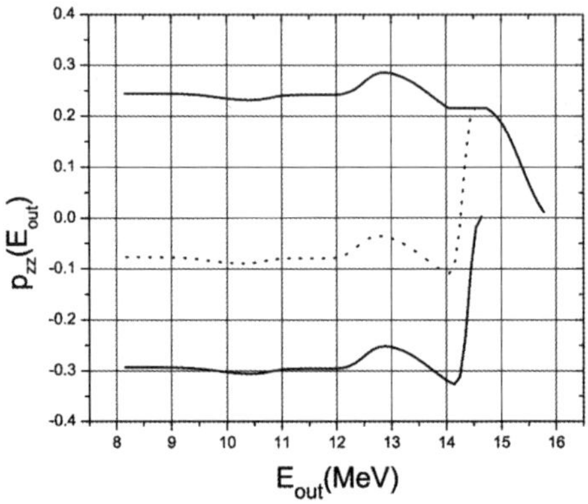

Fig. 4. Tensor polarization p_{zz} produced by graphite targets from unpolarized deuteron beams as function of the energy behind the target, $E_{\rm out}$, depending on the target thickness. The lower full line shows the resulting p_{zz} for $E_{\rm in}$=14.8 MeV. From the upper two lines ($E_{\rm in}$=15.8 MeV), the full line gives p_{zz} for a target with a different sandwiched material suppressing the negative polarization by carbon between 14.8 and 14.0 MeV, whereas the dotted line below $E_{\rm out}$=14.8 MeV is for a pure graphite target.

References

1. V. G. Baryshevsky, Phys. Lett. **A 171**, 431 (1992); J. Phys. G **19**, 273 (1993).
2. V. G. Baryshevsky and A.A. Rouba, Proc. 11th Int. Conf. on Meson-Nucleon Physics and the Structure of the Nucleon, SLAC eConf C070910, 346, `http://arxiv.org/abs/0706.3808` (2008).
3. Göran Fäldt, J. Phys. G **6**, 1513 (1980).
4. L. S. Azhgirey *et al.*, Particles and Nuclei, Lett. (JINR Dubna,Russia) **5**, 728 (2008).
5. R. Engels, Diploma Thesis, Universität zu Köln (Cologne, Germany, 1997), `http://www.ikp.uni-koeln.de/groups/ex/schieck/arbeiten/engels.diplom.pdf`.
6. made by immersing natural flake graphite in a bath of chromic acid, then concentrated sulfuric acid, which forces the crystal lattice planes apart, thus expanding the graphite.
7. performed by Zentralabteilung für Chemische Analysen of Forschungszentrum Jülich GmbH.
8. P. A. Schmelzbach *et al.*, Nucl. Phys. **A264**, 45 (1976).
9. S. A. Tonsfeldt, PhD Thesis, University of North Carolina, 1983, Ann Arbor, MI, USA.
10. M.Bittcher *et al.*, Few-Body Systems **9**, 165 (1990).
11. Philip R. Bevington, Data Reduction and Error Analysis for the Physical Sciences, McGraw Hill (1969).
12. F. Ajzenberg-Selove, Nucl. Phys. A **523**, 1 (1991).
13. B. L. Berman and S. C. Fultz, Rev. Mod. Phys. **47**, 713 (1975).

POLARIZED ELECTRONS AND POSITRONS AT THE MESA ACCELERATOR

K. Aulenbacher* and A. Jankowiak

Institut für Kernphysik der Johannes Gutenberg-Universitt Mainz, D-55118 Mainz, Germany
**E-mail: aulenbac@kph.uni-mainz.de*
www.kph.uni-mainz.de

We suggest starting an accelerator physics project called the Mainz Energy-recovering Superconducting Accelerator (MESA) as an extension to the experimental facilities at the institute for nuclear physics. MESA may allow us to introduce an innovative internal target regime based on the ERL principle. A second mode of operation will be to use an external polarized electron beam for parity violating experiments. Furthermore, MESA could also allow us to establish a CW source of polarized positrons.

Keywords: Polarization in interaction and scattering; electron sources; polarized beams.

1. Introduction

For many years, electron scattering experiments at the institute for nuclear physics in Mainz have been carried out by the CW electron accelerator MAMI (MAinzer MIcrotron). By 2007 a fourth accelerator stage, the so-called harmonic double sided microtron (HDSM), was added which increased the energy from 855 to 1508 MeV.[1] The four staged cascade is called MAMI-C, its floor plan is shown in figure 1.

MAMI-C produces up to 100 μA of external beam current at a maximum energy of 1.508 GeV. It delivers beams for about 7000 hours a year with about 90 % availability for the experiments which are located at sites A1, A2 and A4. More than 50 % of the run time is with polarized electron beams. The excellent and reliable performance of MAMI-C allows us to direct a considerable amount of work force towards new accelerator developments in order to increase the physics achivements of our experimental program. Typically such developments aim at increasing the beam energy but inspection of figure 1 demonstrates that space restrictions are prohibitive.

Another possibility is to improve the precision of experiments in already accessible regions of scattering kinematics. An innovative experimental regime must then be found which allows us to overcome the limitations of the machines already available. Therefore two innovations, namely the ERL (Energy Recovery LINAC) principle and CW operation of superconducting structures at gradients of up to 20 MV/m will enter into our project which is called MESA (Mainz Energy recovering Superconducting Accelerator). The operating principle of MESA will be presented in the following section.

We have identified important hadron and particle physics experiments which may be realized at the MESA beam energy of about 100 MeV. Here, MESA could result in decisive advantages if compared to conventional setups. A significant part of the program is devoted to parity violating (PV) electron scattering. Extreme requirements - even in comparison with the recently realized experiments at JLAB and MAMI - are characteristic for these "next generation" PV-experiments. Finally, we will discuss a potential application of MESA as a CW source of polarized positrons.

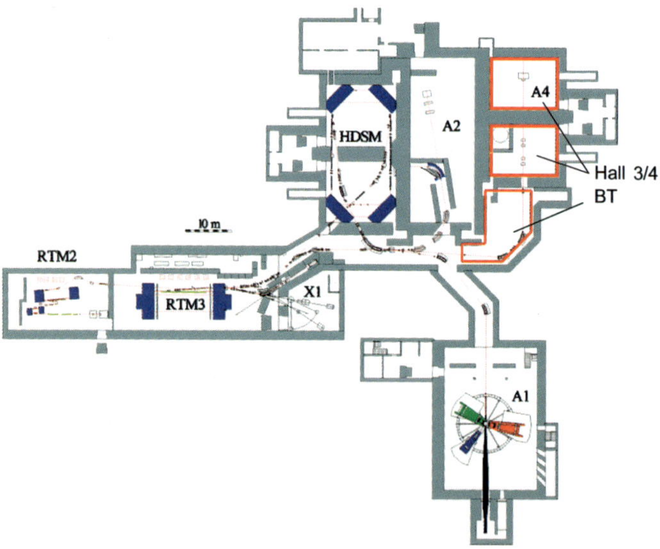

Fig. 1. MAMI-C floor plan with experimental halls. The Beamline Tunnel (BT) and halls 3 and 4 will be available for the installation of MESA and its experiments.

2. MESA layout

2.1. *General considerations, modes of operation*

MESA may be installed in halls 3/4 (see fig. 1), since the experimental program going on there will probably be completed in 2010. A schematic layout of the MESA machine is presented in figure 2. MESA's footprint does not exceed the area of one hall. Since the low beam energy of MESA does not require the large scale installations which are characteristic for experiments at GeV energies the remaining floorspace could provide for two independent experimental areas.

The MESA injector will consist of a 100 keV photosource followed by a chopper and a harmonic buncher. All components of the injector are normal conducting and will essentially be copies of MAMI systems. The injector will receive an additional acceleration structure in order to achieve an output energy of 5 MeV. The photosource will be driven by RF synchronized lasers providing a repetition rate corresponding to the first subharmonic (2.449/2 GHz) of the LINAC RF in order to match the operating frequency of the superconducting main linac. The bunch charge we aim for is 8 pC, yielding 10 mA average current at 5 MeV. Three klystrons (identical to those which operate at MAMI-C) will power the LINAC, allowing to distribute enough RF power to overcome the ohmic losses and for the 50 kW of beam power.[a]

Two resonant structures of ≈ 1.8 m length will be used as the MESA main accelerator. If installed in a suitable cryostat at 2 K, the two cavities are sufficient for an energy gain of 33 MeV per pass at a gradient of 20 MV/m.[2] Two recirculations will then allow for a beam energy of 104 MeV. MESA could be operated in two alternative modes:

2.2. *External Beam (EB) mode*

A third recirculation allows for another pass through the accelerating structure, the resulting 137 MeV beam may then be directed to an external experiment. The beam current will be limited in this mode by the RF power that can be transmitted through the superconducting structure. A power of 25 kW will allow for a beam current of $150\,\mu A$ with some headroom

[a]It may be noted that the attempted performance can be surpassed by using a superconducting injector, as is the case in the JLAB ERL. This, however, would add a considerable amount of investment cost and complexity to MESA. Since one of our primary goals is to start the experiments as fast as possible, we will consider this option only if considerable workforce and additional investment would become available.

for the stabilization systems. The experiments will make use of polarized beam with a degree of polarization of $P = 0.85$ as it is presently routinely achievable for MAMI-C experiments.

2.3. *Energy Recovery LINAC (ERL) mode*

In ERL mode the path length of the third recirculation will be changed by a half integer of the RF wavelength by avoiding a magnetic chicane. This leads to 180° phase shift for the electrons which are recirculated to the accelerating structure, where the beam is decelerated. Since all following orbits are again of integer length, the decelerating process will continue until the beam leaves the structure at injection energy (but at the opposite end of the superconducting structure) and is dumped at 5 MeV. In this mode a windowless target (similar to a storage ring target) can be introduced in the 104 MeV orbit.

The decisive difference compared to storage ring experiments is that each beam particle passes the target only once. Therefore we call it a "Pseudo" Internal Target (PIT). It allows to achieve stationary beam conditions even for the strong scattering that is present at energies of 100 MeV or even lower. The 10 mA average beam current that corresponds to a beam power of 1 MW at the target allows sufficient luminosity even at the low target densities achievable with a windowless gas target. Note that the ERL principle in this setup gives an advantage of a factor ≈ 20 in energy efficiency if compared to a 100 MeV external beam.

3. The MESA source

An already existing copy of the MAMI photosource will be used, which offers an improved vacuum system for even better performance than presently available. It can be set into operation almost without additional cost or development work and will serve for both operation modes of MESA. We now discuss the expected performance considering beam dynamics and cathode lifetime for the different modes.

For EB operation there is no significant difference to MAMI conditions concerning beam dynamics, the excellent emittance conditions of MAMI can be conserved. Even with the presently achieved charge lifetime of 100 Coulomb this would allow for a source availability of > 98 % in continuous experimentation. As in MAMI operation, GaAs/GaAsP superlattice photocathodes[3] will provide highly polarized beams ($P \approx 0.85$).

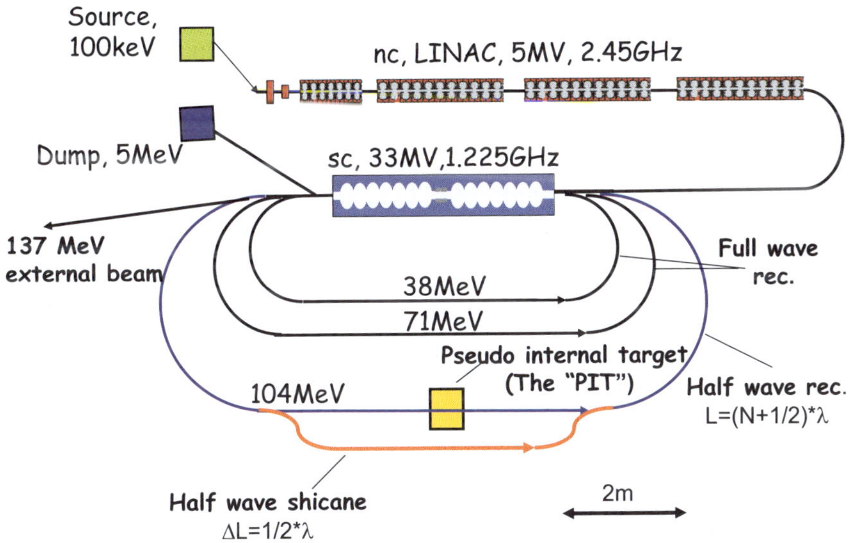

Fig. 2. MESA accelerator layout.

In ERL mode the source will have to deliver almost two orders of magnitude larger average beam current (10 mA). The present source was already tested at average currents of this magnitude.[4] Space charge requires an increase of the spot size on the photocathode by about a factor of ten which leads to a corresponding emittance increase. Experimental experience and computer simulations indicate that normalized transverse emittance is about $10\,\mu$m, leading to $< 50\,nm$ geometrical emittance at 100 MeV. A beam radius lower than 0.5 mm over distance of 1 meter is then achievable, allowing for large clearances in the internal target. A major design issue for MESA is to avoid an increase of effective emittance due to, for example, non linearities of optical elements or space charge effects.

So far only unpolarized experiments are envisaged in ERL mode which allows use of bulk GaAs cathodes. Such cathodes do not require the complex superlattice features which are vital for the production of highly polarized beam, they are therefore cheap and easily available. Due to the presence of the cathode storage and exchange system at our source it is easy to switch between the photocathodes required for the different operation modes. Bulk cathodes offer more than ten times higher quantum efficiency than is available in high polarization mode. Another advantage is that cathode charge lifetime at high photon excitation energies (green light with $\approx 2.4\,\text{eV}$, compared to $1.7\,\text{eV}$ for polarized operation) was found to be increased to 300 Coulomb. In the present state of performance this allows for about one day of continuous operation at 10 mA, which is sufficient to start up MESA and its internal target experiments. However, further improvement is necessary. This could probably be achieved by using KCsSb cathodes, which can operate under much harsher conditions than GaAs.

The ERL mode experiments are concerned with setting limits for the existence of light dark matter particles.[5] Such experiments rely on the best possible signal to background conditions and this requirement is the main reason why operation with an internal target in the ERL mode is considered as superior compared to conventional setups. Since these experiments are done with unpolarized beams we do not go into more detail here.

4. Polarized electron scattering at MESA

The EB experiments which are currently envisaged are parity violating (PV) scattering experiments, which require longitudinally polarized beam on an unpolarized proton (or later deuteron) target. In the case of e/p scattering at 137 MeV the scattering reaction is entirely elastic. Under the assumption of isospin invariance the expected PV-asymmetry can be decomposed into a sum of terms which contain nucleon form factors and the weak charge as observables.[6] The quantities of interest are the nucleon strange magnetic form factor G_M^S and the weak charge $Q_W = (1-4(sin(\theta_W))^2$, θ_W being the Weinberg angle. In the forward direction the contribution from Q_W is dominating the scattering asymmetry, however the absolute value is very small ($\approx$ 100 ppb). Under backward angles, G_M^S contributes by a fraction of a few percent to the total asymmetry of $\approx$ 3 ppm, if a prediction of lattice theory[7] is assumed to be correct.

A measurement of the Q_W induced asymmetry with $\Delta A/A \approx 3$ % accuracy is considered as an important step forward in standard model tests. For the backward angle measurement, if the lattice prediction for G_M^S should be correct, a first detection of a non-zero strange form factor can be obtained if the backward asymmetry is measured with about one percent accuracy.

The PV experiment will require a 20 cm long hydrogen target, yielding a luminosity of almost $8 \cdot 10^{38}$ cm$^{-2}s^{-1}$. A large solid angle detector with a sufficient granularity will be used to measure asymmetries for a wide angular range simultaneously, hence allowing the extraction of the important observables Q_W and G_M^S. Under the envisaged conditions a sufficient statistics can be sampled within several thousand hours of run time.

The asymmetries to be measured range from about 0.1 ppm (Q_W, forward angle) to about 3 ppm (G_M^S, backward angle). For Q_W the systematic errors caused by false asymmetries must be improved by about an order of magnitude if compared to present PV experiments at MAMI. For G_M^S these requirements are less strict, because of the much larger value of the signal. However, since G_M^S contributes only a few percent, an important error source results from the beam polarization accuracy, which requires $\Delta P/P \leq 1\,\%$.

An experiment aiming at the measurement of Q_W (Qweak[8]) will start soon at JLAB, and is operating at about one order of magnitude higher beam energy. The expected advantages for the low energy setup at MESA are the following:

- Theoretical corrections must be applied to the extracted PV signal which are increasing with beam energy. For the low energy experiment the systematic uncertainties introduced by these corrections are minimized.
- Inelastic background is strongly suppressed.
- In contrast to "general purpose" accelerators like MAMI or CEBAF the MESA accelerator can be designed with the primary goal of providing optimum operating conditions for the PV experiments.
- Operating costs for MESA are considerably lower.
- Even if we take into account the other experiments foreseen at MESA, exclusive access to the machine for several thousand hours per year will be possible.

5. Option for polarized positrons at MESA

CW beams of polarized positrons[9] are presently based on radioactive sources and reach intensities of a few 10^5 s^{-1}. With its high intensity polarized electron beam, MESA offers to improve this considerably by transferring helicity from electrons to positrons in the electromagnetic shower: polarized electrons - 1 mA intensity at 38 MeV created by a single pass through the MESA linac - are directed towards a radiator. Since the γ radiation is highly polarized near the endpoint of the bremsstrahlung spectrum, the positrons from subsequent pair production are also polarized. If, as a result of this process, the fraction of positron energy to beam energy is large, a high positron polarization results. Requiring a polarization in excess of 50 %, the efficiency[10] is expected to be $\leq 10^{-3}$. In a further step, one may increase the beam brightness by a moderation technique, which is also used in the existing e^+–sources. This also has a low efficiency of

$10^{-4} - 10^{-3}$, but finally yields 'cold' polarized e^+ with an energy spread of the order of eV or even lower. Of course, the total efficiency for cold $\vec{e}^+$ generation will be tiny, ranging between 10^{-8} and 10^{-6}.

Due to the high intensity of the polarized electron beam it seems feasible to obtain current densities in the range 10^7–10^9 $cm^{-2}s^{-1}$ at polarizations in excess of 50 %. If polarization is not required, the intensity would be increased by at least one order of magnitude. Besides higher brightness, the MESA based source offers fast spin reversal, good time resolution and variable time structure as additional advantages. The e^+ beam emerges from the moderator at low energy ($\approx$ eV), just as in a thermionic e^- source. This offers unique conditions for experiments on spin effects in solids. Alternatively, a reacceleration of e^+ in MESA to the 100 MeV range is conceivable.

6. Conclusion

MESA is an interesting accelerator project that offers unique conditions for several experiments in particle and hadron physics and also in applied science. The compact size and favorable conditions, regarding infrastructure and staff, make the realization of MESA within the given constraints of budget and infrastructure at Mainz conceivable. Future work will concentrate on detailed design studies to be completed within the next two years. We believe that the MESA acellerator could start to operate in 2015.

7. Acknowledgements

This work was supported by the Sonderforschungsbereich 443 der Deutschen Forschungsgemeinschaft (DFG).

References

1. K.H. Kaiser *et al.*, *Nucl. Istrum. Meth. A* **593**, 159 (2008).
2. W. Anders *et al.*, CW Operation of Superconducting TESLA Cavities, in *Proc. SRF 2007*, (Peking University, China, 2007).
3. T. Maruyama *et al.*, Polarized electron emission from strained GaAs/GaAsP superlattice photocathodes, in *Proc. SPIN 2004: Trieste*, ed. K. Aulenbacher and *et al.* (World scientific, Singapore, 2005).
4. R. Barday and K. Aulenbacher, Polarized source operation at currents of several milliampere, in *Proc. SPIN 2006: Kyoto*, eds. K. Imai and *et al.*AIP Conf. Proc. **915** (AIP, New York, 2007).
5. S. Heinemeyer *et al.*, An experiment to search for light dark matter in low energy elastic ep scattering (2007), arXiv:0705.4056.
6. S. Baunack *et al.*, *Phys. Rev. Lett.* **102**, 151803 (2009).

7. P. Wang *et al.*, (2007), arXiv:08071.0944.
8. D. S. Armstrong *et al.*, Qweak: A precision measurement of the protons weak charge, in *8th Conference on Intersection of Particle and Nuclear Physics*, ed. Z. Parsa, AIP Conf. Proc., Vol. 698 (AIP, New York, 2004).
9. J. Van House and P.W. Zitzewitz, *Phys. Rev. A* **29**, 96 (1984).
10. A. P. Potylitsin, *Nucl. Instr. Meth. A* **398**, 395 (1997).

STATUS REPORT OF THE DARMSTADT POLARIZED ELECTRON INJECTOR

Y. Poltoratska*, R. Barday, U. Bonnes, M. Brunken, C. Eckardt, R. Eichhorn, J. Enders, A. Göök, C. Heßler, C. Ingenhaag, M. Platz, M. Roth and M. Wagner

Institut für Kernphysik, TU Darmstadt,
64289 Darmstadt, Germany
**E-mail: yuliya@ikp.tu-darmstadt.de*
www.tu-darmstadt.de

W. F. O. Müller, B. Steiner and T. Weiland

Institut für Theorie Elektromagnetischer Felder, TU Darmstadt,
64289 Darmstadt, Germany

The superconducting electron linear accelerator S-DALINAC in Darmstadt will be extended by a 100 keV polarized electron source. The setup consists of a GaAs polarized gun, a beam line with a Wien filter for spin manipulations, a Mott polarimeter for polarization measurement, as well as many diagnostic elements. We report on the current status of this project and present results of measurements of the beam properties.

Keywords: Polarized electrons; S-DALINAC; Mott polarimeter; beam characteristics.

1. Introduction

The recirculating superconducting electron linear accelerator S-DALINAC[1] is one of very few electron accelerators devoted to nuclear structure physics, including electron scattering experiments at low momentum transfer. In addition to a thermionic 250 keV electron source of unpolarized electrons, a new polarized source is foreseen to facilitate experiments on polarization observables in, for example electron-induced break-up or parity violation. A setup separate from the S-DALINAC has been realized to test all components prior to the installation of the source at the S-DALINAC. We report on the current status of this test stand, the beam characteristics, and the planned installation at the S-DALINAC.

2. S-DALINAC

A schematic drawing of the S-DALINAC with the position of the planned new injector and existing experimental areas is shown in figure 1. The current experimental program at the S-DALINAC includes electron scattering and photon scattering experiments. Nuclear resonance fluorescence experiments are performed behind the superconducting injector at energies between 1 and 10 MeV with average beam currents of up to 60 μA. The experimental setup is described in reference 2. The same experimental site

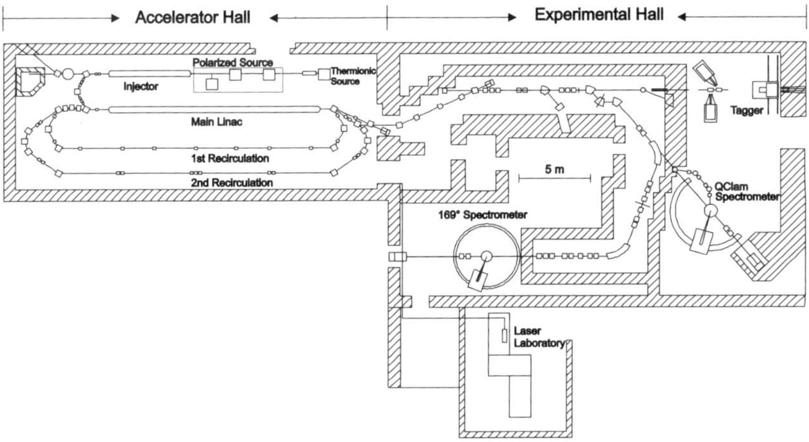

Fig. 1. Layout of the S-DALINAC with its experimental sites. The polarized source position between the thermionic source and the superconducting injector part of the linac is indicated. The laser beam runs through a fiber or an evacuated transport line from the laser lab.

is used for photoactivation experiments of type (γ,n). The electron beam behind the main linac is presently used for electron-scattering experiments. For that purpose, two electron spectrometers – a high-resolution energy-loss system[4] and a large-acceptance spectrometer of QClam type – are available. While at the former mainly form-factor measurements (e.g. ref. 5) are carried out, the latter is used for coincidence experiments or scattering at 180° see for example reference 3.

Two setups provide photon beams behind the main linac section: (i) a site providing about 50–100 MeV electron beams which is prepared for an experiment on the untagged bremsstrahlung (ii) a high-resolution photon tagger for astrophysically relevant photodisintegration and photon scattering studies between 10 and 20 MeV.

The installation of the polarized electron source is planned between the unpolarized thermionic gun and the first superconducting accelerating structure so that polarized electron scattering experiments and experiments with circularly polarized photons will become possible in all experimental areas. The scattering of polarized electrons will allow, for example, one to measure the fifth structure function in the break-up of nuclei at low momentum transfer. This structure function is sensitive to the final state interaction. Polarized gamma-ray beams will be produced at the S-DALINAC by bremsstrahlung. Planned experiments include the search for parity-violating effects in photon scattering and photo-induced fission. Examples for future experiments with polarized beams in Darmstadt are discussed elsewhere.[6]

3. Test stand of the polarized injector

3.1. *Layout*

The basic design of the polarized injector has been adapted from the established source of polarized electrons installed at the Mainz Microtron MAMI.[7] However, due to geometrical restrictions for installing such a source at the S-DALINAC, it was necessary to build the new injector as compact as possible, thus requiring design and development efforts. To test the developed polarized source independently from the operation of the S-DALINAC, a standalone test stand has been built.[8,9] Figure 2 displays the constructed test setup of the new polarized injector with its cathode and preparation chambers and a part of the beamline.

The longitudinally polarized electrons are produced at a GaAs photocathode by photoemission. As cathode material GaAs strained superlattice crystals are used from which electron beam polarizations above 80 % can be achieved. The spin direction is selected by using circularly polarized light from a diode laser system below the electron source focussed onto the photocathode's surface. The photocathode crystal is installed at the front end of a highly polished stainless-steel electrode. The form of the electrode was optimized using numerical simulations.[10,11] To extract the electrons to the vacuum, Negative Electron Affinity (NEA) is achieved through a thin CsO layer on the surface of the GaAs crystal in a separate preparation chamber. With the aid of a load-lock system the activation of the new photocathode can be handled during only a few hours. The emitted electrons are accelerated inside the source electrostatically to an energy of 100 keV and injected by an alpha magnet into the horizontal beamline where the beam properties

Fig. 2. Photograph of the electron source test stand with the part of the beamline.

can be measured. It is essential to create ultra high vacuum conditions in the gun chamber to prevent a rapid degradation of the NEA surface. After an uniform bake out procedure during 12 days at 220 °C, an end pressure of $< 2 \cdot 10^{-11}$ mbar has been achieved using different pumps especially Non-Evaporable Getter (NEG) and ion-getter pumps.

3.2. *Transverse beam properties*

The transverse beam properties have been investigated qualitatively by fluorescent screens and quantitatively by a wire scanner.[12,13] The normalized transverse emittance has been determined from the measurements of the beam radius for different focussing strengths of a solenoid preceding a wire scanner. Values of $\varepsilon_{n,x} = (0.134 \pm 0.012)$ π mm mrad and $\varepsilon_{n,y} = (0.118 \pm 0.004)$ π mm mrad, respectively, have been obtained.

3.3. *Polarization*

To determine the polarization of the beam a 100 keV Mott polarimeter[14] was constructed. The degree of polarization is extracted from count-rate asymmetries in electron scattering on nuclei due to the spin-orbit interaction. The asymmetry occurs only for transverse spin orientation with respect to the electron motion. Thus the initially longitudinally oriented spins need to be rotated by 90°. This is done by a Wien filter acting as a spin rotator. This device is also mandatory for the planned installation at the S-DALINAC in order to adjust the spin orientation for different experimental areas. It can provide 100° degree spin rotation. Using gold foils of different thicknesses, the degree of polarization could be determined to an accuracy of about 3%. Typical values of 34% polarization for Bulk GaAs and up to 86% polarization using a strained superlattice cathode have been demonstrated routinely.

3.4. *Time structure*

For acceleration the produced polarized electron beam needs to be modulated with the frequency of the superconducting cavities of 3 GHz for cw operation. This is achieved by using modulated laser diode as a light source.[9] Pulse length of $<$80 ps have been observed using a fast sampling oscilloscope. At this time, the limited band width of the detection technique prevents shorter pulses from being detected. To verify the time structure of the electron beam an ultra-fast coaxial Faraday cup was constructed. Measurements with a sampling oscilloscope support the assumption that the electron bunch length is comparable to the length of the laser pulses.

4. Implementation of the polarized injector at the S-DALINAC

The new injection scheme needs to fit in between the thermionic gun and the superconducting part of the injector. Furthermore accessability within the accelerator hall has to be taken into account. The present planning status is displayed in figure 3. To meet length requirements, both the source and the Mott polarimeter are located above the beam line. Two differential pumping stages will separate the UHV environment around the photocathode from the beam line.

For capturing beam with the superconducting structures for acceleration the electron-bunch length plays a very important role. A bunch length of 50–80 ps is not sufficient for acceleration. To minimize the background between the electron bunches and to reduce the bunch length to about 5 ps, new chopper and prebuncher systems have been developed, based on a similar setup at MAMI.[15] The new construction will be used both for the polarized and the unpolarized beam. For that purpose, the energy of the thermionic gun has to be reduced from 250 kV to 200 kV.

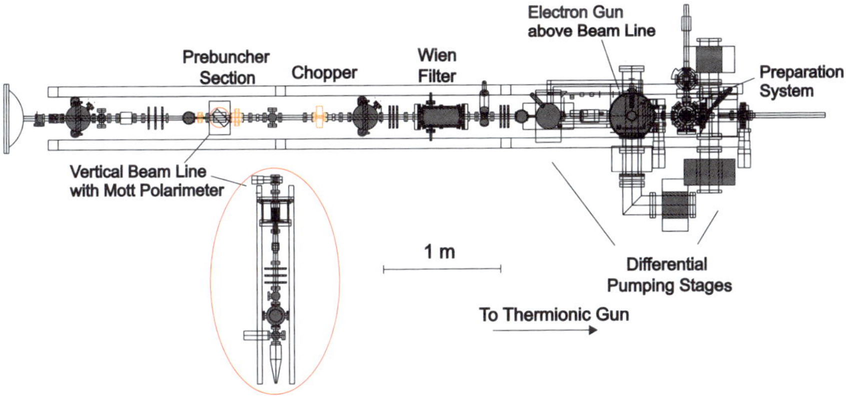

Fig. 3. Technical drawing of the planned installation of the polarized injector at the S–DALINAC. The existing unpolarized thermionic gun is located outside the right edge of the figure; on the far left, the beginning of the cryostat of the superconducting 10 MeV injector linac is shown. The GaAs cathode chamber and the preparation system are located above the beam line as is the Mott polarimeter.

A quantitative analysis of the planned experiments with polarized beams requires the knowledge of the degree of polarization after acceleration close to the experimental sites. Therefore, the polarimeters will be installed at different experimental sites. To support experiments at the injector experimental area of the S-DALINAC one additional Mott polarimeter for 5–10 MeV electrons was developed. For electrons with energies between 30 and 130 MeV, Møller scattering is well suited. A spectrometer for the coincident detection of the recoiling and the scattered electrons is being designed. Further information on this topic may be found in reference 16.

5. Summary and outlook

A source of polarized electrons was developed for the S-DALINAC. It has been set up, characterized, and operated at a test stand. Installation at the S-DALINAC is scheduled starting in January 2010. Experiments with polarized electrons and photons at the S-DALINAC may commence as early as the middle of 2010.

Acknowledgments

Y. Poltoratska thanks C. Eckardt for presentation of the talk at the PST2009 Workshop. This work was supported by Deutsche Forschungsgemeinschaft through SFB 634.

References

1. A. Richter, Operational experience at the S-DALINAC, in *EPAC'96, Sitges, 1996*, 110.
2. P. Mohr *et al.*, *Nucl. Instr. Meth. A* **423**, 480 (1999).
3. N. Ryezaeva *et al.*, *Phys. Rv. Lett.* **100**, 172501 (2008).
4. T. Walcher *et al.*, *Nucl. Instr. Meth.* **153**, 17 (1978).
5. O. Burda *et al.*, *Phys. Rev. Lett.* **99**, 092503 (2007).
6. C. Eckardt *et al.*, Polarized electrons for experiments at low momentum transfer SPIN @ S-DALINAC, in *SPIN'08, Charlottesville 2008*, 919.
7. K. Aulenbacher *et al.*, *Nucl. Instr. Meth. A* **391**, 498 (1997).
8. C. Heßler, Konzeption, Aufbau und Test einer Quelle spinpolarisierter Elektronen für den supraleitenden Darmstädter Elektronenlinearbeschleuniger S-DALINAC, Doctoral Dissertation D17 (TU, Darmstadt, 2008).
9. C. Heßler, Commissioning of the offline-teststand for the S-DALINAC polarized injector SPIN, in *EPAC'08, Genoa, 2008*, 1476.
10. B. Steiner *et al.*, Recent Simulation Results of the Polarized Electron Injector (SPIN) of the S-DALINAC, in *EPAC'06, Edinburgh, 2006*, 2188.
11. B. Steiner Strahldynamik-Simulation einer polarisierten Quelle fr den S-DALINAC (SPIN), Doktoral Thesis D17 (TU, Darmstadt, 2008).
12. C. Eckardt, Emittanzmessung an der Quelle spinpolarisierter Elektronen am S-DALINAC, Diploma Thesis (TU, Darmstadt, 2007).
13. C. Ingenhaag, Bestimmung der Strahlemittanz an der Quelle polarisierter Elektronen bei verschiedenen Batriebsparametern, Bachelor Thesis (TU, Darmstadt, 2009).
14. Y. Poltoratska, Design and setup of a compact Mott polarimeter for the future S-DALINAC polarized injector, Diploma thesis, (Kharkiv Karazin National University and TU, Darmstadt, 2005).
15. V. I. Shvedunov *et al.*, Design of a Prebuncher for Increased Longitudinal Efficiency of MAMI, in *EPAC'96, Sitges, 1996*, 1556.
16. R. Barday, these proceedings.

THE MOTT POLARIMETER AT MAMI

V. Tioukine*, K. Aulenbacher and E. Riehn
Institut für Kernphysik der Johannes Gutenberg-Universität Mainz,
D-55118 Mainz, Germany
**E-mail: tioukine@kph.uni-mainz.de*
www.kph.uni-mainz.de

We have developed a Mott electron polarimeter for the MAMI accelerator in Mainz. Two double focusing magnet spectrometers collect elastically backscattered electrons from gold targets. In spite of the small solid angle a high statistical efficiency is achieved if typical MAMI-beam currents are being used. Background discrimination of inelastically scattered electrons and photons by spectrometers allows for a good signal to background ratio. The measured asymmetries for different target thicknesses have been extrapolated to zero foil thickness using various approximation functions yielding an accuracy of extrapolation better than 2% in an energy range between 1 and 3.5 MeV. The results of asymmetry measurements are independent of the primary beam current in a range from 0.005 to 45 microamperes at a level of $< 1\%$ relative variation.

Keywords: Mott polarimeter; polarized electrons; nuclear targets.

1. Introduction

Mott polarimetry at MeV-scale energies is restricted to comparatively high beam intensities mainly because the Mott cross section scales approximately like $1/E^2$. The statistical Figure Of Merit $FOM = S_{eff}^2(I_s/I_p)$ is therefore much smaller than for standard Mott polarimeters which operate at energies $\leq 120\,\text{keV}$. S_{eff} is the "effective Sherman function" or "analyzing power" of the scattering apparatus; I_s, I_p are scattered and primary current, respectively. This low FOM is not a major concern at high energy electron accelerators where high beam intensities are standard and MeV beam energies are already provided at the injector stage of such a machine. In the following section we demonstrate advantageous features of a MeV-scale Mott which is installed at the Mainz Microtron (MAMI). These include a dynamic range of almost four orders of magnitude in primary current with high statistical efficiency and good reproducibility. The beam intensity range allows us to perform the polarization measurements at the same current level as it is used for almost all experiments presently in operation at MAMI.

Mott polarimeters determine the beam polarization P_{beam} by measuring the experimental asymmetry $A_{exp} = P_{beam} \cdot S_{eff}$ in elastic scattering of polarized electrons off targets with high nuclear charge, typically gold. In this simplest approach the observable is a single spin scattering asymmetry which is theoretically accessible by calculating the analyzing power S_0. Apparative complications such as multiple scattering in the target will lead to a reduction of the analyzing power towards an effective value S_{eff}.

The absolute accuracy achievable for a Mott polarimeter is therefore limited mainly by the accuracy to which S_{eff} can be determined. Operating at the MeV scale offers two distinct advantages for this task, first because the Sherman function can be calculated more accurately than at lower and higher energies and second, because the effects of multiple scattering can be controlled in a model-independent way by Monte Carlo simulation. In section three we will discuss the potential of this method for several results that were achieved in an energy range between 1 and 3.5 MeV.

2. Dynamic range and reproducibility

2.1. *Apparatus*

The Mott-polarimeter is located behind the injector linear accelerator (ILAC) of MAMI which provides 3.5 MeV kinetic beam energy in standard operation. During the measurements the beam is directed towards gold targets of various thicknesses (0.1, 0.25, 0.5, 1 and 15 μm) which are mounted in vacuum in a scattering chamber (fig. 1). Scattered electrons are detected under an angle of 164°. The solid angle is defined by 4 mm diameter collimators; photons and inelastically scattered electrons are suppressed by double focusing magnetic spectrometers. These provide a one to one imaging of the beam spot from the target to the detection scintillator, which is reached by the electrons after passing a 0.3 mm thick aluminum vacuum-window. During measurements the spin is oriented perpendicular to the scattering plane, the necessary spin rotation is achieved by a Wien filter rotator.

2.2. *Statistical efficiencies*

For the thin targets the spin averaged count rate is $R = k \cdot d \cdot I_{Beam}$ with $k = 1.7\,\text{kHz}/(\mu\text{A} \cdot \mu m)$ and d the foil thickness in μm. This is in reasonable

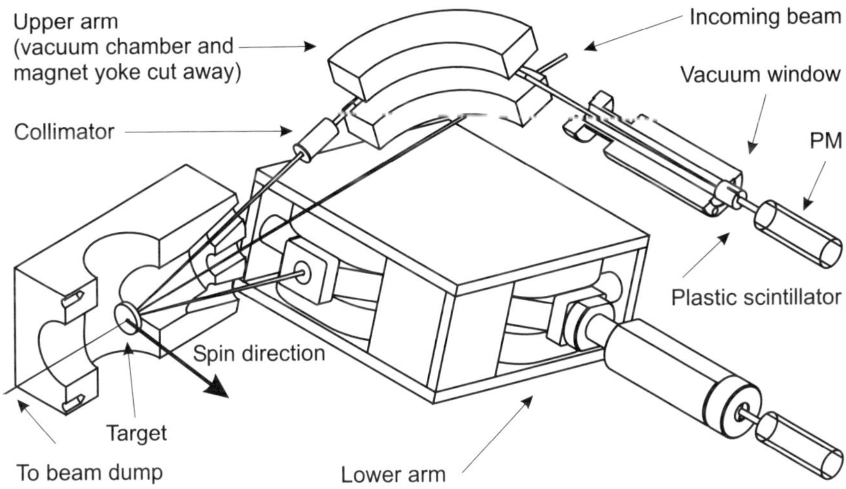

Fig. 1. Artists view of Mott polarimeter with magnetic spectrometers. Upper magnet is partially removed for better visibility of electron trajectory.

agreement with the rate expected from luminosity, solid angle acceptance and the Mott scattering cross section. The parameter k is increasing with target thickness due to multiple scattering, however this effect is only sizable for the 15μm target. Due to the high asymmetries (see figs. 3 and 4) and rates we achieve statistical accuracies of <1% within less than one minute at beam currents of some μA.

We can define a statistical FOM for each target which is given by $FOM = S_{eff}^2(d) \cdot k \cdot d$. We find a continuous decrease of analyzing power with increasing foil thickness with $0.1 < |S_{eff}| < 0.4$ (see fig. 4). The apparent advantage in FOM for thin foils is, however, more than compensated by the factor kd in thick targets, hence yielding the highest FOM for the 15 μm target which still has a relatively large analyzing power of $S_{eff} \approx 0.1$.

The 15 μm target was only used for small beam currents < 1 μA. This restriction was chosen for two conditions, namely first for the count rate to stay below 100 kHz, hence avoiding variations of asymmetries due to dead time effects. Second, there is a certain risk of foil damage to be avoided because of the reduced ratio of beam spot area to foil thickness in the thick targets, which lowers the radiation cooling capability.

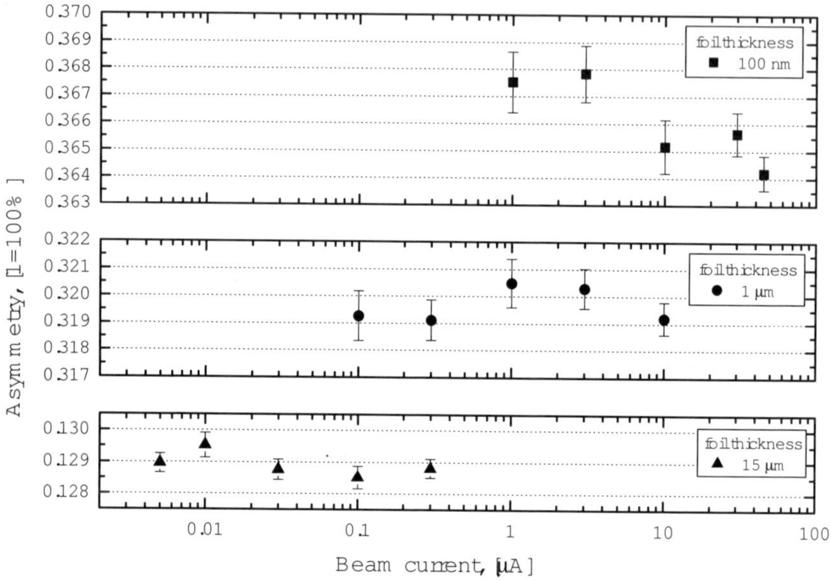

Fig. 2. Asymmetry versus beam current for various targets.(Statistical error bars)

2.3. *Measurements at different beam currents*

The beam currents ranged between 5 nA and 45 μA, the current variation was done exclusively by passively reducing laser beam intensity before it excited the photocathode in order to avoid any change of the laser parameters (e.g. wavelength) which could in turn change the polarization. The electron optical elements of the accelerator were held constant as well. For the reasons mentioned above we decreased the foil thickness with increasing beam current but provided some overlap in the current ranges where this was done (fig. 2). The relative statistical error for each measurement is $< 0.3\%$, which required two hours of run time for the lowest current and correspondingly less for the larger ones.

For the measurements performed on the 1 and 15 μm thick targets no statistically significant variation is observed. This indicates that the reproducibility $\Delta A/A$ of the measurement from one point to the other is better than 0.3 %. For the highest range of currents a statistically significant variation of 0.8 % is observed. A possible explanation is a slight increase of photocathode temperature due to the increased level of laser power which is needed to produce the larger currents.

Long-term variations of asymmetries may either be caused by the well-known changes of beam polarization during aging of photocathodes or by apparative drifts. It is therefore difficult to prove long-term stability without further independent asymmetry measurements. So far, drifts were controlled by periodic comparison with available data from other polarimeters at MAMI. These cross checks indicate a long-term reproducibility of < 1 % but a demonstration at the level of precision demonstrated in figure 2 will require much more frequent cross checks. In order to have even a continuous check in the future we plan to use a Compton asymmetry measurement in the beam dump of the Mott polarimeter which can be performed simultaneously with the Mott measurements.

3. Reduction of systematic errors in absolute calibration

3.1. *Systematic errors due to finite target thickness*

The third section of the MAMI-ILAC provides acceleration from 2 to 3.5 MeV. The beam energy can therefore be varied by either reducing the RF-power in the section or by shifting the RF-phase towards decelerating operation. A combination of both methods allowed for an energy variation between 1 and 3.5 MeV. Figure 3 presents asymmetries for the various targets as observed for different beam energies. The task for calibration is to extrapolate the observed asymmetries $A(d)$ towards target thickness zero - $A(0)$ - for each individual energy. In order to perform the extrapolation a description for $S_{eff}(d)$ must be given which takes into account the physical and apparative effects which change S_{eff}.

The problem of finding the exact form of $S_{eff}(d)$ is a difficult task if performed analytically and it is even more difficult to estimate the model dependent error if a specific method is chosen. However, as was noted by Gay *et al.*,[1] the exact analytical description is unimportant if the curvature of the true function is negligible at target thicknesses that are experimentally accessible. In this case, a model function can either be discarded because it does not fit the data or it will provide the same linear Taylor expansion as the "true" $S_{eff}(d)$, i.e. it will be exact since second order effects are supposed to be negligible by definition. Due to the fact that the lowest thickness available is limited, this "zero curvature" regime is probably easier to achieve at high energies since the reduction of analyzing power with thickness is less fast (see fig. 3).

It can be guessed to which extent the "zero curvature" regime is reached by choosing different fit functions to the data and observing the

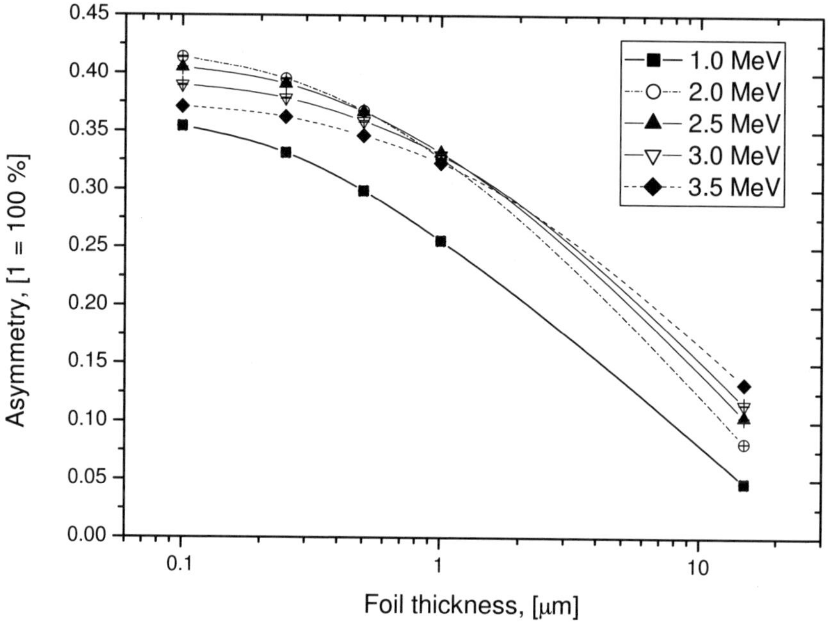

Fig. 3. Dependence of asymmetry on foil thickness for various primary beam energies at a fixed scattering angle of 164 degree, lines are to guide the eye.

variation of the predicted values $A(0)$. In our case these variations are very small: We choose for example $A_{exp}(d) = A_1 + A_2 exp(-kd)$ (a function initially proposed by Gay) and a second one, described by a ratio: $A_{rat}(d) = A_1 + A_2/(1 + kd)$ which implicitly takes into account the polarization dependence of double scattering. Both fit functions have three free parameters (A_1, A_2, k) and are equivalent with respect to their ability to fit the data (fig. 4). They yield variations of the predicted $A(0) = A_1 + A_2$ of less than 0.5 % at energies $\geq$ 2 MeV (fig. 5). The increased difference at energies $<$ 2 MeV may be a result of increased nonlinearity of $S_{eff}(d)$ for the given thickness. The "zero curvature" regime seems therefore to be reached for thicknesses of 100 nm at energies $\geq$ 2 MeV. The ultimate proof for this statement can be given by Monte Carlo simulation, as was noted by Kakhoo *et al.*,[2] a task that is technically feasible also at MeV energies.[3] This will finally eliminate the need for a fit altogether.

The fit uncertainties for the extrapolations are of the order 1.5 % (error bars in fig. 5). They probably result from uncertainties in the (relative) thickness of the foils. A future improvement may be obtained along the following line: We assume that the initial slope of $S_{eff}(d)$ can be estimated

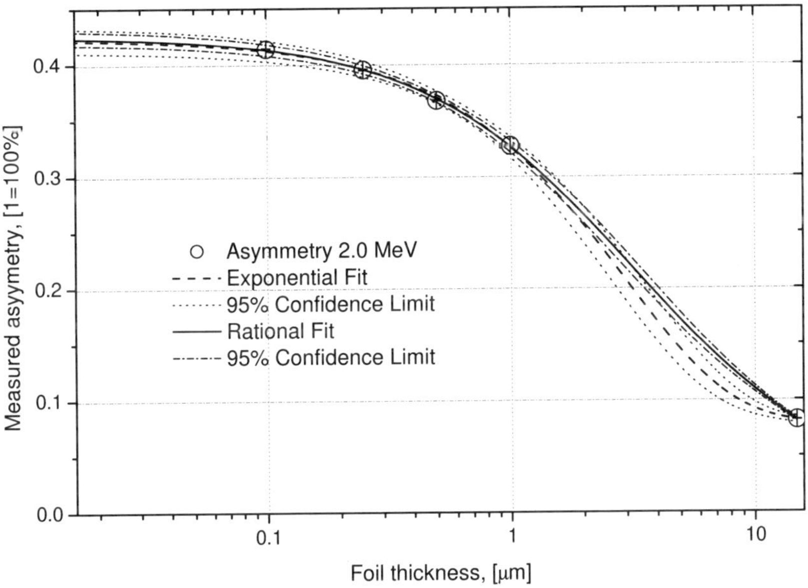

Fig. 4. Fit to the observed Asymmetries at 2.0 MeV by two different functions.

by *ab initio* Monte Carlo to an accuracy of < 5 % - which seems achievable since all input parameters for MC are known to about 1 % accuracy. As long as nonlinear contributions are under control this leads to a small systematic error since the absolute reduction of S_0 for our thinnest target is small due to the high energy: $\Delta S = (S(0) - S_{eff}(100\text{nm})) \approx 0.03$. Even if we assume a 5 % relative error for both the slope and the target thickness the error in ΔS is about 0.002 which is a 0.4 % fraction of $S_{eff}(100\ \text{nm})$.

3.2. *Systematic errors due to calculation of S_0*

We have already pointed out in an earlier paper[3] that the calculation of S_0 is particularly accurate in the energy range between 1 and 5 MeV, since the uncertainty that is introduced by effects of the atomic electron cloud and the structure of the nucleus are minimized. In figure 5 we extract the beam polarization from the extrapolated $A(0)$ and the value of S_0 taken from our calculations. The results are independent of beam energy with a variance of much less than 1 %. These results may also serve for a better estimation of the so-called "radiative" effects, which probably have a pronounced energy dependence. Our results in figure 5 give strong constraints for such an effect, hence supporting the statement that radiative effects are "small".

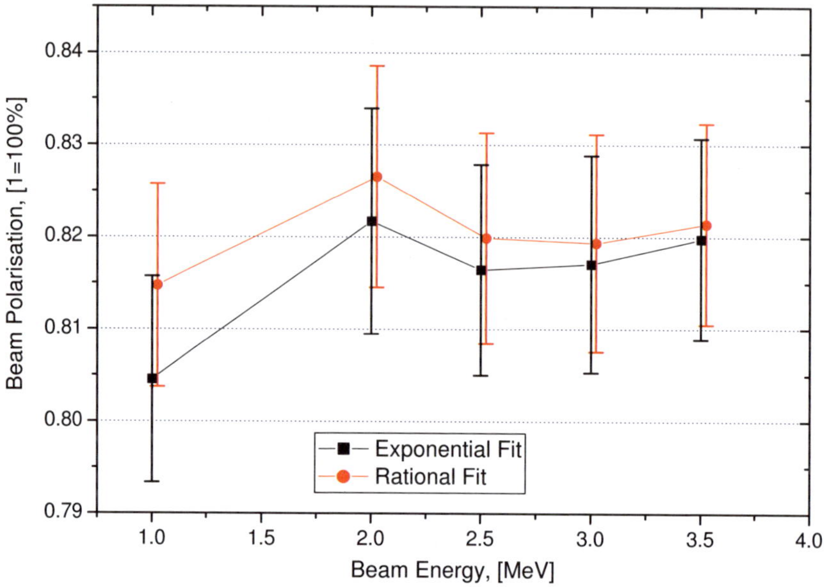

Fig. 5. Polarization determination for several energies. Errors are fit uncertainties only.

4. Conclusion

The 3.5 MeV Mott polarimeter at MAMI offers high statistical efficiency and good reproducibility of results. This is the case for almost all beam intensities that are presently in use at MAMI. The demonstrably small dependence of analyzing power on target thickness and the consistent results that are achieved over a wide energy range support the idea that an absolute accuracy well below 2 % will be available in the near future.

5. Acknowledgments

This work was supported by the SFB 443 of the Deutsche Forschungsgemeinschaft.

References

1. T.J. Gay *et al.* , *Rev. Sci. Instr.* **63 (1)**, 114 (1992).
2. M. A. Khakoo *et al.* , *Phys. Rev. A* **64**, 052713 (2001).
3. K. Aulenbacher and V. Tioukine, *Ab initio* calculation of effective Sherman functions at MeV energies, in *Proceedings 18th international Spin physics symposium SPIN 2008*, ed. D. G. Crabb and *et al.* (AIP, New York, 2008).

PROTON POLARIMETRY AT THE RELATIVISTIC HEAVY ION COLLIDER

Y. Makdisi*, E. Aschenauer, G. Atoian, A. Bazilevsky, G. Bunce, R. Gill, H. Huang, B. Morozov, S. Rescia, M. Sivertz, K. Yip and A. Zelenski
Brookhaven National Laboratory, Upton, NY 11973, USA
**E-mail: makdisi@bnl.gov*

S. K. Lee
State University of New York at Stony Brook

X. Li
Shandong University, China

I. Alekseev and D. Svirida
ITEP, Moscow

The RHIC polarized proton collider employs polarimeters in each of the blue and yellow rings that utilize the analyzing power in p-carbon elastic scattering in the Coulomb Nuclear Interference (CNI) region to measure the absolute beam polarization. These are calibrated by the polarized hydrogen jet target that measures the absolute beam polarization in pp elastic scattering in the CNI region. This paper describes the status and performance of these polarimeters in the FY09 run which included both a 250 GeV/c and 100 GeV/c physics data taking period. We will describe some of the difficulties encountered and the efforts underway to improve the performance in better energy resolution, rate handling capability, and reduced systematic uncertainties.

Keywords: Silicon detectors; polarimeters; beam polarization; polarized hydrogen jet.

1. Introduction

The RHIC polarimeter system relies on the analyzing power in both proton-proton as well as proton-carbon elastic scattering in the Coulomb Nuclear Interference (CNI) region where the predictions without the presence of a hadronic spin-flip amplitude reaches a maximum of the order of 5 and 4 %, respectively, and falls with increasing 4-momentum transfer[1,2] (fig. 1). The presence of this amplitude is difficult to estimate and is likely to change the predictions especially at lower beam energies. The analyzing power is predicted to change slowly with beam energy.

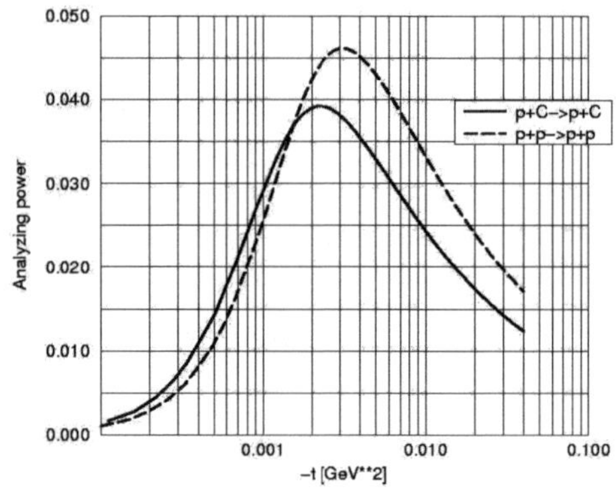

Fig. 1. The analyzing power in pp and p-carbon elastic scattering in the CNI region.

A polarized hydrogen jet target[3–5] with high polarization that measures the analyzing power in pp elastic scattering *in situ* and then uses the same data sample to measure the beam polarization over the respective store. This in turn calibrates the relative and faster beam polarimeters that measure the asymmetry from the p-carbon elastic scattering process.

The RHIC polarimeters utilize several thin carbon targets mounted on a special target drive operated by a motor to select horizontal or vertical targets for scanning across the beam to measure both the beam polarization and beam polarization profile. The targets are surrounded by three pairs of silicon detectors positioned 18 cm away with one pair at 90 ° in the horizontal plane and the other two at ±45 ° above and below. (fig. 2).

The silicon detectors, manufactured by the BNL instrumentation division, comprise 12 strips each independently measuring the scattered recoil carbon energy and time of arrival. The left-right scattering asymmetry determines the degree of the beam polarization. The large p-carbon elastic scattering cross section provides a fast measurement with 2 % statistical accuracy in less than one minute. The polarized jet target at an interaction region at 100 meters from the polarimeters and measures the absolute beam polarization to better than 10 % in an 8-hour store. It takes a few stores to calibrate each polarimeter to an accuracy of 5 %.

The silicon detectors polarimeters are calibrated using americium sources that emit alphas with energy of 5.5 MeV. This is large compared

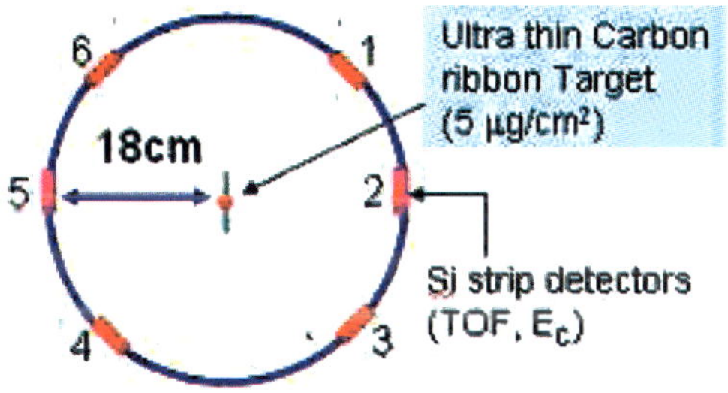

Fig. 2. A beam view of the silicon detectors in the RHIC polarimeters.

to that of the carbon recoils from 300 keV to about 1 MeV. This is a starting point followed by an involved process that relies on fitting the data to the time energy relation as well as data of the energy deposition of carbon in silicon.[6] This measured effective dead layer in the silicon results three times larger than the physical one. Another issue facing the polarimeters is increasing rate as the RHIC bunch intensity moves upwards.

The polarized jet silicon energy calibration utilizes both Am and Gd sources, the latter emits 3.2 MeV alpha particles at energies overlapping those of the recoil protons of interest from 1 to 5 MeV rendering a more comparable calibration. The jet measurements on the other hand are limited by the unpolarized molecular hydrogen background that at this stage is estimated at 3 % and contributes a 2 % systematic accuracy. In what follows we will describe the efforts to overcome some of these difficulties.

2. The silicon energy and rate response

The BNL tandem complex has the capability to accelerate various ion species at prescribed energies and intensities. We used the tandem to systematically scan energies of interest with varying intensities up to 4.10^6 ions/cm^2 to study the BNL manufactured silicon detectors response:

- The beam energy will span from 0.3 to 5 MeV wider than the recoil carbon range to understand detector response to the alpha energy from the americium source.
- Use beam carbon charges of +1, +2, and +3 to assess the detectors response and how soon charge equilibration occurs.
- Study any non-linearity to the energy response to help the calibration process.
- Check the energy resolution vs. beam energy.

- Use a foil to simulate the recoil carbon energy loss as it traverses the target.
- The tests utilized the readout system currently employed by the polarimeters in addition to alternate electronics to assess differences, if any.

3. New polarimeter detectors and electronics R&D

We used the tandem carbon beams to test several Hamamatsu silicon photodiode detectors of several thicknesses, a 5 μm S9724-005 (1x1 cm^2) detector, a 300 μm S3590-19 (1x1 cm^2) detector and a strip-array S4114-35 without window-35 strips (4.4$\times$9 mm^2) in an effort to attain better energy resolution and to develop an understanding of the detector sensitivity to the large number of prompt particles that stream through concurrently when the detectors are responding to the recoil carbon scatters.

- Test new photodiode detectors and array strip detectors under similar conditions to the BNL silicon detectors.
- Test a dual-silicon detector system
  (a) with a thin 5 μm followed by the 300 μm detectors to assess the charge equilibration process.
  (b) Identify the carbon charge with the thin detector at 10 MeV. This then drives the energy loss (dead layer) determination from the earlier data.
  (c) Assess the thin detector's ability to provide a trigger as it is likely to be blind to minimum ionizing prompts, so less rate dependent.
  (d) Test the existing charge amplifier with a lower shaping time.
  (e) Test a new low capacitance cable between the detector and preamp.
- Test a current amplifier concept which is better for high capacitance detectors (thin detectors) and high rate environment, and whether the associated noise level is acceptable.
- Study the effect of reduced silicon volume current by reducing the detector area and thickness.

The results of the measured energy resolution with carbon beams are shown in figure 3. The Hamamatsu detectors seem to indicate significantly better energy resolution. Using such detectors will allow us to reach a lower t value especially in the p-carbon polarimeters.

4. *In situ* beam tests at RHIC

The upgraded polarimeter vessels allowed for dual polarimeter setups in each ring. In the blue ring we installed a pair of photodiode strip detectors at 90 ° in the horizontal plane, instrumented with the current readout

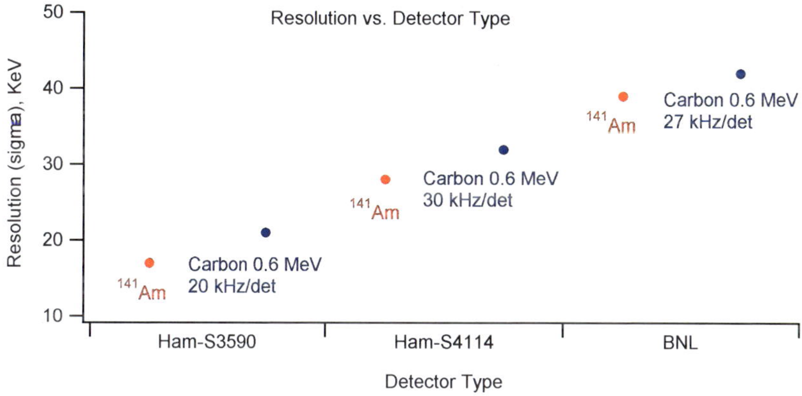

Fig. 3. Measured energy resolution for various types of silicon detectors using carbon beams from the BNL tandem

system. With a vertical carbon target, this polarimeter response will be compared and contrasted with the two 45 ° BNL silicon polarimeters looking at the same target during the normal data taking. Similarly, one of the yellow ring polarimeters utilized one 90 ° pair equipped with Hamamatsu single photodiode detectors (two to a side) to carry out *in situ* comparisons. Dedicated studies compared the responses to varying the number of bunches and bunch intensities. Varying target thicknesses provided rate dependences. Another venue is to assess changes the detectors suffer due to radiation damage in a prolonged run. Run 9 provided a good test bed for our set up in that we ran for the first time with 250 GeV polarized proton beams for several weeks in addition to a ten week run with 100 GeV beams. We also added a scintillation counter mounted on the outside of a thin 2 mm flange to detect the arrival time of prompt particles in an attempt to better define t_0 and better decouple the time and energy measurements. In practice this only served to count the overall rate seen at 90 °.

The dual polarimeters in each beam allowed the use of horizontal and vertical targets independently scanned across the beam and provided both a vertical and horizontal beam polarization profile. A fast target scan also provided beam emittance measurements. These are used to help with machine tuning and more importantly to assess the actual beam polarization as seen by the experiments.

With the increased beam bunch intensities, the carbon polarimeters suffered measurable instabilities that, due to high beam rates, were further exacerbated at 250 GeV due to the smaller effective beam size. Dedicated beam studies were carried out with different target thicknesses to determine the cause of these problems. While data analysis continues, one potential culprit appears to be a 24 times amplification of a shaper stage on the preamp board. Parallel work was also carried out on the new Hamamatsu photodiode detectors using new amplifier shaper boards as well as a separate ADC and TDC system. The early analysis appears to indicate no rate issues and minimal increase in dark current.

5. The polarized jet target operation

Towards the end of run 8, we tested the idea of having the two RHIC beams vertically separated by 5 mm and simultaneously impinging on the jet target. This was quite successful in that the data analysis did not reveal any more background under the elastic signal compared to running with one beam on target.

For run 9 we realigned the jet with respect to the RHIC beam line, and along with an overhaul of the RHIC beam position monitors, we successfully ran with simultaneous measurement of both beams vertically separated by about 3.5 mm, consistent with acceptable beam-beam conditions to increase our vertical acceptance. An online plot showing the jet operating with two beams simultaneously on axis is shown in figure 4.

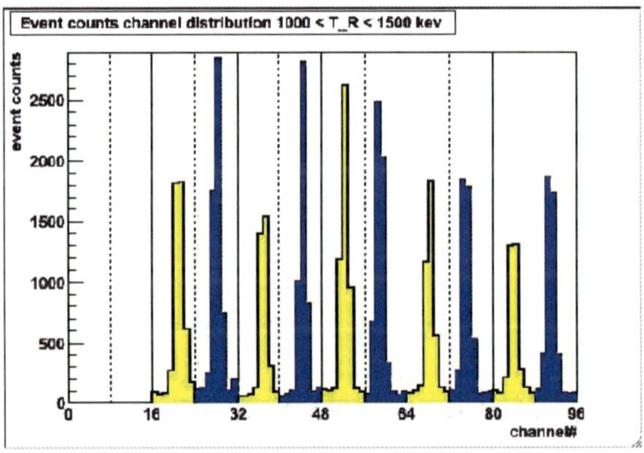

Fig. 4. An online plot of the activity on the six jet silicon strip detectors. The alternating peaks indicate the scattering due to the yellow and the blue, respectively. The level in-between indicated the attained background. Silicon detector 1 was off.

The jet took data in two running periods at 250 and 100 GeV respectively, with continuous monitoring and calibration of both RHIC polarimeters and doubling the statistics. At 250 GeV and with nine hour stores, the jet provided a beam polarization measurement with 5 % statistical accuracy. At 100 GeV, the statistical accuracy per five hour store was 7 %.

So far, the polarized hydrogen target was able to measure the analyzing power in pp elastic scattering in the CNI region at four beam energies of 24 GeV, 31.2 GeV, 100 GeV and 250 GeV. The results are compiled in figure 5. It is interesting to note that the analyzing power appears to be the same over such a wide range spanning RHIC injection to top energy. The data at 24 GeV and 100 GeV with larger t coverage have been published[7,8] in an attempt to measure the hadronic spin-flip amplitude. While there is a discerned contribution at 24 GeV, it appears to be quite small at 100 GeV. Such plots show the capability of the jet system to self-calibrate.

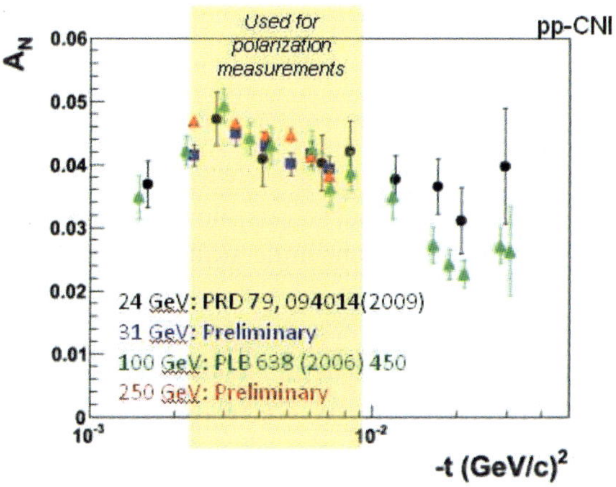

Fig. 5. The analyzing power in pp elastic scattering vs. four momentum transfer. The data at 250 GeV is preliminary.

An effort to monitor and understand the molecular hydrogen jet component *in situ* using luminescence monitoring was not successful. We continue to pursue ideas to quantify this background.

6. The path forward

There is no planned polarized proton running in run 10. We plan to utilize the polarized proton beam at the AGS and its p-carbon polarimeter to test our ideas for a path forward. We plan to use the 45 ° pairs to carry out the following studies:

(a) New Hamamtsu silicon photodiode detectors configured in strip geom-

etry with new amplifier/shaper and separate TDC and ADC read out through VME. The preamp can be configured with a fall time down to 5 μsec which can increase its dynamic range
(b) A new current-sensitive preamplifier with a short rise and fall time response to below 4 ns in an attempt to reduce the pile-up. These will plug into existing preamplifier boards.
(c) Should time allow we will also test a board with the 24$\times$ amplification reduced to 6$\times$ in an attempt to reduce saturation effects.
(d) While the AGS does not present the same environment as RHIC, namely the bunch-to-bunch timing as low as 114 ns, we will use as high as 12 bunch fills and push on the intensity front and use different target thicknesses to approximate the RHIC rate conditions.
(e) Finally, at RHIC we will likely test moving the shaper and waveform digitizers into the RHIC tunnel to avoid the pulse spreading due to the 200 m long cables.

7. Summary

We described the ongoing program to provide reliable beam polarimetry at RHIC and the efforts to improve our understanding of the energy responses, calibration, and rate capabilities of the current RHIC polarimeters. R&D is underway to test an improved set of silicon detectors that will provide better energy resolution, rate capabilities, and allow access to higher analyzing powers. Work continues to improve the polarized jet target calibration of the RHIC polarimeters. Of note is the maturing data analysis process and the improved ability to provide the RHIC experiments with results in a timely fashion.

8. Acknowledgements

We are grateful for significant technical help and support from D. Steski, G. Mahler, J. Ritter, T. Curcio, D. Lehn, and T. Russo and the BNL Tandem operations crew. Thanks to H. Okada and V. Dharmawardane for diligent data analyses of run 8. This work is performed under Brookhaven Science Associates, LLC, under contract No. DE-AC02-98CH10886 with the U.S. Department of Energy. Funding is also provided from the RIKEN BNL Research Center.

References

1. B. Z. Kopeliovich and L. I. Lapidus, in *Sov. J. Nucl. Phys.* **19**, 114 (1974).
2. N. H. Buttimore, E. Gotsman, and E. Leader, *in Phys. Rev. D* **18**, 694 (1978).

3. T. Wise *et al.*, in *Proc. 16th Int. Spin Phys. Symp. (SPIN 2004)*, eds. K. Aulenbacher *et al.*, 757 (World Scientific, Singapore, 2005).
4. A. Zelenski *et al.*, in *Proc. 16th Int. Spin Phys. Symp. (SPIN 2004)*, eds. K. Aulenbacher *et al.*, 761 (World Scientific, Singapore, 2005).
5. A. Nass *et al.*, in *Proc. 16th Int. Spin Phys. Symp. (SPIN 2004)*, eds. K. Aulenbacher *et al.*, 776 (World Scientific, Singapore, 2005).
6. I. Nakagawa *et al.*, in *Proc. 12th International Workshop on Polarized Ion Sources, Targets, and Polarimetry, Upton, NY, 2007*, 370.
7. H. Okada *et al.*, *Phys. Lett. B* **638**, 450 (2006).
8. I. G. Alekseev *et al.*, *Phys. Rev. D* **79**, 094014 (2009).

POLARISATION AND POLARIMETRY AT HERA

B. Sobloher
for the POL2000 collaboration
Deutsches Elektronen–Synchrotron DESY,
Notkestr. 85, 22607 Hamburg, Germany
E-mail: blanka.sobloher@desy.de

Longitudinal polarisation of the lepton beam is a key ingredient in the success of the world's unique $e^{\pm}p$ ring collider HERA. This article aims at providing a brief introduction to the physics motivation for deep-inelastic scattering of polarised electrons or positrons off protons, the basic mechanisms to establish lepton polarisation in the high-energy storage ring and to describe briefly the three different polarimeters, which measured both the transverse and the longitudinal polarisation.

Keywords: Polarised electron storage ring; polarisation; polarimetry; high-energy beams; HERA; Compton scattering.

1. Introduction

The unique HERA facility in Hamburg, Germany collided leptons with protons at centre-of-mass energies of 300 and 318 GeV between 1991 and 2007, incorporating radiative polarisation of the lepton beam. Spin rotators installed around the interaction points of the experiments HERMES, H1 and ZEUS transformed the natural transverse polarisation of the lepton beam to logitudinal polarisation, which is in deep inelastic scattering a powerful tool to study the internal structure of the nucleus.

The polarisation was measured routinely with two polarimeters. Using polarisation dependent Compton scattering, the transverse polarimeter TPOL detected the tiny up–down asymmetries associated with vertical polarisation, while the longitudinal polarimeter LPOL utilised the energy asymmetry caused by longitudinal polarisation. The third polarimeter, employing a Fabry–Perot cavity to provide a high laser photon density to measure longitudinal polarisation, started to collect significant amounts of data towards the end of 2006 and in 2007.

2. The HERA collider

The Hadron Elektron Ring Anlage HERA is the first and only electron–proton or positron–proton storage ring, located at the Deutsches Elektronen–SYnchrotron DESY laboratory in Hamburg, Germany. First $e^{\pm}p$ collisions were achieved in October 1991, with the colliding beams experiments ZEUS and H1 taking first physics data shortly afterwards.[1] The fixed target experiments HERMES and HERA–B went into operation in 1995 and 2000 respectively, the latter taking data till March 2003. The collider went through an ambitious luminosity upgrade in 2000-2001 and has been shut down finally after a successful second running period on 1 July 2007. A schematic of the HERA accelerator complex is shown in figure 1 (left).

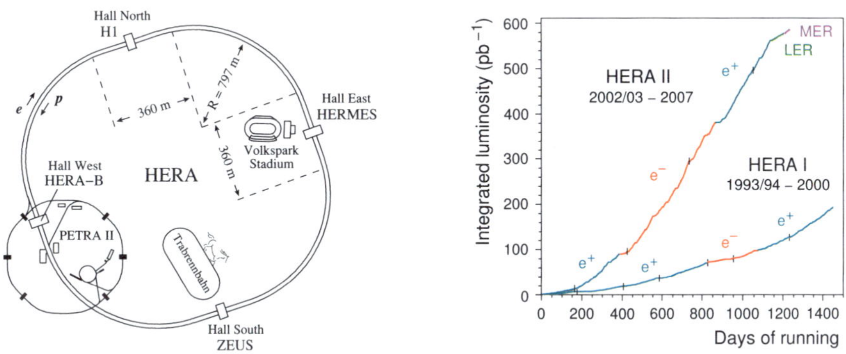

Fig. 1. The HERA accelerator complex (left) and the integrated luminosity at HERA I and HERA II (right).[2] Labels LER and MER: Low and Middle (proton) Energy Runs, black marks: approximate change of year.

At HERA electron or positron beams were accelerated to the energy of 27.5 GeV and the counter-rotating proton beams to energies of 820 GeV or 920 GeV. Colliding at the interaction points of ZEUS and H1 the beams yielded a centre-of-mass energy an order of magnitude larger than conventional fixed target experiments. From the beginning HERA was designed to incorporate a polarised lepton beam and the establishment of transverse polarisation in the storage ring was an important prerequisite for the HERMES experiment. Longitudinal polarisation was delivered to HERMES from its beginning and also since the upgrade to H1 and ZEUS. Till the final shutdown each of the colliding-beams experiments collected an integrated luminosity of $\approx 0.5\,\mathrm{fb}^{-1}$ as can be taken from figure 1 (right).

2.1. *Physics case for longitudinal polarisation*

At HERMES, the longitudinal polarised lepton beams scattered off gas targets, to study the nucleon spin structure by determining the spin-dependent structure functions of various nuclei. The helicity distributions of individual quark flavours inside the nucleon could be determined and further information on the spin structure of the nucleon has been obtained by investigating whether the gluons inside of the nucleon are also polarised.[3]

Being delivered with a longitudinally polarised lepton beam, the physics programs of H1 and ZEUS were extended by a large electroweak program, allowing the study of, for instance, the chirality of Charged Current (CC) interactions. The CC Deep Inelasting Scattering (DIS) cross section depends on the polarisation of the lepton colliding with the proton.[4,5] According to the Standard Model only left handed fermions couple to the W–boson, the cross section should therefore vanish for fully right-handed polarised electrons and fully left-handed positrons. By setting an upper limit on a non-vanishing cross section a lower limit on the mass of a hypothetical right-handed W–boson can be set.

There is also sensitivity to some electroweak parameters of the Standard Model like the W–boson mass M_W. The ratio of Neutral Current (NC) to CC cross sections constrains the W–boson mass in the (M_W, M_t) plane. The sensitivity to the W–boson mass and the electroweak mixing angle $\sin^2\theta_W = 1 - M_W^2/M_Z^2$ provides also a test for electroweak universality.[6]

Other examples are given by the measurement of the light quark (u, d) neutral current couplings and the γZ^0 interference structure functions F_2 and xF_3 by detailed comparison of polarised NC and CC cross sections[7] and potential for new physics in, for example, searches for leptoquarks or R–parity violating supersymmetry.

2.2. *Radiative polarisation*

Acceleration in a circular machine like HERA implies crossing of many depolarising resonances, making it difficult to sustain an initial polarisation at a high degree.[8] Instead, the inevitable emission of synchrotron radiation is used to polarise the stored beam after acceleration has finished. By means of spin-flips caused by a small fraction of synchrotron emissions in the magnetic field of bending dipoles, the spin of emitting particles align parallel or anti-parallel with the transverse magnetic field, leading to a gradual build-up of polarisation. This radiative polarisation was first described by Sokolov and Ternov in 1964.[9]

In an ideal machine the build-up of radiative polarisation P proceeds exponentially with

$$P(t) = P_{\mathrm{ST}}\big(1 - e^{-t/\tau_{\mathrm{ST}}}\big) \tag{1}$$

with the asymptotic polarisation limit and build up time given by $P_{\mathrm{ST}} = 8/(5\sqrt{3}) \approx 0.924$ and $\tau_{\mathrm{ST}} \approx 100\,\mathrm{s} \cdot \rho^3/E^5 \cdot \mathrm{GeV}^5/\mathrm{m}^3$.

In a real storage ring the polarisation build-up is counteracted by several depolarising effects, causing the maximal achievable polarisation P and build-up time τ to be smaller than given by the Sokolov–Ternov effect alone. Spin diffusion in the presence of misalignments, field errors and horizontal magnetic fields along the ring weaken the polarisation build-up and careful alignment and organisation of the quadrupole strengths, known as spin matching, along with harmonic orbit corrections are needed to minimise the influence.[10] In addition, depolarising resonances are avoided by choosing a half-integer spin-tune, i.e. the number of precessions a spin performs per turn in the ring $\nu := a\gamma$ with $a = (g-2)/2$ being the electron gyromagnetic anomaly. HERA operated at $E = 27.5\,\mathrm{GeV}$ had a spin-tune of $\nu = 62.5$.

2.3. *Spin rotators*

In order to obtain longitudinal polarisation at the interaction points of the experiments, pairs of spin rotators were installed around each interaction region, exploiting spin precession arising from deflection in transverse magnetic fields. As the spin-tune ν is large, small commuting orbit deflections ϕ can be used to generate large non-commuting spin precessions $\psi = \nu\phi$.

At HERA the so-called mini–rotator design of Steffen and Buon has been adopted, consisting of a series of six alternating vertical and horizontal bends without any focussing quadrupoles within, due to its relatively small length of 56 m.[11] These dipole spin rotators allow us to have either sign of electron helicity in the longitudinal spin state, though with a maximum orbit distortion of ±22 cm, the bends had to be moved vertically upon a helicity change. The first rotator pair was installed in 1993-1994 for the HERMES experiment, followed by further pairs for H1 and ZEUS during the HERA upgrade in 2000-2001.

2.4. *Polarisation at HERA*

A typical fill of the HERA electron ring could last for more than 12 hours. Initially filled with currents of about 40 mA in 180 − 190 bunches with a spacing of 96 ns, the bunch current during a fill decreased over time due to collisions and other losses.

The polarisation at HERA was monitored independently by two fast polarimeters with a very high availability and providing real-time polarisation values to the machine and the experiments. During the HERA II running period over 99 % of all physics fills had at least one polarimeter operational. The maximum polarisation ever achieved was about 0.76 in the HERA I period before the upgrade. As spin rotators represent a source of spin diffusion, the typical equilibrium polarisation at HERA II with three spin rotator pairs was lower with $\approx 0.4-0.5$ and rise times of about 40 min. The polarisation varied from fill to fill and even within a single fill, as it was subject to the tuning of the machine. In addition, colliding and non-colliding bunches had different asymptotic polarisation values due to beam tune shifts of the colliding bunches caused by beam–beam interactions with the proton bunches.

3. The HERA polarimeters

The requirements of polarimetry at HERA were challenging. The polarimeters should measure the polarisation of the stored lepton beam continuously in a non-invasive manner, providing real-time values to the experiments and the HERA machine control with a statistical accuracy of few percent per minute measurement. In addition, the devices required the ability to handle the frequent changes in the lepton orbit. The systematical uncertainties should be small with $\delta P/P < 2.5\,\%$, if measurements relying on polarisation values shall not be dominated by polarimetry.

One polarimeter measured the transverse polarisation and two more measured the longitudinal polarisation within the HERMES straight section. The basis for all three devices is given by Compton scattering laser photons of the high-energetic lepton beam and observing the backscattered photons. The cross section for Compton scattering is sensitive to the transverse and longitudinal components P_y and P_z of the lepton beam polarisation, provided that the laser photons are circularly polarised:

$$\frac{d^2\sigma}{dEd\phi} = \Sigma_0(E) + S_1\Sigma_1(E)\cos 2\phi + S_3P_y\Sigma_{2y}(E)\sin\phi + S_3P_z\Sigma_{2z}(E) \quad (2)$$

with S_1 and S_3 being the linear and circular components of the laser light polarisation.[12] The polarisation is then measured using the asymmetry of the cross sections, when switching the laser polarisation between left and right helicity states, i.e. $S_3 = \pm 1$ with $S_1 \approx 0$:

$$\mathcal{A}(y, E_\gamma) = \frac{\sigma_L(y, E_\gamma) - \sigma_R(y, E_\gamma)}{\sigma_L(y, E_\gamma) - \sigma_R(y, E_\gamma)} \quad (3)$$

3.1. *Transverse polarimeter TPOL*

The Transverse POLarimeter (TPOL) operated throughout the complete HERA I and II periods in the straight section west near the HERA-B experiment. It measured the tiny spatial asymmetry between the left and right laser helicity states caused by the transverse polarisation.[10,13]

A green argon–ion laser, operated at 10 W in CW mode, was made circularly polarised by means of a Pockels cell, switching the helicity at a frequency of ≈ 80 Hz. The laser beam was transported by an optical system over more than 300 m into the HERA tunnel and was brought into collision with the lepton beam under a vertical angle of 3.1 mrad. The degree of light polarisation was regularly monitored behind the Compton interaction point using a rotating Glan prism, with typical polarisation values $S_3 > 0.99$.

The backscattered Compton photons were detected 65 m downstream of the Compton interaction point in a compact, $19\,X_0$ deep electromagnetic scintillator–tungsten sampling calorimeter, read out using wavelength shifter bars from all four transverse sides. To achieve sensitivity to the vertical position of the incident photon, the scintillator plates were optically decoupled along the central horizontal plane, thus dividing the calorimeter effectively into independent upper and lower halves. Information about the energy and the vertical impact position y of an incident photon is then obtained from the sum of the two halves $E = E_{\rm up} + E_{\rm down}$ and the energy asymmetry η between them:

$$\eta := \frac{E_{\rm up} - E_{\rm down}}{E_{\rm up} + E_{\rm down}}. \tag{4}$$

The operation of the transverse polarimeter relied on the single–photon mode with on average $\bar{n} = 0.01$ backscattered photons per bunch crossing, so that only one photon will be detected. While this requires relatively low photon rates of < 100 kHz, it allows use of the known kinematical endpoint of the Compton scattering process, called Compton edge, for the absolute calibration of the detector. The main background is bremsstrahlung generated along the 7.3 m short straight section which is in the line of sight of the detector. The measurements of backscattered photon distributions for the two laser helicity states were interspersed regularly with measurements where the laser light was blocked by a chopper. This allowed us to measure the background and to subtract it from the Compton data on a statistical basis.

The polarisation of colliding and non-colliding bunches is measured separately and since the upgrade in 2000-2001 also a bunchwise measurement is made possible by means of a new faster DAQ. During this upgrade, a position sensitive detector in the form of crossed silicon strip detectors, including the necessary preradiator of $1\,X_0$ thickness, has been added in front of the calorimeter. These detectors should allow for an *in situ* measurement of the intrinsic calorimeter response, the non-linear transformation between the spatial impact point y and the energy asymmetry η called $\eta(y)$–transformation.

The polarisation is then calculated from the shift of the mean energy asymmetry distributions for left and right laser helicity states using an analysing power Π:

$$\bar{\eta}_{\rm L} - \bar{\eta}_{\rm R} := \Delta S_3 P_y \Pi \tag{5}$$

as illustrated in Fig. 2 (left). At HERA I the analysing power was deter-

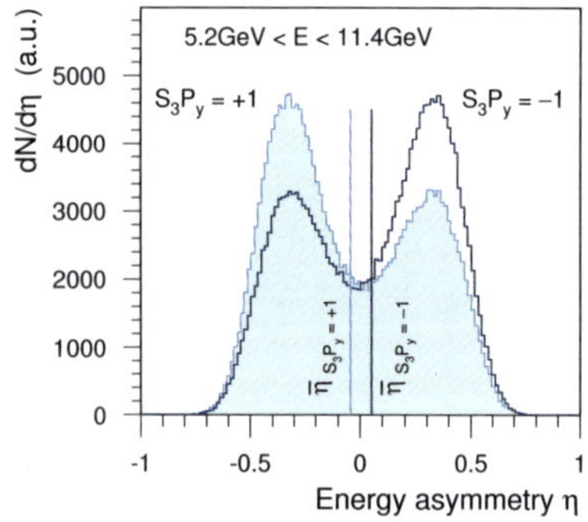

Systematic uncertainty	$\Delta P/P$
Electronic noise	$< \pm 0.1\,\%$
Calorimeter calibration	$< \pm 0.1\,\%$
Background subtraction	$< \pm 0.1\,\%$
Laser light polarisation	$\pm 0.1\,\%$
Compton beam centering	$\pm 0.4\,\%$
Focus correction	$\pm 1.0\,\%$
Interaction point region	$\pm 0.3\,\%$
Interaction point distance	$\pm 2.1\,\%$
Absolute scale	$\pm 1.7\,\%$
Total syst. uncertainty	$\pm 2.9\,\%$

Fig. 2. Illustration of the polarisation dependent shift of the mean energy asymmetry distributions (left) and the preliminary list of contributions to the fractional systematic uncertainty of the TPOL measurement (right).

mined from simulations and from rise time measurements in a flat machine, where the intrinsic relation between the asymptotic polarisation value and the rise time constant as given by the Sokolov–Ternov effect is exploited for the calibration of the absolute polarisation scale. At HERA II the beam conditions as well as the detector have changed. Both the lepton beam size and divergence as well as the longitudinal position of the Compton interaction point became more variable, influencing the photon distribution at the calorimeter surface and thus the analysing power. Also, the exchange of the calorimeter and the added dead material in front are likely to change the analysing power with respect to the HERA I running period.

The statistical uncertainty of the polarisation measurement amounts to about 2–3 % per minute of data, for single bunches to about 10 % per ten minutes of measurement. The current, preliminary estimation of systematic uncertainties amounts to 2.9 % as is shown in the table in figure 2 (right).[14] The dominant contribution is given by the analysing power Π, in the current breakdown of sources divided into three contributions given by the influence of the intrinsic beam width and divergence, the distance of the Compton interaction point and the absolute scale. While the first of the three has been corrected for since 2004,[15] for the second only an upper limit from geometrical acceptances is known. The three mentioned dominant contributions are correlated and have to be evaluated thus in a correlated fashion, using a detailed realistic simulation of the magnetic beam line and a precise modelling of the calorimeter response including $\eta(y)$–transformation and energy resolution.

3.2. *Longitudinal POLarimeter LPOL*

The second polarimeter (LPOL) measured the longitudinal lepton beam polarisation within the HERMES spin rotator pair, downstream of the HERMES gas target.[16] It went into operation in 1997 and used the sizable asymmetries in the energy distributions of the backscattered Compton photons when switching between the left and right laser helicity states.

The polarimeter operated in multi–photon mode, where on average $\bar{n} \approx 10^3$ photons are backscattered per bunch crossing. In this mode background like bremsstrahlung becomes less important. With most of the long straight section east of the line of sight of the calorimeter, bremsstrahlung background would be too high to operate in single–photon mode.

The key ingredient to such high backscattering probabilities are high power laser pulses. Generated by a frequency-doubled green Nd:YAG laser, pulsed at 100 Hz, each laser pulse had a fixed power of 100 mJ and a length of 3 ns. The laser was synchronized with the lepton bunches and a trigger for readout at twice the laser pulse frequency allowed us to measure background every second event. The circular polarisation, achieved by a Pockels cell, flipping helicity for every pulse, was regularly measured using a Glan–Thompson prism with $S_3 > 0.99$. The laser was transported with an optical system over 70 m into the HERA tunnel and collided with the lepton beam at a vertical crossing angle of 8.7 mrad. The backscattered Compton photons were detected 54 m downstream by a compact electromagnetic Cerenkov calorimeter, consisting of four 19 X_0 deep $NaBi(WO_4)_2$ crystals (NBW), read out separately. The crystals were optically decoupled and arranged in a rectangular 2×2 array to allow for a positioning of the calorimeter in the photon beam.

In the multi–photon mode, the detector signal is proportional to the integral of the energy-weighted Compton cross section:

$$I_{S_3 P_z} := \int_{E_\gamma^{\rm min}}^{E_\gamma^{\rm max}} r(E_\gamma) E_\gamma \frac{d\sigma_{\rm C}}{dE_\gamma} dE_\gamma, \tag{6}$$

with $r(E_\gamma)$ being the single–photon relative response function, a constant for a perfectly linear detector. The energy-weighted Compton cross section is shown in figure 3 (left). The energy dependent asymmetry then becomes

$$\mathcal{A} := \frac{I_{S_3 P_z < 0} - I_{S_3 P_z > 0}}{I_{S_3 P_z < 0} + I_{S_3 P_z > 0}} = \Delta S_3 P_z \Pi_z \tag{7}$$

The statistical uncertainty of the measurement is about 1–2 % per minute and about 6 % per five minutes measurement, clearly limited by the repetition rate of the laser.

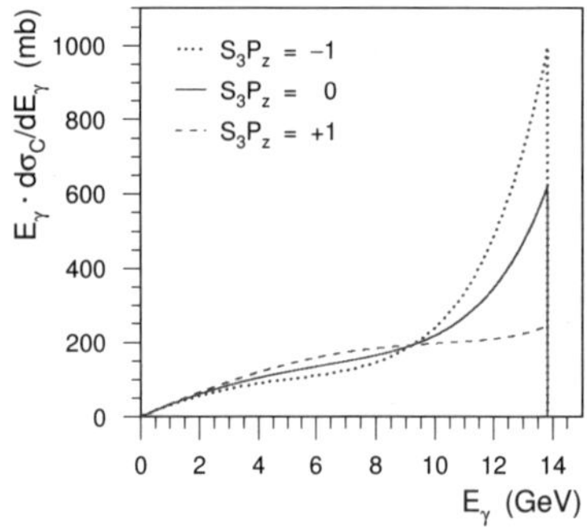

Systematic uncertainty	$\Delta P/P$
Analysing power	$\pm 1.2\,\%$
long-term stability	$\pm 0.5\,\%$
Gain matching	$\pm 0.3\,\%$
Laser light polarisation	$\pm 0.2\,\%$
Helicity dep. luminosity	$\pm 0.4\,\%$
Interaction region stability	$\pm 0.8\,\%$
Total (HERA I)	$\pm 1.6\,\%$
Extra (new calorimeter)	$\leq \pm 1.2\,\%$
Total (HERA II)	$\pm 2.0\,\%$

Fig. 3. The energy-weighted single differential Compton cross section $E_\gamma d\sigma_{\rm C}/dE_\gamma$ (left) and the list of contributions to the fractional systematic uncertainty of the LPOL measurement (right).

The current estimation of systematic uncertainties is shown in the table in figure 3 (right).[14] The dominant systematic uncertainty is given by the analysing power $\Pi_z = 0.1929 \pm 0.0017$.[17] Its main contributions are given by the shape of the single–photon response function as measured with test beam data and the extrapolation from single– to multi–photon mode. The latter was validated by attenuating the signal over three orders of magnitude using neutral density filters and monitoring the polarisation value in comparison with the independent measurement of the TPOL. After the replacement of the calorimeter crystals in 2004 the performance of the new calorimeter was ascertained in alternating measurements with a sampling calorimeter. From this an upper limit of 1.2 % systematic uncertainty due to the new calorimeter was estimated, increasing the formerly quoted HERA I systematic uncertainty to 2 %.[14]

3.3. *Cavity longitudinal polarimeter*

A third polarimeter project has been started in the early HERA II running phase employing a Fabry–Perot cavity to stock laser photons with a very high density at the Compton interaction point. Working in continuous few–photon mode, backscattering on average $\bar{n} \approx 1$ photons per bunch crossing, it combines the virtues of both existing operational methods. While providing a very high statistics with scattering rates in the order of MHz, it can make use of the Compton and bremsstrahlung edges for the calibration of the calorimeter.

The cavity polarimeter measured the longitudinal polarisation within the HERMES spin rotator pair, located about 10 m downstream of the LPOL interaction point and utilising the same detector location for the measurement of the backscattered Compton photons. After installation of the Fabry–Perot cavity in spring 2003, the first Compton events were observed in March 2005 with a much increased operation till the end of HERA. Over 500 hours of efficient data could be collected.

The cavity is driven by an infrared Nd:YAG laser with an intial power of 0.7 W, located together with all optical components on an optical table close to the cavity. Circular polarisation of the laser light is achieved by rotating quarter wave plates, flipping the helicity every few seconds, and monitored behind the cavity.[18] The cavity mirrors are located inside the vacuum vessel at 2 m distance from each other, providing a vertical crossing angle of 3.3 °. With a finesse of $\approx 3 \times 10^4$ the initial laser power is amplified by means of constructive interference with an effective gain of ≈ 5000 to about 3 kW.[19]

The measurement of the longitudinal polarisation proceeds by an overall fit of a parametrised model to the energy distributions for the two laser helicity states collected separately. Absolute calibration is done using the known Compton and bremsstrahlung edge positions. The description of the energy spectra besides the Compton spectrum also includes contributions of background like synchrotron and Compton scattered black-body radiation, the bremsstrahlung spectrum as well as detector resolution and non-linearity parameters. Detailed simulations of the calorimeter response were needed, e.g. for a precise description of the synchrotron radiation peak. The statistical uncertainty with about 3 % per bunch and 10 s doublet is unprecedented at HERA.

Based on more than 500 hours of data including dedicated data samples, most of it taken during the final stage of the HERA operation, detailed systematic studies have been performed. The preliminary list of systematic uncertainties includes the modelling of the detector response and of the synchrotron radiation peak, electronic pile-up, detector parameter fitting, a varying HERA beam, the calorimeter position and the laser polarisation inside the cavity. Whereas the contributions of parameter fitting and HERA beam variations are found to be negligible, the other contributions are of approximately the same size, adding to a total of $\delta P/P = 0.9\,\%$.[20]

4. Conclusions

The running of HERA was efficiently covered with measurements of the lepton beam polarisation. At HERA II over 99 % of all the physics fills had at least one polarimeter operational.

The preliminary estimation of the systematical uncertainties for TPOL amounts to about 2.9 % and for LPOL to 2 %. However, the agreement of the two polarimeters shows a varying behaviour over time which is not yet understood. To cover these discrepancies an additional systematical uncertainty of 3 % had been assigned, raising the uncertainty of the combined measurement to about 3.4 %.[14] Currently, efforts are under way to validate and improve the polarisation analyses of both polarimeters to decrease the systematical uncertainty of the combined measurement and final results are expected within the next few months.

The polarisation measurement with a high finesse Fabry–Perot cavity at HERA has been established, successfully operating with increasing data taking frequency till the end of HERA. The analysis of the systematical studies is nearly finished, indicating that the goal of a sub-percent systematic precision has been achieved.

Acknowledgments

Special thanks are due to T. Behnke, R. Fabbri and Z. Zhang for giving me invaluable advice for the preparation of the talk and this article. I am particularly indebted to the organisers of the PST2009 conference, whose support made my contribution possible.

References

1. G. A. Voss and B. H. Wiik, *Ann. Rev. Nucl. Part. Sci.* **44**, 413 (1994).
2. Picture adapted from ZEUS collaboration.
3. G. van der Stoonhoven, *Prog. Part. Nucl. Phys.* **55**, 181 (2005).
4. B. Antunovic, H1prelim-06-041 (2006), presented at DIS 2006.
5. S. Glazov, in *Proc. 24th Int. Symposium on Lepton and Photon Interactions at High Energies (LEP09), Hamburg, Germany, 2009.*
6. R. Beyer, E. Elsen, S. Riess, F. Zetsche and H. Spiesberger, in *Proc. Workshop on Future Physics at HERA, Hamburg, Germany, 1996.*
7. M. Klein, in *Proc. Ringberg Workshop on New Trends in HERA Physics, Ringberg Castle, Germany, 2003.*
8. E. Steffens, in *Proc. Workshop PST2007*, eds. A. Kponou and *et al.*, AIP Conf. Proc., Vol. 980 (AIP, New York, 2008).
9. A. A. Sokolov and I. M. Ternov, *Phys. Dokl.* **8**, 1203 (1964).
10. D. P. Barber *et al.*, *Nucl. Instr. Meth. A* **338**, 166 (1994).
11. J. Buon and K. Steffen, *Nucl. Instr. Meth. A* **245**, 248 (1986).
12. F. Lipps and H. A. Tolhoek, *Physica* **20**, 85 and 385 (1954).
13. D. P. Barber *et al.*, *Nucl. Instr. Meth. A* **329**, 79 (1993).
14. A. Airapetian *et al.*, POL2000-2007-001 (2007), `http://www.desy.de/~pol2000`.
15. F. Corriveau, V. Garibyan, O. Ota and S. Schmitt, internal note (2004), `http://www.desy.de/~pol2000`.
16. M. *et al.*. Beckmann, *Nucl. Instr. Meth. A* **479**, 334 (2002), DESY-00-106.
17. A. Airapetian *et al.*, HERMES Internal Report 05-47 (2005), `http://www.desy.de/~pol2000`.
18. Z. Zhang, LAL-01-87, PRHEP-HEP2001-261, hep-ex/0201033 (2001).
19. F. Zomer, LAL 03-12, Habilitation thesis, LAL (Orsay, France, 2003).
20. M. Beckingham *et al.*, & Jaquet, M.*et al.*, in preparation.

POLARISATION MEASUREMENT AT THE ILC WITH A COMPTON POLARIMETER

C. Bartels* and J. List

Deutsches Elektronensynchrotron, DESY,
Notkestrasse 85, 22607 Hamburg, Germany
**E-mail: christoph.bartels@desy.de*
www.desy.de

This article provides an overview of the conceptual design of the ILC polarimeters, the analysing power calibration and data-driven polarisation measurement. A Cherenkov detector prototype for Compton polarimetry is presented.

Keywords: ILC; polarimetry; Compton scattering; Cherenkov detector; analysing power.

1. Polarimetry at the ILC

The International Linear Collider (ILC) is planned to collide electrons and positrons at center of mass energies in the range of $\sqrt{s} = 200$–500 GeV. Longitudinal polarisation is foreseen for both the electron and positron beams. The beams are structured in bunch trains with a repetition rate of 5 Hz. For the nominal set of beam parameters, each train consists of 2800 bunches in intervalls of about 370 ns.[1] To fully exploit the ILC's potential for precision physics, it will be crucial to know the initial state of the colliding beams as precisely as possible. It is expected that the beam energy can be measured with an accuracy of 1–$2 \cdot 10^{-4}$. Polarimeters are planned both up- and downstream of the main e^+e^- interaction point (IP), allowing for fast polarisation measurement, giving feedback to the machine, reducing systematic uncertainties and adding redundancy to the entire system. For the polarisation measurement it is planned to achieve a precision of $\delta P/P = 0.25$ %.[2] This will clearly be limited by systematic effects. To reach this ambitious goal, extensive research is ongoing for the design of possible sources, polarimeters and the performance evaluation of measurement schemes. Just recently a Cherenkov detector prototype for an ILC polarimeter has been designed and constructed at DESY and put into operation at test facilities at DESY and the ELSA ring at Bonn university.

2. Sources

The ILC baseline configuration as described in the Reference Design Report (RDR)[1] already provides polarised electron and positron beams ($P_{e^-} = 80\ \%, P_{e^+} \geq 30\ \%$). The electron source is a DC photocathode gun. The electron polarisation can be flipped fast (train-by-train) by changing the helicity of the source laser with pocket cells. A measurement of the electron polarisation close to the source will be done with a Mott-polarimeter. The positrons will be produced using a 150 m helical undulator placed in the main electron linac at 150 GeV. Circularly polarised photons from the undulator hit a thin target, producing polarised e^+e^--pairs, from which the positrons are extracted. This design is expected to deliver a positron polarisation of $P_{e^+} \geq 30\ \%$.[1,3]

3. Compton polarimeters

Compton polarimetry will be used for the polarisation measurement at high beam energies. The process of Compton scattering is well known from QED,

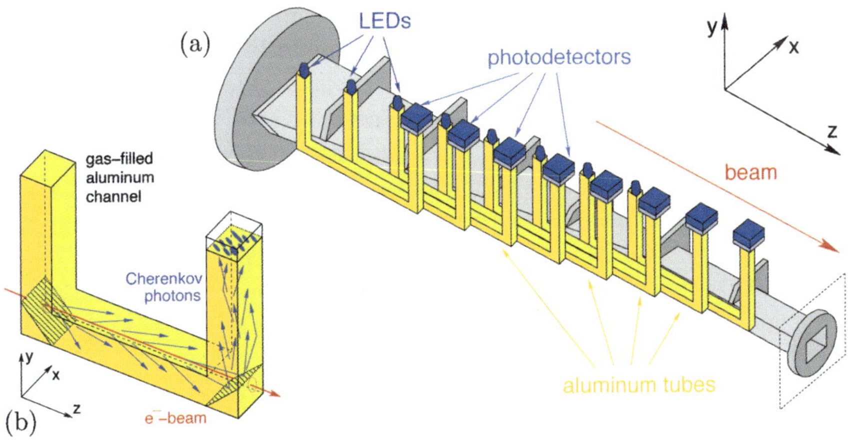

Fig. 1. (a) Conceptual drawing of an ILC Cherenkov detector with an array of U-shaped aluminum channels (yellow) covering the tapered exit window and (b) electrons traversing the basis of one U-shaped channel producing Cherenkov light.

and furthermore Compton polarimetry is relatively non-invasive, leaving the beam undisturbed and thus allowing for a polarisation measurement even during collisions.[4] The incoming electrons are scattered under an angle of $\approx$ 10 mrad with a circularly polarised laser. The cross section of Compton scattering shows a large asymmetry near the Compton edge, depending on the product of electron polarisation P_e and laser helicity λ. Rapid flipping of the laser helicity allows us to measure this asymmetry, and thus the beam polarisation. The scattered e^- and e^+ are separated from the main beam line via a magnetic chicane and the energy spectrum is sampled with a segmented Cherenkov detector (fig. 1). One Cherenkov detector consists of up to 20 staggered U-shaped aluminum channels covering the tapered chicane exit window. Each channel is filled with a high threshold Cherenkov gas (for background reduction) and has a base length of approximately 15 cm. Light produced by traversing electrons is reflected upward to photomultipliers mounted on top of one U-leg. A second leg houses a calibration system based on LEDs or allows us to couple laser light into each channel.

4. Upstream chicane

The polarimeter chicane upstream of the IP shown in figure 2, is located about 1800 m before the IP within the beam delivery system. It has a length of about 75 m and is made of four dipoles with $\approx$0.1 T field strength.[4] The Compton IP is located between dipoles two and three. A precise polarisation measurement at the main IP requires the beam at the Compton IP to be aligned with the main IP to about 50 μrad. Dipoles three and four guide the scattered particles to the Cherenkov detector located behind dipole four. The chicane is operated with a fixed field strength, keeping the position of the Compton edge and the spread of the spectrum stationary on the Cherenkov detector for a wide range of beam energies. However, due to the fixed field, the Compton IP moves laterally depending on the beam energy. This movement of up to 10 cm has to be compensated by a movable stage for the laser beam. The upstream polarimeter benefits from the clean environment and allows for fast polarisation measurement. A laser for an ILC polarimeter should be able to mimic the ILC bunch structure, and allow us to measure every bunch. A suitable laser is already in operation at the TTL/Flash source at DESY. Per bunch roughly 1000 e^- are scattered and distributed over the channels of the Cherenkov detector. The upstream chicane allows for instant recognition of intra-train variations and helps

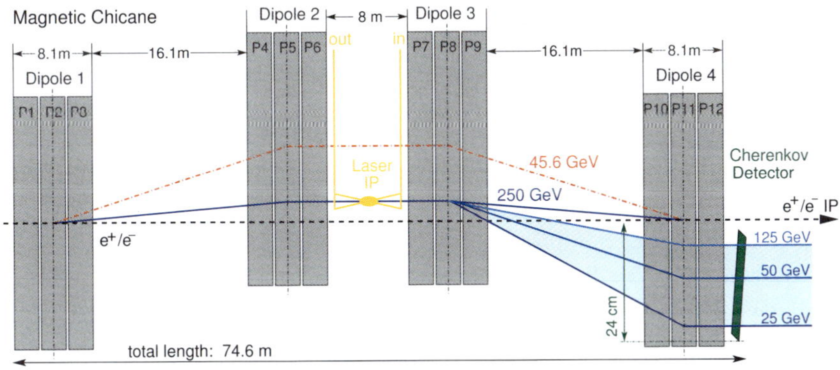

Fig. 2. The upstream magnetic chicane at 1800 m before the main collision point. The Compton IP is between dipoles two and three, the position of the Cherenkov spectrometer is indicated near dipole four.

to monitor time dependent effects in-between trains. Due to high rates of scattered electrons a statistical precision of 3 % can be expected for any bunch position after two measurements with opposite laser helicity. The average statistical precision over two entire trains will be $\Delta P/P = 0.1$ %.[3]

5. Downstream chicane

The downstream chicane 150 m after the main IP in the extraction line is conceptually very similar to the upstream chicane, one difference being a six dipole layout compared to the four dipoles of the upstream chicane. However, since the backgrounds are expected to be higher than at the upstream position (due to disrupted beams and additional diagnostic instruments in its proximity) a high power laser is required, limiting the sampling frequency. A possible setup utilises three 5 Hz lasers firing at three different bunch positions for up to a minute, and then scanning all the other bunches in a predetermined pattern. The statistical uncertainty is less than 1 % for one bunch position after one minute, and will be comparable to the upstream polarimeter after 17 hours. The downstream chicane gives a handle on depolarisation effects during beam collisions by measuring the polarisation with and without collisions. Additionally both polarimeters ought to calibrate each other outside collisions.[3]

6. Analysing power calibration

The dominant contributions to the systematic uncertainties of both polarimeters up- and downstream are the Analysing Power (AP) calibration (0.1–0.2 %), the detector linearity (0.1–0.2 %) and the laser helicity (0.1 %), resulting in an overall systematic uncertainty of 0.25 %. The AP calibration is dominated by time-independent effects like the chicane geometry and the employed magnetic fields. Fast simulation studies have been performed to evaluate the impact of a possible misalignment of the detector with respect to the beamline. To limit this contribution to the overall systematics to 0.1 %, the detector has to be aligned with a precision of 0.4 mm.[5] This seems to be possible, however it has to be pointed out that an overall analysing power calibration to a precision of 0.2 % has not yet been demonstrated. In fact, the most precise Compton polarimeter to date, which is the SLD polarimeter had a systematic uncertainty contribution due to the analysing power calibration of 0.4 %.

7. Measurement schemes

While the polarimeters give the beam polarisation on short timescales, i. e. measurements in seconds to hours, the absolute polarisation scale has to be obtained from annihilation data. Usually, well-known SM processes like Z^0 production or the decay of W^+W^- pairs are used.[6] There are currently two possible measurement schemes under investigation. The first method is the measurement of the absolute cross section in the Blondel scheme, which has been succesfully used for polarisation measurement at LEP,[7,8] and the second uses the angular distribution of semileptonic decaying W-pairs. Both schemes require that data is taken for all possible helicity configurations of the incoming beams, e. g. $++, --, +-$ and $-+$. Furthermore, it has to be assumed that the absolute values for positive and negative polarisation are the same, $|P^+| = |P^-|$. Corrections to the absolute polarisation scale are given by the polarimeters. Both schemes have their benefits and drawbacks, the Blondel scheme being applicable to a larger class of channels, while for the fit method of the W angular distribution, the luminosity spent on the physically less interesting $++, --$ configurations could be better reduced. It shows that for both methods a statistical uncertainty of less than 0.2 % can be achieved with approximatly 500 fb^{-1} for the electron polarisation, which correponds to roughly the first four years of ILC operation.[6] Overall, the polarisation measurement from the angular W distribution fit seems to require less luminosity. A great improvement is obtained when the option of a higher positron polarisation of ≥ 60 % is realised, reducing the required measurement time to achieve the same precision by more than a factor of two, with respect to the $P_{e^+} = 30$ % baseline design.

8. Cherenkov detector prototype

A two-channel Cherenkov detector prototype was constructed and put into operation by the DESY polarimetry group in 2009. It is concieved as a small version of an ILC-like detector with non-staggered channels. The prototype is build entirely of aluminum, and flooded with C_4F_{10}, which has a high Cherenkov threshold of 10 MeV. The Cherenkov length in the basis of the U-shaped channels is 150 mm, and the entire structure is housed in an aluminum box of dimensions in cm $23 \times 9 \times 15$ (L×W×H, fig. 3). Data

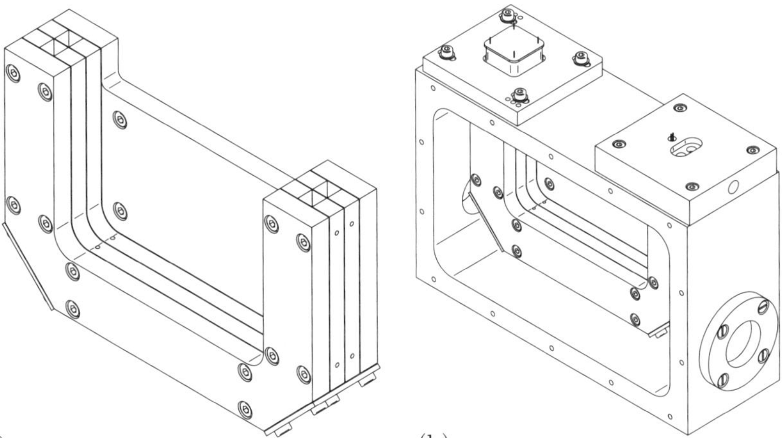

(a) (b)

Fig. 3. Technical Drawing of the inner U-shaped two-channel structure (a), and (b) the channel structure located in the outer surrounding box, with a photomultiplier mounted on the left-hand leg.

have been taken with the prototype at DESY II and at the ELSA ring in Bonn which offers higher rates and multi-electron events. The setup has been tested with four different photomultipliers, three from Hamamatsu[9] and one from Photonis.[10] First analysis results show that the data is in agreement with a detailed GEANT4 simulation accompanying the testbox construction and operation.

Due to its geometry, the light distribution on the photocathode depends on the incident position of the electrons entering the Cherenkov channels.

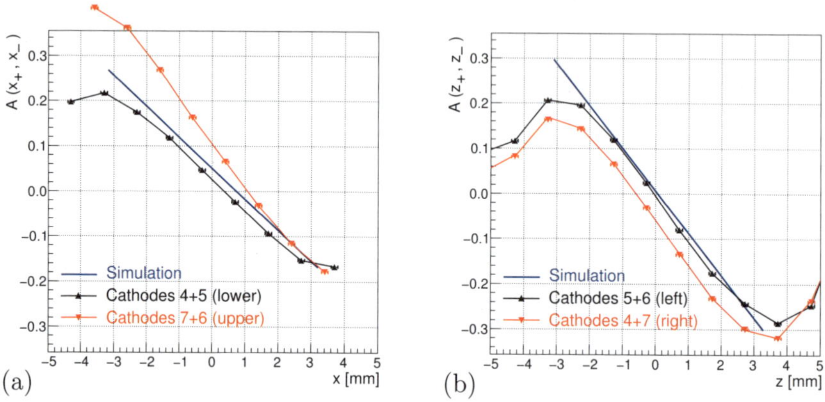

Fig. 4. Comparison of light yield asymmetry as function of (a) horizontal and (b) vertical beam position x and z between simulation and data.

By measuring the light distribution inside a channel with a segmented photomultiplier one can calculate the light yield asymmmetry regarding the left-hand side and right-hand side of the photocathode as a function of the incident beam position. Figure 4 (a) shows this asymmetry from simulation (blue line) compared to an actual measurement. Here the left-right light yield asymmetry was measured separately in the upper and lower part of the channel (denoted as cathodes 4+5 and 7+6 respectively). As can be seen, the measurement for the lower part of the photomultiplier is in good agreement with the simulation, while the second measurement shows a small offset and tilt. An interesting feature is that due to a lower reflectivity of the inner wall, which separates the two channels, an offset at the (0,0) central position occurs which could also be seen in the data. Figure 4 (b) shows the results for the top-bottom light asymmetry, here measured separately for the left-hand side and the right-hand side of the channel. Again the data is close to the simulation. A similar offset at the central position is not expected and not seen in the left-hand side of the channel. The asymmetry measured on the right-hand side of the channel deviates from the ideal expectation. A detailed and quantitative understanding of the results is subject to future studies.

References

1. N. Phinney *et al.*, LC Reference Design Report Volume 3 - Accelerator, arXiv:0712.2361 [physics.acc-ph].
2. G. A. Moortgat-Pick *et al.*, The role of polarized positrons and electrons in revealing fundamental interactions at the linear collider, *Phys. Rep.* **460**, 131 (2008). [arXiv:hep-ph/0507011]

3. B. Aurand *et al.*, Executive Summary of the Workshop on Polarization and Beam Energy Measurements at the ILC, arXiv:0808.1638 [physics.acc-ph].
4. S. Boogert *et al.*, Polarimeters and Energy Spectrometers for the ILC Beam Delivery System, in *J. Instrum.* **4**, 10015 (2009). [arXiv:0904.0122 [physics.ins-det]]
5. J. List, Presentation: Analyzing Power Calibration, https://indico.desy.de/contributionDisplay.py?contribId=23\&sessionId=8\&confId=585.
6. P. Bechtle *et al.*, Measurement of the beam polarization at the ILC using the WW production, LC-DET-2009-003.
7. A. Blondel, A Scheme To Measure The Polarization Asymmetry At The Z Pole in LEP, *Phys. Lett. B* **202**, 145 (1988). [Erratum-ibid. **208**, 531]
8. K. Mönig, The use of positron polarization for precision measurements, LC-PHSM-2000-059.
9. Hamamatsu Photonics, http://www.hamamatsu.com.
10. Photonis USA, http://www.photonis.com.

TIME EVOLUTION OF GROUND MOTION-DEPENDENT DEPOLARISATION AT LINEAR COLLIDERS

I. Bailey[a],C. Bartels[b], M. Beckmann[b], A. Hartin[b*], C. Helebrant[b], D. Käfer[b], J. List[b], and G. Moortgat-Pick[b]

[a]*Physics Department, Lancaster University, Lancaster LA1 4YB, UK*

[b]*DESY FLC, Notkestrasse 85, Hamburg 22607, Germany*

**Email: anthony.hartin@desy.de*

Future linear colliders plan to collide polarised beams and the planned physics reach requires knowledge of the state of polarisation as precisely as possible. The polarised beams can undergo depolarisation due to various mechanisms. In order to quantify the uncertainty due to depolarisation, spin tracking simulations in the International Linear Collider (ILC) Beam Delivery System (BDS) and at the Interaction Point (IP) have been performed. Spin tracking in the BDS was achieved using the BMAD subroutine library, and the CAIN program was used to do spin tracking through the beam-beam collision. Assuming initially aligned beamline elements in the BDS, a ground motion model was applied to obtain realistic random misalignments over various time scales. Depolarisation at the level of 0.1 % occurs within a day of ground motion at a noisy site. Depolarisation at the IP also exceeds 0.1 % for the nominal parameter sets for both the ILC and for the Compact Linear Collider (CLIC). Theoretical work is underway to include radiative corrections in the depolarisation processes and simulation of the depolarisation through the entire collider is envisaged.

Keywords: Depolarisation; spin-tracking; ILC.

1. Introduction

The precision physics program of the ILC requires precise knowledge of the state of beam polarisation. To that end, the Compton polarimeters intended for the ILC (one upstream and one downstream of the IP) will have to measure the polarisation with error a factor of two smaller than the previous best measurement at the SLAC SLD experiment.[1] A prototype of a high precision Cherenkov detector to record Compton scattered electrons from the interaction of a longitudinal laser and the charged beams has been developed and tested at the ELSA test beam in Bonn.[2] Further sources of uncertainty in the beam polarisation come from depolarisation processes in the accelerator. The depolarisation is due to misaligned elements along beamlines and from beam-beam processes at the interaction point (IP) of the collider. It is crucial to understand these uncertainties as a limiting factor in the overall precision of the polarisation measurement.

In general, two effects influence the spin motion in electric and magnetic fields: a) spin precession governed by the Thomas–Bargmann-Michel-Telegdi (T-BMT) equation and b) the spin-flip Sokolov-Ternov (S-T) effect via synchrotron radiation emission. Usually the spin precession effect is dominant in the beam-beam interaction at the interaction point of a collider unless the magnetic fields of the bunches are an appreciable fraction of the Schwinger critical field (4.4×10^{13} G). However for beam parameters of planned future linear colliders, the magnetic fields at collision are significant, and quantum spin-flip effects lead to depolarisation. The precision requirements for physics processes with polarized beams require then a review of the simulation of beam-beam effects at collision which is achieved by the program CAIN.[3]

For passage of polarised beams through beamlines, the field strengths of the beamline magnetic elements are much lower and the S-T effect can be neglected entirely. It is only required to simulate the spin precession and such a simulation is implemented as part of the BMAD library of beam dynamics subroutines.[4] One aim of this paper is to apply BMAD to simulations of the International Linear Collider's (ILC) Beam Delivery System (BDS) as described in the machine's Reference Design Report (RDR).[5]

Since depolarisation is a cumulative effect it is necessary to link up the simulation of the various parts of the accelerator. Assuming an intial distribution of polarisation vectors of individual charges within a bunch, the bunch can be tracked through the linac, BDS (which includes the upstream polarimeter to measure its state), through the IP collision, and in the extraction line to the downstream polarimeter. In this paper, the program PLACET[6] is used to track the bunch through the linac. Since PLACET has no polarisation implementation, no depolarisation is assumed to occur in the linac. BMAD is employed for the BDS and planned orbit correction feedbacks at the end of the linac and at the IP are implemented as PID controllers within OCTAVE.[7] A block diagram representing the general program flow is shown in figure 1.

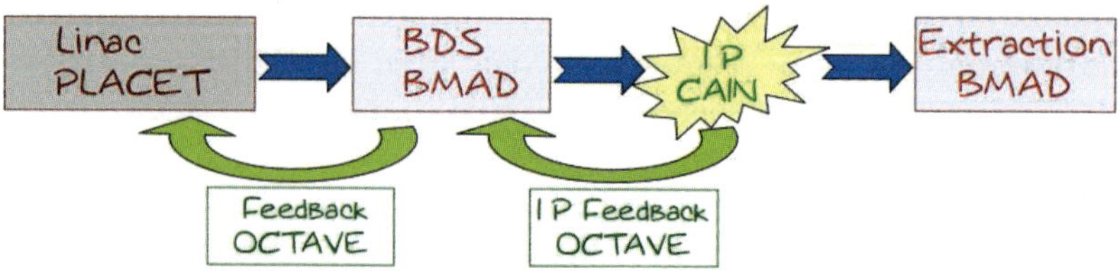

Fig. 1. Software block diagram of the spin tracking in a linear collider.

2. BDS spin precession and time dependent depolarisation

The BDS of the ILC as described in the RDR is 2226 metres long and consists of a skew correction/diagnostics section (including the upstream polarimeter), a betatron collimation section, and energy collimation section and final focus. With a single particle on the design orbit of the optical lattice of the BDS, particle spin at the IP matches with the upstream polarimeter location, and significant precession takes place in the latter half of the lattice (fig. 2).

The real orbit of the beam will not be ideal and consequently the spin precession will not exactly match at the polarimeters and IP. If the orbit

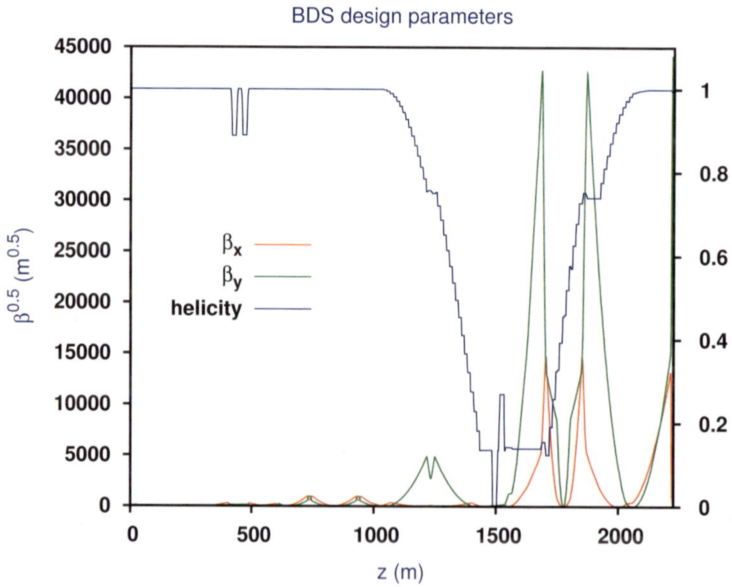

Fig. 2. Beta functions and spin precession in the BDS of the ILC.

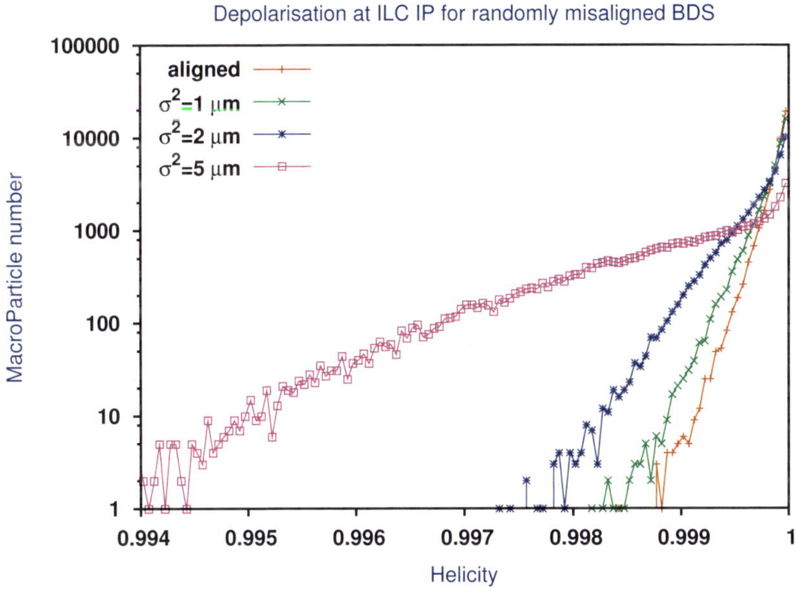

Fig. 3. Depolarisation in a bunch due to random misalignment of BDS beamline elements.

randomly varies within some distribution, the spin precession will likewise vary and depolarisation will result. Orbit variation (from the ideal) can occur because of random misalignments of magnetic elements in the beamline. The misalignments are both static in the less than perfect intial alignment, and dynamic due to natural ground motion and environmental noise.

Assuming that the initial beamline survey and results in micron level alignment, BMAD can be employed to investigate depolarisation in a bunch of 50,000 macroparticles. Defining 0.1 % depolarisation as significant within the total required precision of the ILC polarisation measurment of 0.5 %, a random misalignment of magnetic elements of up to 5 μm RMS is significant (fig. 3).

In order to know the extent of beamline misalignment between surveys, ground motion studies have been performed at potential facility sites around the world. Using broadband Streckeisen STS-2 seismometers and piezosensors, RMS amplitudes of vibration in different frequency bands and power spectra can be obtained.[8] In this study, ground motion data for a "noisy" site – the so-called ground motion model C was used.[9]

In order to apply ground motion power spectra to a beamline, correlated displacement in beamline elements over longitudinal distance and time was

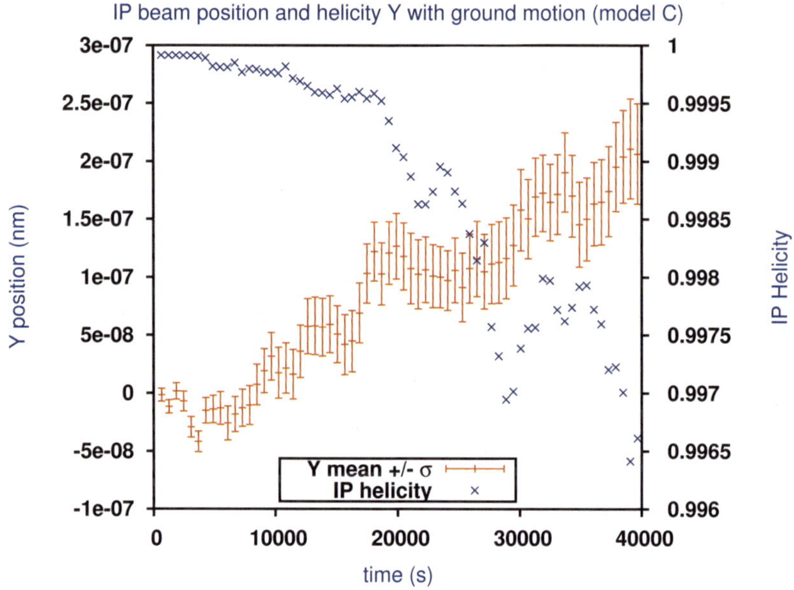

Fig. 4. Depolarisation growth due to ground motion induced misalignment.

required. Such a correlation is obtained by convoluting random offsets in the frequency domain with the measured power spectra and transforming back to the time domain. A coherency function is then used to correlate vertical motion with longitudinally separated beamline elements.[10]

Using these methods, time dependent sets of y-offsets were applied to beamline elements within the BMAD simulation of the ILC BDS. The offsets were applied only in the y direction since the beam profile is narrower in y, and consequently the orbit is more sensitive to misalignment in this direction. The net effect of the time dependent vertical displacements is a random offset in beam orbit and a corresponding increasing depolarisation over time. Within a day of ground motion induced misalignment, depolarisation becomes significant and means to recover the polarisation will need to be investigated (fig. 4).

3. Depolarisation at the IP

The CAIN program models both classical and quantum depolarization effects in beam-beam collisions and is used here to simulate the IP depolarisation for two linear collider models, the ILC and CLIC. CAIN has been modified slightly to include full polarisation of all pair producing processes in the beam-beam interaction, however the overwhelming contribution to the depolarisation is from the classical precession and from the beamstrahlung spin-flip process. The depolarisation is more significant for the aggressive set of CLIC parameters for which the magnetic field associated with the charge bunch is so high (of order of the Schwinger critical field) that the quantum effects dominate (tab. 1).[11]

Table 1. Comparison of the luminosity-weighted depolarizing effects in beam-beam interactions for the ILC and CLIC.

Parameter set	Depolarization ΔP_{lw}		
	ILC 100/100	ILC 80/30	CLIC-G
T-BMT	0.17 %	0.14 %	0.10 %
S-T	0.05 %	0.03 %	3.4 %
incoherent	0.00 %	0.00 %	0.06 %
coherent	0.00 %	0.00 %	1.3 %
total	0.22 %	0.17 %	4.8%

Since depolarisation at the IP is a significant fraction of the overall budget (i.e. it again exceeds 0.1 %) then, in the interests of precision, any variaion obtained by including radiative corrections is of concern. Even classical spin precession, as described by the T-BMT equation,

$$\frac{\mathrm{d}\vec{S}}{dt} = -\frac{e}{m\gamma}[(\gamma a + 1)\vec{B}_T + (a+1)\vec{B}_L - \gamma(a + \frac{1}{\gamma+1})\beta\vec{e}_v \times \frac{\vec{E}}{c}] \times \vec{S}, \quad (1)$$

is subject to radiative corrections by the symbol a which describes the anomalous magnetic moment of the electron in the bunch magnetic fields. The anomalous magnetic moment is only included to first order, in the approximation of ultra-relativistic electrons, and on the mass shell. The S-T equation is also subject to higher order radiative corrections. The theoretical, experimental and simulation aspects of just such studies were the topic of a recent workshop[12] and are the subject of ongoing work.

4. Conclusion

The precision requirements of physics with polarised beams requires a detailed understanding of the spin transport in all parts of a planned future linear collider. Details have been provided here of the spin transport in the BDS and during the bunch collisions at the IP, both of which contribute significant depolarisation.

The studies need be extended to further parts of the machine in order to obtain a full understanding of the spin transport. For the polarised sources, an extension of Geant is available that includes polarised particle transport.[13] The various feedbacks for orbit correction are implemented and can add to the understanding of the time evolution of the depolarisation. Spin transport in the linac can be modelled using either BMAD or Merlin[14] and the spin transport in damping rings is comprehensively studied using the SLICKTRACK program.[15]

Once all components of the simulation process are linked together, an overall understanding of the luminosity weighted polarisation at physics collision can be developed. Further work is then required to understand the value of the polarisation measurement at the upstream and downstream polarimeters.

References

1. B. Aurand *et al.*, Executive Summary of the Workshop on Polarisation and Beam Energy Measurements at the ILC, ILC-NOTE-2008-047 (2008).
2. C. Bartels *et al.*, Precision Polarimetry at the ILC: Concepts, Simulations and experiments, in *Proc. TIPP09, 2009.*
3. K. Yokoya, CAIN, `http://www-acc-theory.kek.jp/members/cain/default.html` .
4. D. Sagan, BMAD Subroutine Library for Relativistic Charged-Particle Simulations, `http://www.lns.cornell.edu/~dcs/bmad/`.
5. ILC Reference Design Report, Vol 3- Accelerator, ILC-REPORT-2007-001, (2007).
6. D. Schulte, PLACET: A program to simulate drive beams, in *Proc. of EPAC, 1402, Vienna, 2000.*
7. J. W. Eaton, GNU Octave Manual, `http://www.gnu.org/software/octave/` (2002).
8. A. Seryi *et al.*, Ground Motion Studies and Modeling for the Interaction Region of a Linear Collider, in *Proc. 20th International LINAC Conference, Monterey, California, 2000.*
9. A. Seryi, Ground Motion Studies, `http://www.slac.stanford.edu/~seryi/gm` .
10. Y. Renier and P. Bambade, Description of PLACET compatible ground motion generator, CARE/ELAN document-2007-004 (2007).
11. I. R. Bailey, A.F. Hartin *et al.*, Depolarization and Beam-Beam Effects at the Linear Collider, EUROTeV-Report-2008-026 (2008).
12. *Proc. of the Advanced QED methods for future colliders Workshop, J Phys Conf Series* **198** (2009).
13. A. Schälicke, Polarised Positron Source - Simulation, `http://pps-sim.desy.de/` .
14. N. J. Walker, Merlin `http://www.desy.de/~merlin` .
15. D. P. Barber and G. Ripken, *Handbook of accelerator physics and engineering,* eds. A. W. Chao, M. Tigner (World Scientific, Singapore, 2002).

ELECTRON BEAM POLARIMETRY AT LOW ENERGIES AND ITS APPLICATIONS

R. Barday[a*], S. Tashenov[b,c], T. Bäck[b], B. Cederwall[b], C. Eckardt[a], J. Enders[a], A. Göök[a], A. Khaplanov[b], Y. Poltoratska[a], K.-U. Schässburger[b], A. Surzhykov[d] and M. Wagner[a]

[a] *Institut für Kernphysik, Technische Universität Darmstadt, D-64289, Darmstadt, Germany*
** E-mail: barday@ikp.tu-darmstadt.de*

[b] *Nuclear Physics Department, Royal Institute of Technology, SE - 106 91, Stockholm, Sweden*

[c] *Atomic Physics Department, Stockholm University, SE - 106 91, Stockholm, Sweden*

[d] *Physikalisches Institut Heidelberg, University of Heidelberg, D-69120, Heidelberg, Germany*

Low energy ($E_k \sim 100$ keV) Mott scattering polarimetry is a widely established technique to measure the polarization of an electron beam. We analyze the feasibility of Mott scattering at energies up to 20 MeV. For further studies of the electron spin dynamics in the scattering process a correlation between the linear polarization of bremsstrahlung radiation and the electron beam polarization has been measured for the first time using a planar HPGe Compton polarimeter at the 100 keV source of polarized electrons at TU Darmstadt.

Keywords: Polarimetry; polarization; Mott scattering; polarization correlation.

1. Introduction

For experiments with polarized electron and photon beams the precise measurement of the absolute degree of the beam polarization is mandatory. Different methods to measure the electron beam polarization at low energy have been applied so far. At an energy of about 100 eV there are, for example, polarimeters based on spin polarized low-energy electron diffraction[1] and low-energy diffuse scattering spin polarimeters.[2] "Traditional" Mott polarimeters are used at beam energies typical for electron guns (50–

120 keV) and low-energy injectors (up to 10 MeV). Since Mott scattering at these energies has a very high cross section and simple, well understood kinematics, it can be used for measuring the absolute degree of the beam polarization. In this contribution, we will discribe experiments for determing the degree of polarization at the source of polarized electrons for the superconducting Darmstadt electron linear accelerator S-DALINAC.[3]

2. Source of polarized electrons

A longitudinally polarized electron beam at a kinetic energy of 100 keV is produced by illumination of the GaAs/GaAsP strained superlattice photocathode with circularly polarized light.[4] The degree of the circular polarization of the laser light illuminating the photocathode is determined by measuring the intensity of the light

$$P_{circ} = \frac{2\sqrt{I_{min} I_{max}}}{I_{min} + I_{max}} \approx 99.9\ \%, \tag{1}$$

where I_{min} and I_{max} are minimum and maximum light intensity, respectively, passed through a linear polarizer. The helicity of the light may be switched from positive to negative by changing the polarity of a Pockels cell. We use an external cavity diode laser (ECDL) in Littrow configuration to shift the wavelength of 785 nm. To measure the wavelength we use a self made Czerny-Turner spectrometer[5] with an accuracy of 0.3 nm. The laser spot diameter on the cathode can be varied between 140 and 520 μm. With the help of two mirrors the spot position on the cathode can be adjusted.

3. Mott scattering

The scattering of relativistic electrons from the bare nucleus has been considers by Mott[6]. The cross section of a transversely polarized electron beam has a right-left asymmetry due to coupling of the electron spin to its orbital motion. The cross section for elastic scattering can be written as:

$$\frac{d\sigma}{d\Omega} = \left(\frac{d\sigma}{d\Omega}\right)_{unpol} \left(1 + S(E, Z, \theta)\vec{P} \cdot \vec{n}\right), \tag{2}$$

where S is the analyzing power, $\vec{P}$ the incident electron polarization and $\vec{n}$ the axial vector which is normal to the scattering plane. The analyzing power S is large at low energy, hence Mott scattering is most useful for studying the polarization near the gun. The asymmetry function increases with increasing Z. For this reason heavy elements are often favorable as targets. Gold ($Z = 79$) is the most used target because it has high Z, can be made in a thin foil, is nonreactive and does not oxidize[a]. Thorium ($Z = 90$) and uranium ($Z = 92$) provide higher sensivity and can be used for polarization analysis, too,[7] but it is difficult to fabricate thin foils.

3.1. *100 keV Mott polarimeter*

We use as analyzing targets self supporting gold targets with thicknesses between 42.5 and 500 nm to measure the beam polarization at 100 keV. Elastically scattered transversely polarized electrons are detected by four silicon surface barrier detectors, located at azimuthal angles of 45°, 135°, 225° and 315°. The detectors are 250 μm thick, which is sufficient to completely absorb 100 keV Mott electrons. The scattering angle of 120° is defined by an aluminium collimator with a hole of 2 mm. Because of high Mott scattering probability at 100 keV, the electron current is limited to about 1 nA, in order to avoid pile-up effects. For eliminating instrumental asymmetries,[8]

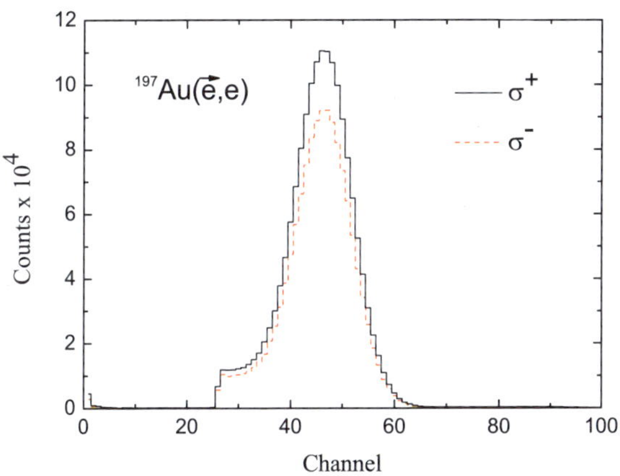

Fig. 1. Energy spectra for electrons scattered on a 122 nm gold target.

two measurements with opposite beam helicities are performed (fig. 1). But the most substantial systematic error arising in Mott polarimeters comes from multiple and plural scattering. This error is reduced by measuring the polarization with targets of several thicknesses and extrapolating the

[a]Its analyzing power for single atom scattering at 120° for 100 keV electron energy amounts to -0.391.

results to zero thickness (fig. 2). Table 1 summarizes the results of different fit functions. The average value of the asymmetry is $A_0 = 0.34$. This corresponds to a beam polarization of $P = 87$ %. Since the analytical function for the fit procedure is unknown, the accuracy of the Mott polarimetry at 100 keV is not better than 3 %.

Table 1. Extrapolated values of the asymmetry for the infinitely thin target and the fit parameters.

Fit function	a	b	c	A_0	χ^2/d.o.f.
$A(t) = a - bt$	0.328	0.00137	-	0.328	-
$A(t) = a/(1 + bt)$	0.380	0.00865	-	0.380	4.7
$A(t) = a/(1 + bt)^2$	0.321	0.00236	-	0.321	5.4
$A(t) = a + b\exp(-t/c)$	0.057	0.276	169,457	0.333	3.5

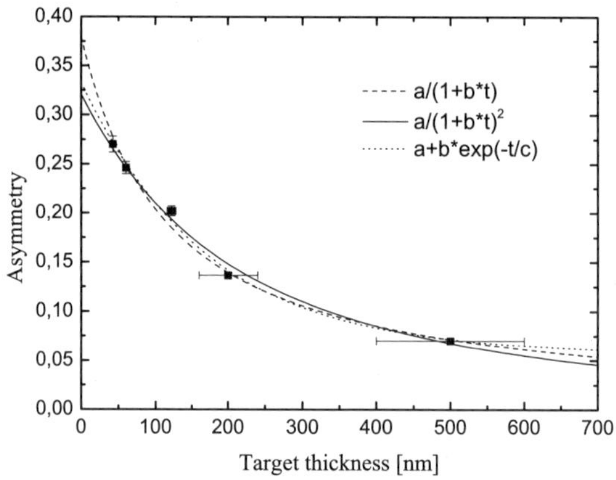

Fig. 2. Experimental asymmetry as a function of the gold foil thickness and foil thickness extrapolation.

In order to reduce the error by the determination of the beam polarization considerably, targets with thickness comparable with the elastic mean free path are desirable. For gold and 100 keV beam energy the mean free path amounts to $\lambda_{\rm mfp} \sim 7$ nm. It is extremely difficult to produce such thin targets and to install them at the accelerator. Silver, on the other hand, has $\lambda_{\rm mfp} \sim 15$ nm. Thus, silver targets may lead to a smaller systematic error of the polarization measurement, although the asymmetry is lower.

3.2. *5–10 MeV Mott polarimeter*

The superconducting injector of the S-DALINAC provides beams with energies between 5 and 10 MeV, which are used for photo-induced reactions.[9] We foresee to measure the absolute degree of polarization at these energies using a Mott polarimeter. The polarimeter design is shown in figure 3. The backward scattering angle of 165° seems to be a good compromise between the signal-to-noise ratio and the maximum of the analyzing power. The electrons scattered from gold/silver targets pass through a copper collimator within the vacuum chamber, and exit the scattering chamber through a 25 μm thick stainless steel window. The primary beam is deflected by a dipole magnet into a Al-Cu beam dump, angled at 40°. The lower total scattering probability should allow us to measure the beam polarization at microampere beam current. The dilution of the analyzing power by plural and multiple scattering is lower than at 100 keV. This makes the uncertainty due to the foil-thickness extrapolation much smaller and the measurement of the beam polarization more precise. Above beam energies of about 20 MeV the scattering angle where the analyzing power reaches its peak becomes impractically close to 180°. The Mott scattering probability becomes very small at this angle. Therefore the Mott polarimetry at higher energy is unfavorable.

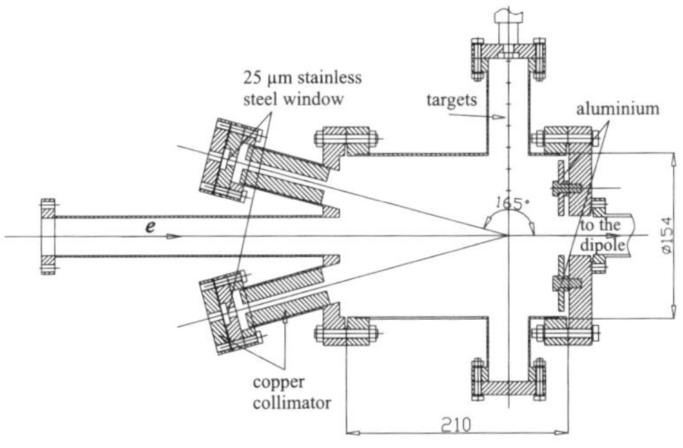

Fig. 3. Prototype of the 5–10 MeV Mott polarimeter scattering chamber.

4. Orientation of the beam polarization

As mentioned above, Mott polarimetry requires that the beam has transverse polarization with respect to the scattering plane. Furthermore at different experimental areas the electron spin should have a certain orientation relative to the beam momentum, and the polarization vector precesses during the beam transport in magnetic fields. Therefore a spin rotator is required. We manipulate the electron spin using a Wien filter.[10] The polarization vector is rotated with respect to the beam momentum to an angle of

$$\varphi = \frac{eLB}{mc\beta\gamma^2}, \tag{3}$$

where e and m are electron charge and mass, and L is the effective length of the Wien filter. Because of the strong energy dependence the effectiveness of the Wien filter falls rapidly with increasing energy. For a 100 keV electron beam a 90° spin rotation requires $B = 5.4$ mT and $E = 0.97$ MV/m. The present Wien filter has been adopted from a SLAC design; it was tested at $E = 1.1$ MV/m. This field provides ±100° spin rotation (fig. 4). By additional reversing the laser light, the spin orientation can be flipped. As a result the spin can be rotated up to 360° and any spin orientation within the rotation plane of the Wien filter can be obtained.

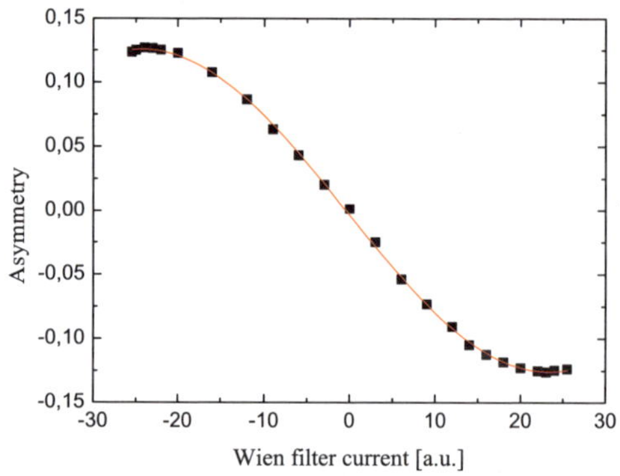

Fig. 4. Wien filter calibration data and a fit using $A = A_0 \cdot \sin(aI_{Wien} + \phi)$.

5. Bremsstrahlung of polarized electrons

Atomic field electron bremsstrahlung is a dominant radiative process in the electron-atom collisions in the 100 keV energy range. It has been predicted that the polarization of the emitted photons is correlated with the spin orientation of the incoming electrons.[11] As a first application of the recently constructed polarized electron source the dependence of the linear polarization of bremsstrahlung radiation on the polarization of the electron beam has been studied experimentally. The linear polarization of bremsstrahlung photons is described in terms of the Stokes parameters P_1 and P_2

$$P_1 = \frac{I_0 - I_{90}}{I_0 + I_{90}}, \quad P_2 = \frac{I_{45} - I_{135}}{I_{45} + I_{135}}, \tag{4}$$

where I_φ denotes the bremsstrahlung intensity component with the electric vector oriented at angle φ with respect to the emission plane. The degree and the angle of the linear polarization are then defined as

$$P_L = \sqrt{P_1^2 + P_2^2}, \quad \tan(2\chi) = \frac{P_2}{P_1} \cdot P_e, \tag{5}$$

where P_e is the degree of the electron beam polarization. It has been predicted that in the case of bremsstrahlung from polarized electrons the Stokes parameter P_2 may become non zero and thus the polarization of the emitted radiation is rotated out of the emission plane by a finite angle (fig. 5). This situation is contrary to the case of an unpolarized electron beam which can only produce bremsstrahlung polarized in the reaction plane. At the

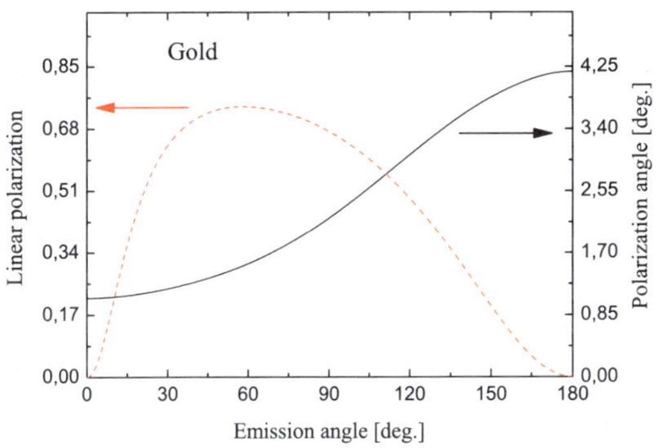

Fig. 5. Bremsstrahlung polarization from longitudinally polarized beam at the short wavelength limit.

electron beam energy 100 keV the angle χ is small and this effect hitherto has not been observed. We have addressed it by application of Compton scattering polarimetry.[12] Here the angular asymmetry of Compton scattering of linearly polarized photons is used to deduce the degree and the angle of the photon polarization. Figure 6 shows the measured scattering asymmetries of bremsstrahlung photons produced by longitudinally and transversally polarized electrons at 90° emission angle. The change in the angular distribution indicates the polarization-polarization correlation effect. The analysis of this phenomenon can bring further insight into the relativistic dynamics of the electron and its spin in the strong field of heavy nuclei. Better understanding of this dynamics may on the other hand bring further advance to the field of relativistic electron beam polarimetry.

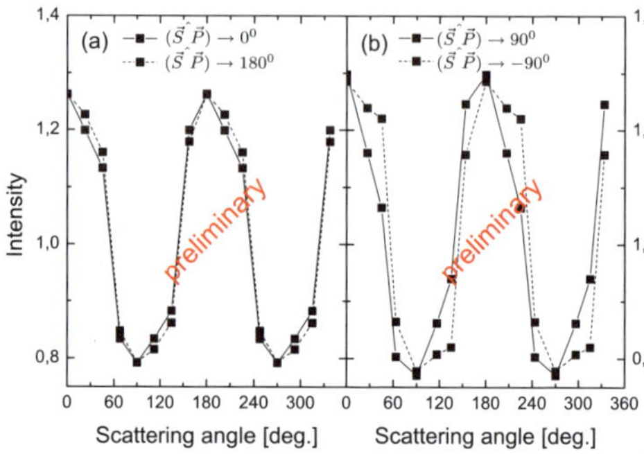

Fig. 6. Intensity distribution for Compton scattering as a function of the azimuthal scattering angle for (a) longitudinal and (b) transverse polarized electron beams.

6. Acknowledgments

The authors want to thank Th. Walther for helpful discussion concerning the laser system and K. Aulenbacher for discussions on the Mott polarimeter. This work was supported through SFB 634 of the Deutsche Forschungsgemeinschaft.

References

1. J. Sawler and D. Venus, *Rev. Sci. Instrum.* **62(10)**, 2409 (1991).
2. J. Unguris *et al.*, *Rev. Sci. Instrum.* **57(7)**, 1314 (1986).
3. A. Richter, in *Proc. EPAC'96, 1996*, 110.
4. Y. Poltoratska *et al.*, these proceedings.
5. F. Schneider, Aufbau eines Spektrometers für Wellenlängen zwischen 700 und 950 Nanometern, Bachelor Thesis (TU, Darmstadt, 2009).
6. N. F. Mott, *Proc. Royal Society (London)*, **A124**, 425 (1929), and N. F. Mott, *Proc. Royal Society (London)*, **A135**, 429 (1932).
7. J. J. McClelland *et al.*, *Rev. Sci. Instrum.* **60(4)**, 683 (1989), D. P. Pappas and H. Hopster, *Rev. Sci. Instrum.* **60(9)**, 3068 (1989).
8. A. Gellrich *et al.*, *Rev. Sci. Instrum.* **61(11)**, 3399 (1990).
9. P. Mohr *et al.*, *Nucl. Instr. Meth. A* **423**, 480 (1999).
10. M. Salomaa and H. A. Enge, *Nucl. Instr. Meth.* **145**, 279 (1977).
11. H. K. Tseng and R. H. Pratt, *Phys. Rev. A* **7**, 1502 (1973).
12. S. Tashenov *et al.*, *Phys. Rev. Lett.* **97**, 223202 (2006).

POLARIZED SOLID TARGETS: RECENT PROGRESS AND FUTURE PROSPECTS

C. D. Keith*

Thomas Jefferson National Accelerator Facility,
Newport News, VA 23606, USA
**E-mail: ckeith@jlab.org*
www.jlab.org

Polarized targets, both solid and gas, are in ever-increasing demand for nuclear scattering experiments. On top of this, the technology for producing these targets is now being applied in other fields, such as materials science and medical research. In this article the author will review recent advances that have been made in the arena of solid polarized targets.

Keywords: Polarized target; frozen spin target; dynamic nuclear polarization; HD Nuclear polarization.

1. Introduction

Nuclear-spin polarized targets have been used in experiments around the world to study a variety of subjects, including the structure of both nuclei and nucleons, the spin-dependence of the strong interaction, and fundamental symmetries such as parity and time-reversal invariance. The use of these targets in scattering experiments dates back to the 1950s, when polarized, thermal neutrons were used with statically polarized targets such as ^{55}Mn and ^{155}In to determine the angular momentum of various nuclear states.[1,2] The invention of Dynamic Nuclear Polarization (DNP) in the same decade by Abragam, Jeffries, and others signaled the advent of highly polarized proton targets suitable for use with low intensity beams of charged particles. The first of these targets were used at Saclay and at Berkeley in the early 1960s.[3,4]

In the ensuing decades, considerable technological progress has been made in every facet of solid polarized targets. New target materials for DNP with a greater percentage of polarizable nucleons and with better resistance to ionizing radiation were developed in the 1970s. More powerful refrigeration techniques, microwave sources and superconducting magnets became available at about the same time. As a result, dynamically polarized proton polarizations approaching 100 % were achieved along with beam-target luminosities of about 10^{35} cm^{-1} s^{-1}. Owing to its lower magnetic dipole moment, deuteron polarizations lagged behind, with a maximum achievable polarization of about 50 %.

Solid polarized targets can be used in either one of two operational modes. In the continuously-polarized mode, the conditions necessary for producing the target polarization are maintained while beam strikes the target. The alternative is the so-called frozen spin mode: the target is initially polarized, one or more polarizing conditions are then relaxed (e.g. a higher temperature or lower magnetic field), and the scattering data is obtained while the polarization slowly decays. In this case, the experiment must be periodically paused in order to replenish or reverse the target polarization.

While polarized *solid* targets are the subject of this review, it should be noted that two techniques for polarizing ^{3}He gas were also developed in the 1950s, Spin-Exchange Optical Pumping (SEOP)and Metastability-Exchange Optical Pumping (MEOP). However, it would be three decades before sufficiently powerful light sources were available to make polarized ^{3}He scattering targets viable instruments for nuclear physics. Since that time though, both SEOP and MEOP targets have been utilized routinely, almost always as a substitute for a polarized neutron target.

Nowadays polarized targets are used in experimental programs at nearly all nuclear/particle physics labs: Brookhaven, CERN, DESY, ELSA, Jefferson Lab (JLab), MAMI, SPRING8, etc. The importance of these targets can be illustrated with the observation that three polarized targets were operated simultaneously at JLab during the winter months of 2009: a polarized ^{3}He gas target in experimental hall A, and two polarized solid targets in halls B and C. It is also noteworthy that the technology originally developed for building these targets has expanded into other research arenas. Hyperpolarized ^{3}He and ^{129}Ze gases have been used for medical imaging for more than a decade, and DNP is now being used to hyperpolarize organic samples for similar purposes.

In this talk I review the basics of both the static and dynamic polarization methods of polarizing solid targets and briefly describe the current "state-of-the-art" for both. I also describe a recent attempt to dynamically polarize solid HD.

2. Statically polarized targets

Any nucleus with a nonzero magnetic dipole moment μ may be polarized statically by cooling the target material to a very low temperature and exposing it to a high magnetic field. The nuclear Zeemann energy levels will experience large population differences if the ratio $\mu B/kT \geq 1$. The resulting vector polarization of nuclei of spin I in the sample is given by the Brillioun function:

$$P = \frac{1}{I}\frac{\sum_m m e^{mx}}{\sum_m e^{mx}} \tag{1}$$

$$= \frac{2I+1}{2I}\coth(\frac{(2I+1)x}{2I}) - \frac{1}{2I}\coth(\frac{x}{2I}) \tag{2}$$

where $x = \mu B/kT$. Higher orders of orientation such as tensor polarization, or alignment, can be defined for nuclei with spin $I > 1/2$, but these will not be addressed here.

In certain cases the magnetic field that aligns the nuclear spins is generated internally, either by a strong hyperfine interaction or a ferromagnetic phase within the material. More commonly though, an external magnet is used, for which the term "brute-force polarization" has been coined. The degree of polarization is then limited by the size of the nuclear moment, the strength of the applied field, and the ultimate temperature to which the sample can be cooled. A partial list of nuclei polarized in this manner is given in table 1.

Table 1. Examples of brute-force polarized targets.

Isotope	Sample	B(T)	T(mK)	P(%)	Ref.
^{1}H	TiH_2	9.0	12	78	5
^{2}H	ZrD_2	7.5	40	6.7	6
^{3}He	solid	7.0	12	38	7
^{27}Al	metal	9.0	10	49	8
^{93}Nb	metal	7.0	9.4	56	9

For most nuclei μ/k is in the range of a few millikelvin per Tesla, and so fields of several Tesla and temperatures of a few millikelvin are desired. While these conditions can be met with modern-day superconducting magnets and dilution refrigerators, the low temperature requirement restricts this method to very low luminosity experiments with neutral beams, while the superconducting magnet restricts the acceptance of scattered particles. Furthermore, weak coupling between the nuclear spins and phonons in the sample material can result in unacceptably long polarizing times for non-metallic samples. Finally, the target polarization can only be reversed by reversing the magnetic field. For these reasons, static polarization is not a widely applied technique.

One statically polarized target material that has received continued interest in recent decades is solid hydrogen deuteride, HD. The ground state configurations of both molecular hydrogen and deuterium (para-H_2 and ortho-D_2) are magnetically inert and can not be polarized in solid form. Molecular HD does not have this restriction and has long been viewed as an attractive material for polarized targets due to its ideal dilution factor f

$$f = \frac{\#\ \text{polarizable nucleons}}{\text{total}\ \#\ \text{nucleons in sample}}. \tag{3}$$

Honig[10] first proposed that this material could be polarized via brute force, and utilized in beam experiments as a frozen spin target due to the extremely long spin-lattice times of H and D nuclei in the HD molecule. The nuclei are polarized with the aid of a small quantity of ortho-H_2 and para-D_2 added to the HD sample. These magnetic species act as a "relaxation switch", initially promoting the relaxation of the H and D spins to the lattice temperature. After a sufficient time has elasped, the o-H_2 and p-D_2 convert to their ground-state, nonmagnetic counterparts. This removes the primarily relaxation path for the H and D nuclei, and the polarization is thus "frozen" at a high value.

Over the last two decades a viable target based on Honig's ideas, *HDice,* has been constructed by a team originally based at Brookhaven National Laboratory and now at JLab. To enhance cooling, aluminum wires are embedded into the HD sample and reduce its dilution factor from 100 % to about 80–85 %. While this is still better than any other solid proton or deuteron polarized target, drawbacks to Honig's scheme do exist, including:

(1) A long time is required to polarize an HD sample – up to six months. Much of this time is spent waiting for the o-H_2 and p-D_2 molecules to convert to the corresponding nonmagnetic species.
(2) The cryogenics required for the target are challenging. The sample is transferred between no fewer than four separate cryostats before it is placed in the beam.
(3) Because it operates in the frozen spin mode, the target is very susceptible to beam heating and radiation damage. This limits its use to low luminosity experiments.

Further details about the HDice target and its future use at JLab are discussed in these proceedings by Wei.[11]

3. Dynamically polarized targets

Dynamically polarized targets offer both higher and faster proton and deuteron polarizations compared to their statically polarized counterparts. In addition, the field and temperature requirements are less strict, allowing use in higher luminosity and/or charged particle experiments. The primary drawback is that the DNP technique is successful with only on a handful of materials, and none with a dilution factor higher than 50 %.

To realize DNP, the material must contain paramagnetic centers (i.e. free or quasi-free electrons) with concentrations up to approximately 10^{19} cm^{-3}. These centers are added to the material via chemical doping or by ionizing radiation, and can be fully polarized under brute-force conditions of $B/T \gtrsim 5$. This electronic polarization can be transferred to nearby nuclei via the off-center saturation of the centers' ESR line with microwave irradiation. The polarization transfer can occur due to one or more mechanisms (solid effect, cross effect, thermal mixing, ...) depending on the properties and density of the paramagnetic centers.[12] A brief of list of modern-day DNP materials and their properties is presented in table 2 below. A more thorough list can be found in the review of Goertz, Meyer, and Reicherz.[13]

Table 2. A sampling of modern-day DNP target materials. The dilution factor f and achievable polarizations P for each material are given.

Name	Butanol	Ammonia	Lithium Hydride
Formula	C_4H_9OH	NH_3	^7LiH
Dopant	Chemical	Irradiation	Irradiation
f (%)	13.5	17.6	25.0
P (%)	90–95	90–95	90
Name	d-Butanol	d-Ammonia	Lithium Hydride
Formula	C_4D_9OD	ND_3	^6LiD
Dopant	Chemical	Irradiation	Irradiation
f (%)	23.8	30.0	50.0
P (%)	70-80	50	55
Comments	Easy to prepare	Polarizes well at 5 T / 1 K	Slow polarization and relaxation

Historically, the polarization of deuterons has lagged behind that of protons by about a factor of two. Within the framework of thermal mixing this can be understood in terms of the deuteron's smaller magnetic moment. In the high-temperature limit a simple, qualitative expression for the maximum polarization can be written in terms of the nuclear magnetic moment μ and the ESR linewidth of the paramagnetic centers D,

$$P_{\max} = \frac{I+1}{3}\sqrt{\frac{T_D}{T_Z}\frac{\mu B}{\hbar D}}P_e \tag{4}$$

Here T_D and T_Z are the characteristic temperatures describing the Boltzmann population distributions of the Dipolar and and Zeemann energy levels populated by the paramagnetic centers, while P_e is the thermal equilibrium polarization of the centers. For additional information the reader is directed to an introductory article by Goertz *et al.*[14] and references therein. From the above equation it is apparent that a narrow ESR linewidth is essential for maximizing the nuclear polarization, particularly for nuclear with a small dipole moment like the deuteron. In recent years a substantial improvement in deuteron polarization was reported by the Bochum group for d-butanol and d-propanediol.[15] In lightly irradiated samples of d-butanol, a polarization in excess of 70 % was obtained at 5 T and $\sim$ 200 mK, and 80 % could be reached at 2.5 T in samples of d-butanol and d-propanediol doped with trityl radicals,[16] recently synthesized for medical imaging purposes.

High cooling power ^{4}He evaporation refrigerators permit the use of targets continuously polarized by DNP with relatively intense beams of charged particles. At JLab electron beam currents up to 120 nA have been utilized with 3 cm long targets of NH_3 and ND_3. Ammonia is the most highly utilized material for intense beams due to its resistance to radiation damage. This damage, which is detrimental to the DNP process, can largely be repaired by annealing the ammonia at 80–100 K for a short period of time. Proton (deuteron) polarizations up to 95 % (50 %) have been achieved at 1 K and 5 T.

A drawback to targets of this type is the large superconducting magnet used to generate the homogeneous field ($\Delta B/B \lesssim 10^{-4}$) necessary for DNP. The geometry of the magnet can severely limit the solid angle available for observing scattered particles, and for this reason the frozen spin target was invented.[17–19] The operation of the frozen spin target is depicted schematically in figure 1. The target material is periodically polarized with microwaves via DNP in a high field, high homogeneity "polarizing" magnet. The microwaves are switched off, and the target is transferred to a second "holding" field where the polarization slowly decays while the experimental scattering data are acquired. In order to realize long depolarizing times, the material must be cooled to a temperature of 50 mK or lower using a ^{3}He-^{4}He dilution refrigerator. The same refrigerator cools the target during the DNP process, warmed to 200–300 mK by the microwaves.

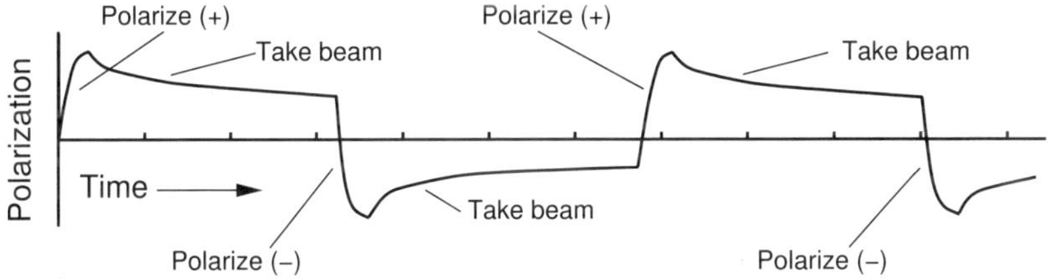

Fig. 1. Schematic representation of the operation of a frozen spin polarized target.

In the first generation of frozen spin targets the holding field was generated by the fringe of the polarizing magnet, by a magnetic spectrometer, or by a separate, dedicated holding magnet located outside the target target cryostat. Recently constructed targets have utilized a superconducting coil attached to a heat shield inside the target cryostat and thin enough (~1 mm) to permit scattered particles to pass through with an acceptably low energy loss. The field produced by the internal coil can be uniform enough to resolve the target's NMR signal and permits polarization measurements while in the holding mode. Internal solenoids were first implemented by Niinikoski[20] and further developed and utilized in scattering experiments by the Bonn group[21] with fields up to 0.4 T. Recent examples have been constructed at JLab[22] and Mainz[23] with fields of 0.56 T and 1.0 T, respectively. A four-layer, 0.54 T racetrack-shaped dipole for transverse polarization has also been constructed and successfully tested at JLab.

Butanol and propanediol are the most frequently utilized materials for frozen spin targets, in part because of their ease of handling. TEMPO-doped butanol beads (1.0–1.5 mm diameter) were used in the JLab target and could be polarized up to 95 % at 5 T and 0.3 K. Under a photon flux of 5×10^7 s^{-1}, a 1/e relaxation time of 2800 hours was observed at 30 mK and 0.56 T for positive polarization, and about one-half that value for negative polarization. This difference between the positive and negative decays has been observed before and possibly arises from NMR-induced stimulated emission of the negative spin state. Reversal of the target polarization at

JLab was performed every five to seven days and required six hours from beginning to end, resulting in a beamtime efficiency better than 95 %.

4. Dynamic polarization of solid HD

As mentioned in section 2, solid HD is a very attractive material for polarized targets because of its high dilution factor. Attempts to dynamically polarize HD have been largely unsuccessful. In 1974, Solem[24] reported a proton polarization of about 4 % at 1.24 T and 1.2 K. Paramagnetic centers consisting of trapped H atoms with an estimated density of 3×10^{18} cm^{-3} were created by irradiating the sample with a 60 MeV brehmsstrahlung beam. A small O_2 impurity ($\sim 10^{-4}$) was added to the HD in order to reduce the H atom relaxation time from 95 ms to about 0.1 ms. More recently, Radtke *et. al*[25] irradiated a sample of HD at 1 K using a Sr^{90} source to produce a paramagnetic density of H atoms of about 10^{18} cm^{-3}. No enhancement of polarization was observed at 70 GHz (2.5 T). This was attributed to a very short proton relaxation time due to isotopic impurities in the sample.

At JLab, we have attempted to dynamically polarize HD using TEMPO as the paramagnetic center. The TEMPO was evaporated upon a sample of aerogel at approximately 80 °C. Samples with spin densities ranging from 0.5×10^{19} to 10×10^{19} spins/cm^3 were produced. One sample with 4×10^{19} spins/cm^3 was crushed to a powder and poured into a 1 cm^3 PCTFE container with a small NMR coil wound around the outside. High purity HD gas (0.2 % H_2 and 0.1 % D_2), provided by the JLab HDice group, was then condensed into the container at 1 K and 5 T. Microwave exposure at 1 W and 140 GHz failed to produce any polarization enhancement above the observed thermal equilibrium (TE) signal. The proton spin-lattice time T_{1p} was determined to be less than 1 s by warming the sample with a heater and watching the growth of the TE signal back to its 1 K value after the heater was switched off. A resistance thermometer inside the sample container was used to make bolometric ESR measurements of the TEMPO (fig. 2). Similar measurements were performed on a sample of TEMPO-doped butanol (2×10^{19} spins/cm^3) using the same sample container. Contrary to the HD, this sample could be polarized to approximately 30 %. Additional investigations are planned.

5. Summary

Following more than forty years of development, polarized solid targets have become invaluable and widely-utilized instruments in the field of subatomic

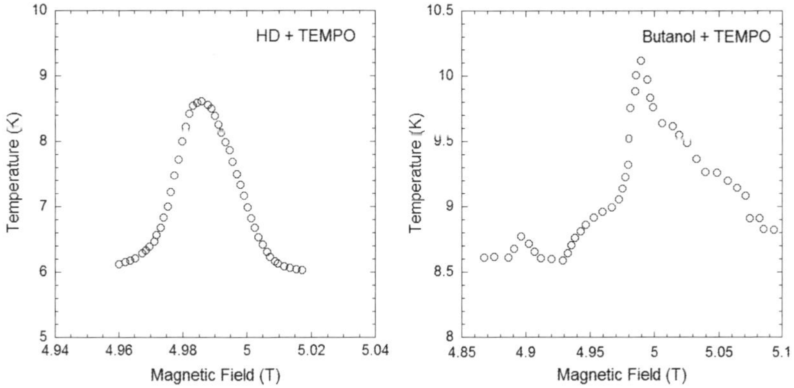

Fig. 2. Bolometric EPR signals of solid HD and butanol samples doped with TEMPO. The temperature measurements are somewhat inaccurate due to microwave absorption of the resistance thermometer.

physics. Dynamically polarized targets are more prevalent than their statically polarized counterparts in part because of their versatility. Targets continuously polarized at 1 K and 5 T can be used with beam intensities approaching 10^{11} particles/s, although the solid angle for detecting scattered particles is limited to about $1/4\ \pi$ or less. On the other hand, frozen spin targets are better suited for less intense beams (10^8 particles/s) but provide scattering angles approaching 4π. In both causes proton polarizations in excess of 90 % and deuteron polarizations up to 50 % are possible. In recent years deuterons polarizations of 70–80 % have been demonstrated in lightly irradiated d-butanol as well as in trityl-doped d-butanol and d-propanediol.

In many respects solid HD is the ideal polarized target material for nuclear physics. Thus far only static methods have produced reasonably polarized samples, limiting its use to low intensity, neutral-particle experiments. However, tests with very low electron-beam currents are planned at JLab in the near future using the HDice target. Attempts to dynamically polarize HD, either lightly irradiated samples or samples doped with the paramagnetic radical TEMPO, have been disappointing. This lack of success can be explained in part by too-high concentrations of ortho-H_2 which causes the proton spin-relaxation time to be too short for DNP. The ortho-H_2 concentration can be reduced either by aging the HD sample for an extended period of time at 4.2 K prior to polarizing, or by using a more highly purified sample.

Finally, I conclude with the observation that nuclear medicine has begun to utilize dynamic polarization techniques invented more than forty years ago for nuclear physics. The use of DNP to hyperpolarize organic compounds for medical imaging is a recent trend that is expected to expand in coming years. As a result both sides will benefit from mutual collaboration between these two communities. The invention of trityl paramagnetic radicals by medical researchers is one such example.

Acknowledgments and Appendices

Authored by Jefferson Science Associates, LLC under U.S.DOE Contract No. DE-AC05-06OR23177. The U.S. Government retains a non-exclusive, paid-up, irrevocable, world-wide license to publish or reproduce this manuscript for U.S. Government purposes.

References

1. S. Berstein *et al.*, *Phys. Rev.* **94**, 1243 (1954).
2. J. W. T. Dabbs *et al.*, *Phys. Rev.* **98**, 1512 (1955).
3. A. Abragam *et al.*, *Phys. Lett.* **2**, 310 (1962).
4. O. Chamberlain *et al.*, *Phys. Lett.* **7**, 293 (1963).
5. R. Aures *et al.*, *Nucl. Instr. Meth.* **224**, 347 (1986).
6. H. Postma, *Hyp. Inter.* **61**, 1261 (1990).
7. C. D. Keith *et al.*, *Nucl. Instr. Meth. A* **357**, 34 (1995).
8. W. Heeringa *et al.*, *Phys. Rev. Lett.* **63**, 2456 (1989).
9. C. R. Gould *et al.*, *Phys. Rev. Lett.* **57**, 2386 (1986).
10. A. Honig, *Phys. Rev. Lett.* **19**, 1009 (1967).
11. X. Wei, these proceedings.
12. M. Goldman, *Spin Temperature and Nuclear Magnetic Resonance In Solids*, (Oxford University Press, London, 1970).
13. St. Goertz *et al.*, *Prog. Part. and Nucl. Phys.* **49**, 403 (2002).
14. S. T. Goertz *et al.*, *Nucl. Instr. Meth. A* **526**, 28 (2004).
15. S. T. Goertz *et al.*, *Nucl. Instr. Meth. A* **526**, 43 (2004).
16. J. Wolber *et al.*, *Nucl. Instr. Meth. A* **526**, 173 (2004).
17. T. J. Schmugge and C. D. Jeffries, *Phys. Rev. A* **138**, 1785 (1965).
18. P. H. T. Banks *et al.*, *Rutherford Laboratory Report A* **81** (1970).
19. T. O. Niinikoski and F. Udo, *Nucl. Instr. Meth.* **134**, 219 (1976).
20. T. O. Niinikoski, *CERN-EP79-19.*
21. H. Dutz et al. *Nucl. Instr. Meth. A* **356**, 111 (1996).
22. C. D. Keith, *Proc. 18th International Spin Physics Symposium*, eds. D. G. Crabb, D. B. Day, S. Liuti, X. Zheng, M. Poelker and Y. Prok, AIP Conf. Proc. **1149**, 886 (AIP, New York, 2008).
23. A. Thomas, *private communication.*
24. J. C. Solem, *Nucl. Instr. Meth.* **117**, 477 (1974).
25. E. Radtke *et al.*, *Nucl. Instr. Meth. A* **526**, 168 (2004).

HD GAS DISTILLATION AND ANALYSIS FOR HD FROZEN SPIN TARGETS

A. D'Angelo*, A. Fantini, C. Schaerf and V. Vegna

Dipartimento di Fisica, Università di Roma Tor Vergata,
and INFN Sezione di Roma Tor Vergata Via della Ricerca Scientifica, 1 I-00133 Roma, Italy
** E-mail: annalisa.dangelo@roma2.infn.it*

B. Buick, S. Del Gobbo and W. Richter

Dipartimento di Fisica, Università di Roma Tor Vergata,
Via della Ricerca Scientifica, 1 I-00133 Roma, Italy

E. Speiser

ISAS - Institute for Analytical Sciences, Berlin Dept.,
Albert-Einstein-Str.9 12489 Berlin, Germany

A. Deur, T. Kageya, M. Lowry, A. Sandorfi and X. Wei

Thomas Jefferson National Accelerator Facility,
Newport News, Virginia 23606, USA

C. S. Whisnant

James Madison University,
Harrisonburg, Virginia 22807, USA

The production of HD targets relies on a longitudinal relaxation time switch mechanism. The longitudinal relaxation time of solid HD samples strongly depends on the concentration of ortho-hydrogen and para-deuterium in pure HD. At low temperatures these contaminants decay into H_2 and D_2 molecular ground states and the reduction of their concentration causes a dramatic increase of the longitudinal relaxation time of H and D in the HD solid. This is obtained by aging the target sample, keeping it at about 10 mK temperature while a 15-17 Tesla magnetic field is applied. The ortho-hydrogen and para-deuterium concentrations in the HD gas to be polarized are therefore critical parameters for the whole polarization process. A careful procedure for distilling commercial HD gas and storing the purified gas has been developed. An accurate technique to analyze the HD gas before and after the polarization procedure, which is based on gas chromatography and Raman scattering, has also been setup to optimize the aging time as a function of the initial concentration of the contaminants.

Keywords: Solid polarized targets; Raman scattering; gas chromatography.

1. Introduction

The use of polarized HD targets is becoming very attractive in nuclear and sub-nuclear physics experiments because of the several advantages provided by this new technology.[1] The fraction of free polarized protons exceeds any other type of polarized target and, considering the contribution from bound protons and neutrons in the polarized deuteron, very high dilution factors for both nucleons are obtained. H and D nuclei may be independently polarized and their polarization may be easily reversed. With present technologies the polarization degree may reach values as high as 95 % and 66 % for H and D respectively. When frozen-spin conditions are met targets may be *cold transported*, stored and used in experiments keeping them at values of temperatures (T) and magnetic fields (B) that are compatible with complex and large solid angle detectors ($B = 1$ T and $T = 0.5$ K).[2]

The price to pay is a long and complicated production cycle, which may still need some research and development activity to be optimized. The whole procedure is based on symmetry properties of molecular hydrogen isotopes. The next section will point out how symmetry properties constrain the possibility of polarizing different molecular hydrogen isotopes. HD gas distillation and analysis will be covered in the following section and details about the Raman scattering technique setup in Rome to analyze the relative content of isotopes in a mixture are finally explained.

2. Hydrogen isotopes properties and nuclear polarization

Homo-nuclear H_2 and D_2 molecules must obey symmetry constraints. Since protons are spin 1/2 fermions, the molecular wave-function must be anti-symmetric under the exchange of identical nuclei. On the contrary the D_2 molecular wave-function must be symmetric under the exchange of spin 1 deuterons.

In the Born-Oppenheimer approximation the molecular wave-function may be separated into the product of the nuclear, electronic, vibrational and rotational functions, each one dependent on the respective degrees of freedom only. Nuclear exchange corresponds to space inversions for electronic and space variables. Therefore, the wave-function symmetry is directly related to its parity. Both vibrational and electronic ($P = (-1)^L$ where L

is the orbital eigen-value) ground state wave-functions are symmetric with respect to space inversion. The rotational levels are symmetric for even values of the eigen-values J, and anti-symmetric for odd values ($P = (-1)^J$). In the case of H_2 molecules the nuclear wave-function is symmetric when the two nuclear spins couple to a total value $I = 1$ (ortho-hydrogen) and anti-symmetric for $I = 0$ (para-hydrogen). The result is that the wave-function symmetry limits the rotational and nuclear states combinations: J odd rotational eigen-values are coupled to ortho-H_2 while J even values are coupled to para-H_2. For D_2 molecules the nuclear wave-function is symmetric when the two $I_d = 1$ nuclear spins couple to $I = 0, 2$ (ortho-deuterium) and it is anti-symmetric for $I = 1$ (para-deuterium). Again J odd rotational eigen-values may be coupled to $I = 1$ nuclear state and the J even values to the $I = 0, 2$ nuclear states, only.

$I = 1$ molecular nuclear states are the only ones that are easily polarizable, but, being coupled to J odd values of rotational states, they are meta-stable. Decays from $I = 1$ and $J = 1$ ortho- to $I = 0$ and $J = 0$ para- states are inhibited ("forbidden") because two transitions must happen simultaneously within the same molecule: an E1 molecular transition to change the rotational state and an M1 nuclear spin-flip to preserve the symmetry of the wave-function. The consequence is that homo-nuclear H_2 and D_2 ground states may not be used to produce polarized targets. For ethero-nuclear HD molecules these constraints do not apply and H and D nuclei may be independently oriented in the molecular ground state. This property makes HD molecules ideal for polarized targets. High magnetic fields (B = 15–17 T) at very low temperatures (T = 10 mK) may align their nuclear spins in the direction of the magnetic field. The maximum degree of polarization that can be obtained at thermal equilibrium is ruled by the Brillouin function:

$$P = B_I(x) = (\frac{2I+1}{2I})coth[(2I+1)x] - \frac{1}{2I}coth(x), \qquad (1)$$

where $x = \mu B/K_B T$ depends on the ratio B/T, the nuclear magnetic moment μ and the Boltzmann constant K_B. The highest obtainable degree of polarization is $P = 0.91$ for hydrogen and $P = 0.30$ for deuterium for $B/T = 15$ T/10 mK.

These extreme environmental conditions, that allow for high nuclear polarization at thermal equilibrium, are not compatible with any particle detector typical of nuclear and sub nuclear experiments. Moreover it has been found[3] that for pure HD, solid direct spin-lattice relaxation mechanisms are suppressed and both longitudinal relaxation time for hydrogen (T_1^H) and deuterium (T_1^D) in HD solid are extremely long.

If *impurities* of ortho-hydrogen molecules are present in HD solid, an important relaxation mechanism is due to cross-relaxation of polarized nuclei in H_2 with neighbor protons in HD, which have the same Larmor frequency.[4] Longitudinal relaxation time T_1^H is strongly dependent upon the concentration of ortho-H_2 *contaminants*: T_1^H is as low as a few minutes for ortho-H_2 concentrations of the order of 10^{-3} and increases to the order of months for concentrations of 10^{-6}.

Since the energy difference between ortho- and para-H_2 molecular states corresponds to $\Delta T = 172$ K, if the HD sample is kept at low temperatures the ortho-H_2 contaminants decay into the para-H_2 state with a decay time of $\tau_H = 6.3$ days. By preparing the initial concentration of ortho-H_2 contaminants in the HD gas to be of the order of 10^{-4}, the relaxation time T_1^H may be kept short enough to reach the equilibrium polarization value in a few days.[2] Leaving the HD at low temperatures and high magnetic field for a period of time longer than four times the ortho-H_2 decay time (one or two months) the concentration of contaminants decreases by two orders of magnitude and the relaxation time T_1^H increases to values of the order of a few months. This spin-lattice relaxation *switch* is the key feature of the whole polarization procedure: by aging the solid HD sample at $T = 10$ mK and $B = 15 - 17$ T for some months a frozen-spin polarized H target is obtained.

The same procedure could be applied to polarize deuterium nuclei by introducing para-D_2 contaminants in the HD gas. However the para-D_2 decay time $\tau_D = 18.6$ days requires very long and impractical aging time and the final thermal equilibrium degree of polarization is quite low. An *adiabatic fast passage*[2] has been developed to transfer the polarization from H to D in the HD solid. Details of the polarization instrumentation, technologies and applications may be found in reference 5.

3. HD gas distillation and analysis

The polarization procedure critically depends on the initial concentration of ortho-H_2 in the HD gas. Commercial HD gas is 98 % pure and contains concentrations of H_2 and D_2 contaminants at the level of 1.5 % and < 0.5 %, respectively. Concentrations useful for a frozen-spin target are two order of magnitudes smaller. A distillation procedure as been developed at the James Madison University to purify the HD gas and optimize the ortho-H_2 concentration. The principle of operation is based on the fact that at low temperatures the vapor pressures of H_2, D_2 and HD differ. The distiller operates at $T = 20$ K. It is cooled by a Gifford-McMahon refrigerator and a temperature gradient is set along the distiller tube between the lower still pot, connected to a boil-up heater, and the upper cold finger. A Stedman packing, made of 30 double layers of stainless steel screen mesh and located along the distillation tube, amplifies the vapor pressure difference among the different isotopes and enables the purification of contaminants by one order of magnitude for each distillation cycle. The gas is distilled in batches of 12 moles. Once the system reaches a steady state the gas stratifies in the distiller column and may be extracted at a rate of 1 mole/day. First three extracted moles consist of H_2 enriched HD gas, six moles of purified HD gas follow while the three moles remaining in the the still pot consist of D_2 enriched HD gas. The extracted gas is stored in tanks each containing two moles of gas. A double-distillation process allows the reduction of contaminants to the required level of a few hundred parts per million.

A Residual Gas Analyzer (RGA) is part of the system. It uses an electric quadrupole field to momentum analyze ionized particles flowing at a fixed velocity, to determine their mass. Since for hydrogen isotopes the molecular dissociation energy is lower than ionization energy, introducing pure HD gas in the RGA results in some recombination of dissociated H and D atoms into H_2 and D_2, and a small fraction of H_2 and D_2 is always observed. The sensitivity of this device is limited to some percents and it is used only to monitor the extraction of the first H_2 enriched moles of HD gas.

To quantify the content of hydrogen isotopes in a mixture at sensitivities higher than few percents, gas chromatography and Raman spectroscopy may be used. A commercial instrument for gas chromatography has been used at JMU to analyze the distilled gas. The technique consists in the measurement of the thermal conductivity difference of the gas to be analyzed with respect to a neon gas carrier, as a function of the retention time in a capillary column. A sensitivity of the order of 10^{-3} has been obtained for the hydrogen isotopes separation. This encouraging result is, however, not enough to measure the ortho-H_2 concentration at the level of 10^{-4}, useful for HD polarized targets.

4. Raman spectroscopy of hydrogen isotopes mixtures

Raman spectroscopy offers a very interesting alternative to analyze the relative content of hydrogen isotopes in a mixture. The laser light is scattered by the molecules, which may change their rotational state by $\Delta J = \pm 2$. Raman scattering therefore corresponds to transitions among ortho-H_2 states

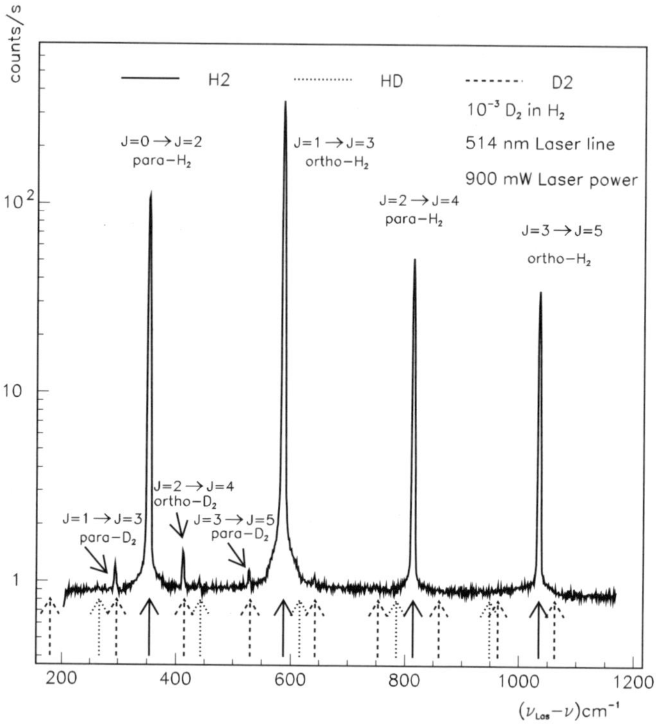

Fig. 1. Raman spectrum of a mixture of 10^{-3} concentration of D_2 in H_2 gas at 2 atm pressure on a logarithmic scale. The incoming light is the 514 nm green line of an ion-argon laser. Laser power was 900 mW. The arrows point to the position of the Raman peaks of the different hydrogen isotopes.

(or among para-H_2) and enables a direct measurement of ortho-H_2 content. In the Rome setup the light spectrum is measured by means of a triple monochromator and a charged coupled device (CCD). Figure 1 shows the Raman spectrum of a mixture of 10^{-3} concentration of D_2 in H_2 gas at 2 atm pressure on a logarithmic scale, obtained using the 514 nm green line of an ion argon laser, having 900 mW output power. The horizontal axis shows the energy difference between the the incident laser light and scattered light, expressed in cm^{-1} units. The molecular rotational energy levels are given by the relation $E_R = (\hbar^2/2I)J(J+1) = hcb_0J(J+1)$, where I is the molecular moment of inertia and b_0 is the corresponding Raman constant. The Raman peaks positions correspond to the energy differences:

$$\Delta E = hcb_0[(J+3)(J+2) - J(J+1)] = hcb_0(4J+6) \tag{2}$$

which, for each molecular species, are equally separated by $4hcb_0$. The intensity of the Raman peaks is a function of the gas mixture temperature and it is given by the following relation:[6]

$$I(J,T) = I_0 A(\nu)\nu^3 f(J)\gamma^2 \frac{45\pi^4}{7}\frac{N}{Q(T)} g_s(J)\frac{3(J+1)(J+2)}{2(2J+3)} e^{-\frac{hcb_0 J(J+1)}{K_B T}} \tag{3}$$

where I_0 is the laser intensity, $A(\nu)$ is the spectral response function of the experimental setup, $f(J)$ is the an-harmonicity correction, γ is the anisotropic matrix element, N is the total number of molecules of the species, $Q(T)$ is the partition function given by:

$$Q(T) = \Sigma_J \; g_s(J)(2J+1)e^{-\frac{hcb_0 J(J+1)}{K_B T}} \tag{4}$$

and $g_s(J)$ is the nuclear spin multiplicity. For a single hydrogen isotope, neglecting the an-harmonicity dependence upon the rotational state and the frequency dependence of the spectral response function, the product $C = I_0 A(\nu)\nu^3\gamma^2(45\pi^4/7)$ is constant. The intensity of each Raman peak may be expressed by an exponential dependence upon the gas temperature:

$$I(J,T) = \frac{CN}{Q(T)} h(J) e^{-\frac{hcb_0 J(J+1)}{K_B T}} \tag{5}$$

where:

$$h(J) = g_s(J)\frac{3(J+1)(J+2)}{2(2J+3)} \tag{6}$$

is a function of the rotational index J, only. The Raman spectrum has been fitted using gaussian functions for the peaks and a constant value for the background. Measured intensities for each peak $I_{meas}(J)$ may be obtained by integrating the fitted Gaussian functions. The gas mixture temperature T and the constant product $CN/Q(T)$ may be extracted fitting the ratio $I_{meas}/h(J)$ by a linear dependence upon $hcb_0 J(J+1)/K_B$ in a semi-logarithmic scale. Once the temperature T is known, the partition function $Q(T)$ may be explicitly evaluated and it is possible to extract, for each value of J, the product:

$$CN(J) = \frac{I_{meas}(J)}{h(j)} Q(T) e^{-\frac{hcb_0 J(J+1)}{K_B T}} \tag{7}$$

which, at thermal equilibrium, should be constant for all peaks of the same hydrogen isotope. An average value of the product of the total number of molecules times the constant factor C may be obtained. Moreover, for H_2 and D_2 gasses, peaks corresponding to transitions among ortho- and para- states may be analyzed as if belonging to different nuclear species, by evaluating the partition function for odd or even values of J separately. Repeating the same analysis procedure for peaks separately connecting even and odd values of J, one may obtain the average values of the products CN_{para} and CN_{ortho}, in addition to CN_{tot}, where N_{para}, N_{ortho} and N_{tot} are the number of molecules of ortho- para- and total molecules of each hydrogen isotope and the constant C assumes the same value for different nuclear species of the same isotope. The relative ortho- or para- content of an isotope may be directly obtained by the ratio of the previous products: $C_{ortho} = CN_{ortho}/CN_{tot} = N_{ortho}/N_{tot}$ and $C_{para} = CN_{para}/CN_{tot} = N_{para}/N_{tot}$.

The determination of the relative content of different isotopes requires the normalization of the C constants corresponding to different isotopes. This may be inferred by the published values of relative Raman intensities.[7] Using the ratio of the $D_2(J = 2 \rightarrow J = 4)$ and the $H_2(J = 1 \rightarrow J = 3)$ peak intensities ($I(2)_{D_2}/I(1)_{H_2} = 0.47$), it is found that $C_{D_2}/C_{H_2} = 0.95$. The result for the measured relative content of the gas mixture of D_2 in H_2 at 2 atm, shown in figure 1, is $N_{D_2}/N_{D_2} = (3.2 \pm 0.3)10^{-3}$, where the error of the measurement is in the 10^{-4} range.

Improvements of the present setup are foreseen to increase the signal to noise ratio by a factor of ten. A more powerful Coherent sabre laser will shortly be available which should allow to increase sensitivity of the analysis based on Raman scattering to the required hundreds parts per million.

5. Conclusions

A systematic study of the dependence of the longitudinal relaxation time T_1^H upon the ortho-H_2 concentration in HD gas to be polarized may be performed to optimize the solid HD target aging time, if precise measures of relative content of hydrogen isotopes mixtures would be possible. After double distillation of commercial HD gas, the gas content may be analyzed by Raman scattering before and after polarization to monitor the quality of the sample.

References

1. S. Hoblit *et al.* (LEGS Collaboration), *Phys. Rev. Lett.* **102**, 172002 (2009).
2. A. Honig *et al.*, *Nucl. Instr. Meth. A* **356**, 39 (1995).
3. W. N. Hardy and J. R. Graines *Phys. Rev. Lett.* **17**, 1278 (1966).
4. M. Bloom, *Physica* **23**, 767 (1957).
5. X. Wei *et al.*, these proceedings.
6. M. Koppitz *et al.*, *J. Chrystal Growth* **68**, 121A (1984).
7. K. Okuno *et al.*, *J. Nucl. Sci. Techn.* **28**, 509 (1991).

ELECTRON SPIN RESONANCE STUDY OF HYDROGEN AND ALKYL FREE RADICALS TRAPPED IN SOLID HYDROGEN AIMED FOR DYNAMIC NUCLEAR POLARIZATION OF SOLID HD

T. Kumada*

Advanced Science Research Center, Japan Atomic Energy Agency, Shirakata-Shirane, Tokai, Ibaraki 319-1195, Japan
**E-mail: kumada.takayuki@jaea.go.jp*

We carried out X-band ESR studies of H, CH_3, C_2H_5, and C_2D_5 radicals trapped in solid normal-H_2, para-H_2, and HD to establish the suitability of these radicals as a polarization source for DNP. Spin-lattice relaxation time T_{1e} of H-atom radicals, which have been used for DNP of solid hydrogens, amounted to the order of 10 minutes in highly purified solid p-H_2 and HD, being much larger than that required for DNP (milliseconds). Moreover, T_{1e} of the H-atom radicals varied with the concentration of ortho-H_2 molecules and temperature in a similar manner as spin-lattice relaxation time T_{1n} of protons. These results suggest that it is very difficult to satisfy both short T_{1e} and long T_{1n} requested for DNP. Instead of the H-atom radicals, we propose to use alkyl radicals, which were cheaply obtained by UV-photolysis of alkyl iodide, and have moderate T_{1e} for DNP.

Keywords: Dynamic nuclear polarization; solid HD; electron spin resonance.

1. Introduction

Solid HD is focused on as a polarized target for particle physics experiments, because, unlike other targets, all nuclear species in a HD molecule are polarizable. Until now, a proton polarization $P_H = 70$ %[1] and a deuteron polarization $P_D = 30$ %[2] have been achieved by a "Brute Force" (BF) method where the H and D nuclei are statistically polarized. However, BF requires extremely low temperature (10 mK), high magnetic field (15 T), long time-periods (2–6 months), and sophisticated cryogenic handlings such as low-temperature transfer between cryostats for scattering experiments. In addition, the target is only useful for low-luminosity neutral-beam experiments. It would be better if such highly polarized HD could cheaply obtained by a Dynamic Nuclear Polarization (DNP) method instead of BF.

Solem[3] in 1974 obtained $P_H = 3.75$ % by DNP of solid HD, but the value is much smaller than $P_H \approx 1$ obtained by DNP of other targets such as NH_3.[4] Why is P_H in solid HD so small? Solem used H-atom radicals produced by radiolysis as a source of polarization for DNP; however, electron spin-lattice relaxation time T_{1e} of the H-atom radicals in pure solid HD is too large to absorb the microwave for DNP. In order to accelerate the relaxation, Solem added paramagnetic species of O_2 in the solid HD sample. However, it is generally known that O_2 also reduces nuclear spin relaxation time T_{1n}, and then limits the achievable P_H.[5]

If only free radicals having better magnetic properties for DNP can be doped, higher P_H of solid HD can be expected. In this paper, we present ESR results of H-atom and alkyl free radicals doped in solid hydrogens to propose the alkyl radicals as a polarization source for DNP of solid HD.

2. Experiment

Highly purified para-H_2 (p-H_2, 99.8 %) was obtained by immersing paramagnetic catalyser FeO(OH) into liquid normal-H_2 liquid for 10 h, whereas HD gas containing normal-H_2 at 2.5 % and normal-D_2 at 1.5 % was used as purchased. Concentration of ortho-H_2 (o-H_2) in solid hydrogen was controlled by adding normal-H_2 gas into the p-H_2 sample. The hydrogen gas at 0.01 mol was sealed in a quartz cell, and then solidified by cooling the bottom tip of the cell in a temperature controller (Scientific Instrument, Model 9650) down to 5 K. H-atom radicals were generated by X-ray (Mac Science, XM590) radiolysis of the solid hydrogen sample in the temperature controller to a dose of about 0.1 kGy, and then measured with a X-band ESR spectrometer (JEOL JES-TE200) at 4 K.

CH_3, C_2H_5, and C_2D_5 radicals were produced by UV-photolysis of the highly purified solid p-H_2 doped with CH_3I, C_2H_5I, and C_2D_5I, respectively. In order to avoid heating and aggregation of these dopants, p-H_2 gas containing the alkyl iodide at 0.2 mol % was very slowly (10^{-5}–10^{-6} mol/s) introduced into the pre-cooled bottom tip. The samples were irradiated with low-pressure mercury lamp until no more alkyl radical was generated ($\sim$ 10 min.). Details of the experimental procedure were described in references 6,7.

3. Results and discussions

3.1. *H-atom radicals*

The sharp doublet in figure 1 shows an ESR spectrum of H-atom radicals produced by radiolysis of solid n-H_2. Roughly $\sim$ 1 ppm/kGy of H-atom radicals were produced, whereas yields of other radicals are much smaller. The peak-to-peak widths ΔH_{pp} of the first derivative lines of the H-atom radicals in isotopic hydrogens are listed in table 1.[8–12] Except at very high concentrations of the radicals ($\geq 10^{20}$ spins/cm^3), ΔH_{pp} in solid HD and D_2 is determined by superhyperfine interaction between the radical and neighboring hydrogen molecules.[8,12] ΔH_{pp} in solid normal-H_2 is 0.02 mT,[6] which is significantly smaller than in HD and D_2, because o-H_2 molecules at

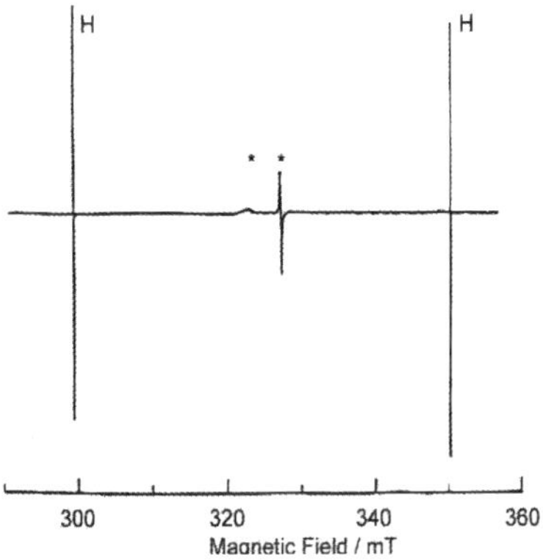

Fig. 1. ESR spectrum of H-atom radicals in solid p-H_2. * are from the quartz cell.

Table 1. Width of H-atom lines in solid HD and D_2. Data marked by * is the linewidth of integrated spectrum. The width divided by 1.18 gives ΔH_{pp} for gaussian-shaped lines.

Solid	Production by	ΔH_{pp}(4 K)	ΔH_{pp}(1 K)	Microwave
HD[8]	γ-ray	0.27	0.27	9 GHz
HD[9]	X-ray	0.27	0.27	9 GHz
HD[10]	UV photolysis of HI	0.28		9 GHz
HD[11]	X-ray	1.1*	2.0*	24 GHz
D_2[8]	γ-ray	0.12	0.12	9 GHz
D_2[9]	X-ray	0.12		9 GHz
D_2[10]	UV photolysis of HI	0.13		9 GHz
D_2[12]	Discharge	0.14*	0.14*	9 GHz

the nearest neighbor of the H-atom radicals are converted into p-H_2 having no nuclear spin moment.[13] Since the superhyperfine interaction is independent of external magnetic field H_0, the width would be independent of H_0 as well. Therefore, we speculate that the linewidth reported by Solem and Rebka,[11] which is much larger than the other values, would not be intrinsic, but due to too fast sweep of H_0 (6 mT/s). In any case, the linewidths are much less than $(\gamma_n / \gamma_e)\ H_0$ (= 1.8 mT at H_0 = 1.2 T), where γ_n / γ_e is the ratio in magnetic moments of proton to electron. Therefore, in the viewpoint of the ESR linewidth, proton spins can be dynamically polarized by the H-atom radicals using the solid effect.[14]

Figure 2 plots inverse of electron spin-lattice relaxation time T_{1e} (H) of the H-atom radicals in solid H_2, HD, and their mixtures as a function of the sum of the concentration of ortho-H_2 (o-H_2) and p-D_2 (p-D_2), which have $J = 1$ rotational quantum number at around 4 K. T_{1e} $(H)^{-1}$ varies in proportional to the square of the sum of concentrations of o-H_2 and p-D_2, [o-H_2 + p-D_2], whereas it is independent of temperature around 4 K (not shown). These results indicate that transfer of Zeeman-energy of electron spins to electric quadrupole-quadrupole (EQQ) interaction between the $J = 1$ hydrogens would be the dominant path for the spin-lattice relaxation in solid hydrogens.[6]

T_{1e}(H) in solid p-H_2 containing o-H_2 at 0.2 % amounts to 10 minutes, which is much larger than that requested for DNP (milliseconds). If T_{1e}

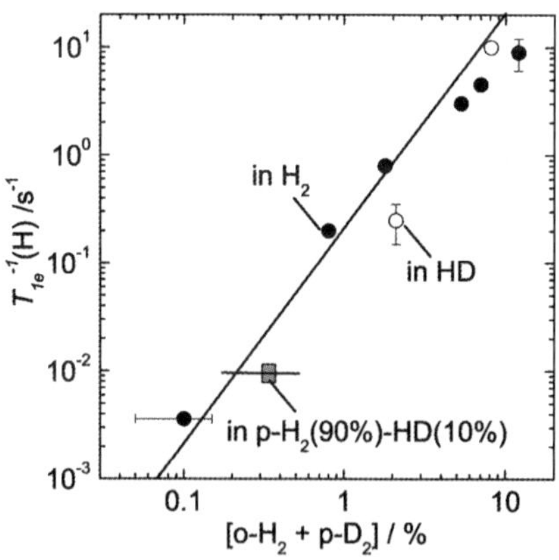

Fig. 2. T_{1e}*(H)* in solid H_2, HD, and their mixtures at 4.2 K. The datum in solid HD for [o-H_2 + p-D_2] = 8 % is reported by Solem and Rebka.[11] The others were obtained by us.[6,9]

is too large, electron spins cannot effectively absorb microwave for DNP. To make matters worse, since $T_{1e}(\mathrm{H})$ increases with increasing H_0,[12] the $T_{1e}(\mathrm{H})$ may amount to hours at DNP conditions ($H_0 = 1$–5 T).

Unlike T_{1e}, long T_{1n} is essential for DNP. However, like $T_{1e}(\mathrm{H})$, T_{1n} was found to be nearly independent of temperature, but remarkably decrease with increasing the o-H_2 concentration up to 1 %.[15] It means that, as long as the H-atom radicals are used, it is difficult to independently optimize T_{1e} and T_{1n} for DNP. As mentioned in the introduction, Solem succeeded at DNP of solid HD by adding paramagnetic species of O_2 to shorten $T_{1e}(\mathrm{H})$; however, O_2 simultaneously shortens T_{1n} as well, and decreases the attainable P_H. Based on these results, we propose to use free radicals other than the H-atom radicals.

3.2. *Methyl radicals*

Figure 3 shows an ESR spectrum of methyl radicals produced by the UV-photolysis of CH_3I in solid p-H_2. ΔH_{pp} of each line was 0.02 mT in solid p-H_2, and 0.12 mT in solid D_2 (not shown). Since the width of the CH_3 lines in solid D_2 is close to that of the H-atom lines in D_2, the width is probably determined by superhyperfine interaction. Like H-atom lines, the linewidth would be broadened up to 0.3 mT in solid HD, and independent of H_0 and temperature. Therefore, the linewidth is small enough for the solid effect as well.

Figure 4 shows microwave-power-saturation behaviors of CH_3 and C_2D_5 radicals in solid p-H_2. T_{1e} of CH_3, $T_{1e}(CH_3)$, was determined using the ESR linewidth and the saturation method for inhomogeneously-broadened system[16] to be

$$0.1\mathrm{ms} \leq T_{1e}(CH_3) \leq 10\mathrm{ms}. \tag{1}$$

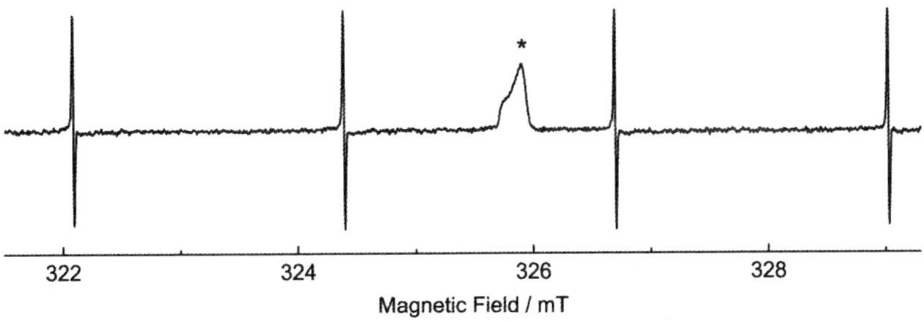

Fig. 3. ESR spectrum of CH_3 in solid p-H_2. * are from irradiated quartz cell.

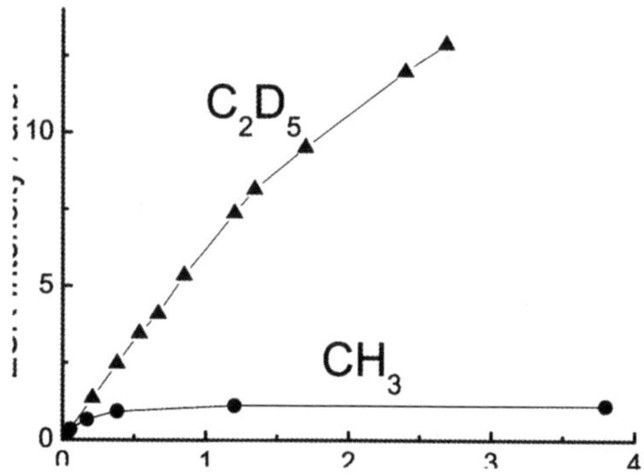

Fig. 4. Microwave-power-saturation behavior of CH_3 and C_2D_5 in solid p-H_2. Note that intensity of the H-atom radicals saturates at a microwave magnetic field of $\ll 0.1$ μT and, when H_0 is held at the H-atom line position, the intensity decreases due to long T_{1e} (not shown).

Similar value is expected for purified solid HD. $T_{1e}(CH_3)$ is much smaller than $T_{1e}(H)$ by a factor of 10^5–10^7. Probably, internal degree of freedom such as vibrational, rotational, or bending motion of CH_3 plays an important role in the dissipation of the Zeeman energy of the electron spin. Because of small T_{1e}, even if O_2 molecules are not added, the CH_3 radicals will effectively absorb microwaves for DNP.

3.3. *Ethyl radicals*

Figure 5 shows ESR spectra of C_2H_5 and C_2D_5 radicals produced by the UV-photolysis of solid p-H_2. Unlike CH_3, since the main axis of the ethyl radical does not rotate, each line is broadened by anisotropic hyperfine interaction. Although the width of each line is 0.02–0.03 mT in solid p-H_2, it would also be broadened up to ~0.3 mT in solid HD due to the superhyperfine interaction. In this case, although the C_2H_5 lines do not, the C_2D_5 lines overlap for each other in an integrated spectrum. The width of the envelope line in the integrated spectrum amounts to ~2 mT, which is close to double the proton Zeeman energy E_{pZee} [$2\ E_{\rm pZee}/(g_{\rm n}\mu_{\rm n}) = 3.6$ mT at $H_0 = 1.2$ T]. Therefore, proton spins can be dynamically polarized using C_2D_5 by thermal mixing[14] rather than the solid effect.

T_{1e} of C_2D_5 is determined by the linewidth and the microwave-power-saturation behavior in figure 4 to be

$$0.01\ \text{ms} \le T_{1\text{e}}(\text{C}_2\text{D}_5) \le 0.1\ \text{ms}. \tag{2}$$

which is short enough for DNP.

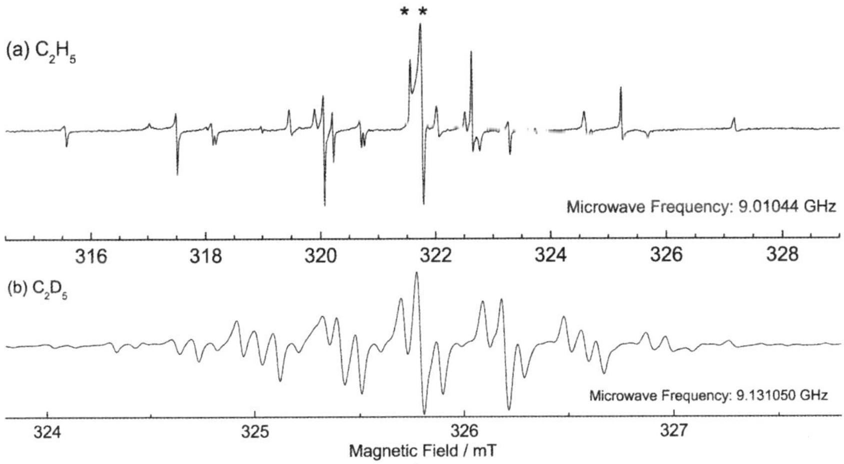

Fig. 5. ESR spectra of C_2H_5 and C_2D_5 in solid p-H_2. * are from irradiated quartz cell.

4. Possible problems for DNP using alkyl radicals

The author lists three possible problems for the DNP of solid HD using alkyl radicals. First, it is empirically known that free radicals at an amount of 2×10^{19} spins/cm^3 are needed for the best performance of DNP, whereas the concentration of these alkyl radicals we have isolated in solid p-H_2 was at most 10^{18} spins/cm^3. However, Fajardo *et al.*[17] succeeded in introducing isolated 10^{19} spins/cm^3 of light molecules such as CH_3OH, CO, and N_2 in millimeter thick and optically transparent solid p-H_2. We believe we can also introduce such high concentration of free radicals by consulting their experimental settings.

Second, longer alkyl radicals may be needed for DNP. Although DNP studies of alkyl radicals produced by radiolysis of polyethylene,[18] and butyl radicals produced by UV-photolysis of solid butanol[19] have been reported, to my knowledge, no DNP study using methyl and ethyl radicals has been reported. The number of ESR lines of the methyl and ethyl radicals may be too small to overlap to form single enveloped line in an integrated spectrum having a linewidth comparable to $2E_{pZee}$, which is needed for the thermal mixing. In order to introduce such longer alkyl radicals, we have to introduce longer alkyl iodide such as pentyl iodide and butyl iodide, which are more difficult to be introduced into solid HD than methyl iodide and ethyl iodide.

Third, I-atom radicals, byproducts of the UV-photolysis of alkyl iodides, may play an important role in depolarization of proton spins. ESR lines of the I-atom radicals have not been observed in UV-irradiated solid hydrogens containing alkyl iodides so far. This result suggests that the I-atom radicals might be recombined to be I_2 by the two-molecule reaction, $2\ CH_3I \rightarrow 2\ CH_3 + I_2$, or the I-atom lines were broadened out due to anisotropic g-value. If the I-atom radicals cause serious depolarization of protons, we should look for other chemicals such as $CH_3N_2CH_3$, which does not generate any by-product radical, $CH_3N_2CH_3 \rightarrow 2\ CH_3 + N_2$. If these problems are overcome, these alkyl free radicals can be used for DNP of solid HD.

References

1. X. Wei, *Nucl. Instr. Meth. A* **526**, 1578 (2004).
2. S. Bouchigny *et al.*, *Nucl. Instr. Meth. A* **544**, 417 (2005).
3. J. C. Solem, *Nucl. Instr. Meth.* **117**, 477 (1974).
4. D. G. Crabb *et al.*, *Phys. Rev. Lett.* **64**, 2627 (1990).
5. T. Kumada *et al.*, *Nucl. Instr. Meth. A* **606**, 669 (2009).
6. T. Kumada *et al.*, *J. Chem. Phys.* **116**, 1109 (2002).
7. T. Kumada *et al.*, *J. Chem. Phys.* **114**, 10024 (2001).
8. T. Miyazaki and H. Morikita, *Bull. Chem. Soc. Jpn.* **66**, 2409 (1993).
9. T. Kumada, unpublished.
10. T. Kumada, *J. Chem. Phys.* **124**, 94504 (2006).
11. J. C. Solem and G. A. Rebka, Jr., *Phys. Rev. Lett.* **21**, 19 (1968).
12. A. S. Iskovskikh *et al.*, *Sov. Phys. JETP* **64**, 1085 (1987).
13. T. Kumada *et al.*, *Chem. Phys. Lett.* **288**, 755 (1998).
14. A. Abragam and M. Goldman, *Rep. Prog. Phys.* **41**, 395 (1978).
15. W. N. Hardy and J. R. Gains, *Phys. Rev. Lett.* **26**, 1278 (1966).
16. C. P. Poole, Jr., in *Electron Spin Resonance* (Dover, New York, 1996).
17. M. Fajardo and S. Tam, *J. Chem. Phys.* **108**, 4237 (1998).
18. D. G. Crabb, *Nucl. Instr. Meth. A* **526**, 56 (2004).
19. T. Kumada *et al.*, to be published in *J. Magn. Reson.*

CHANGE OF ULTRAFAST NUCLEAR-SPIN POLARIZATION UPON PHOTOIONIZATION BY A SHORT LASER PULSE

T. Nakajima

Institute of Advanced Energy, Kyoto University
Gokasho, Uji, Kyoto 611-0011, Japan
E-mail: nakajima@iae.kyoto-u.ac.jp

Y. Matsuo and T. Kobayashi

RIKEN, 2-1 Hirosawa, Wako, Saitama 351-0198, Japan

Recently we have proposed a novel scheme to polarize nuclear-spin using a short laser pulse where nuclear-spin polarization is realized among a hyperfine manifold of an atomic bound state. In this work we investigate how much the degree of nuclear-spin polarization changes if we apply a time-delayed second laser pulse to ionize atoms. We find that the nuclear-spin dynamics upon photoionization can be quite different, depending on the pulse duration of the second laser pulse.

Keywords: Nuclear-spin; polarization; short laser pulse.

1. Introduction

There is a great demand to efficiently produce spin-polarized nuclei.[1] A few well-known methods such as nuclear fragmentation, optical pumping,[2] and its variant combined with spin-exchange collisions[3] are being used to polarize different nuclei.

In the last few years we have been working on ultrafast spin polarization of electrons (nuclei) using pulsed lasers,[4–10] which is entirely different from any of the existing methods mentioned above. Our method is based on the transient spin dynamics and utilizes spin-orbit (hyperfine) interactions of coherently excited fine structure (hyperfine) manifolds by short laser pulses. There are two main advantages in our transient scheme. The first advantage is that spin-polarization can be realized in the ultrafast ($\ll \mu$s) time scale. This is a very nice feature if one is to polarize spin of unstable nuclei. The second advantage is that by choosing a different timing between the pump (excitation) and probe (ionization) laser pulses, one can prepare nuclei in states with different *hyper-spin-polarizability*, which can be defined beyond the conventional up-and-down spin polarization.

After a recent experimental demonstration of ultrafast electron-spin polarization of Sr ions upon photoionization,[9] we are currently working on the next step, that is, an experimental demonstration of ultrafast nuclear-spin polarization of alkaline-earth elements we have proposed in recent papers.[8,10] Since our current investigation is not yet with an accelerated beam but with a thermal beam, we employ a photoionization technique to remove a *single electron* and *optically* monitor nuclear-spin polarization of photoions with a narrow-band ns laser pulse.

Since the hyperfine interactions are still active before and after photoionization, unless all valence electrons are removed through ionization, it is important to investigate how much change of nuclear-spin polarization we will have upon photoionization.

In this paper we investigate the change of ultrafast nuclear-spin polarization upon photoionization by a short laser pulse. Our theoretical results indicate that, although the hyperfine interactions of photoions destroy nuclear-spin polarization to some extent, the remaining nuclear-spin polarization is still significant (> 30 %), and under some conditions it can be as much as ~ 70 % for I=1/2 alkaline-earth isotopes. We also give a brief description of the on-going experiment in our group to realize ultrafast nuclear-spin polarization.

2. Scheme

The scheme we study in this paper is shown in figure 1. We assume that alkaline-earth atoms with nuclear-spin $I = 1/2$ are initially in the ground state, ns^2 $^1S_0$, with a total angular momentum of $F = 1/2$. They are excited to the nsmp $^3P_1$ or $^1P_1$ state by the right-circularly polarized pump pulse with a duration of τ_{pump}. The excited state has a hyperfine doublet structure with total angular momenta of $F = 1/2, 3/2$, and an energy separation of E. If the duration of the pump pulse satisfies the condition of $\tau_{pump} \ll E^{-1}$, we can create a coherent superposition of the hyperfine doublet, and thus realize ultrafast nuclear-spin polarization.[8,10] By choosing the origin of time at the moment of the pump pulse, we turn on the circularly or linearly polarized probe pulse at t_1 to induce photoionization, where its duration is τ_{probe}. After photoionization, there is only one valence electron left, and hence the total electronic angular momentum of the pho-

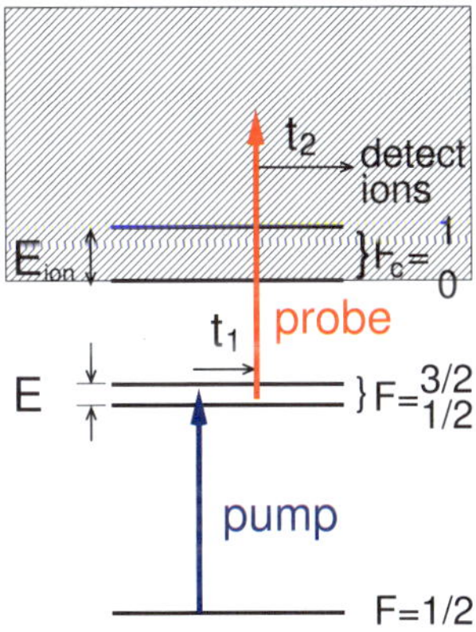

Fig. 1. Level scheme. The shaded area indicates a continuum.

toion is $J_c = 1/2$, which results in the ionic hyperfine doublet with total angular momenta of $F_c = 0$ and 1. Finally we detect photoions at $t_1 + t_2$. If we wish, instead of detecting photoions, we can further remove the remaining valence electrons at $t_1 + t_2$ by a thin carbon foil for the case of a fast atomic beam experiment.

3. Numerical results and discussions

In order to analyze the change of nuclear-spin dynamics for the scheme described in the previous section, we employ the transition-rate approximation which is valid for the pump and probe laser intensities well below the saturation. The use of transition-rate approximation significantly simplifies the complexity of the time-dependent analysis of nuclear-spin for the excitation and ionization processes. A more rigorous way that is valid from the low to high laser intensities is to use time-dependent Schrödinger equations, as we have demonstrated for the ultrafast electron-spin polarization of photoions and photoelectrons.[6]

In figure 2 we show the change of nuclear-spin polarization through the singlet excited state, $nsmp\ ^1P_1$, as a function of time t_1, which is actually a delay time between the pump and probe pulses. Solid, dashed and dot-dashed lines in figure 2 correspond to the results by the linearly and right-/left-circularly polarized probe pulses. Similar results through the triplet excited state, $nsmp\ ^3P_1$, are presented in figure 3. Clearly the presence/absence of spin-orbit interaction makes this difference.

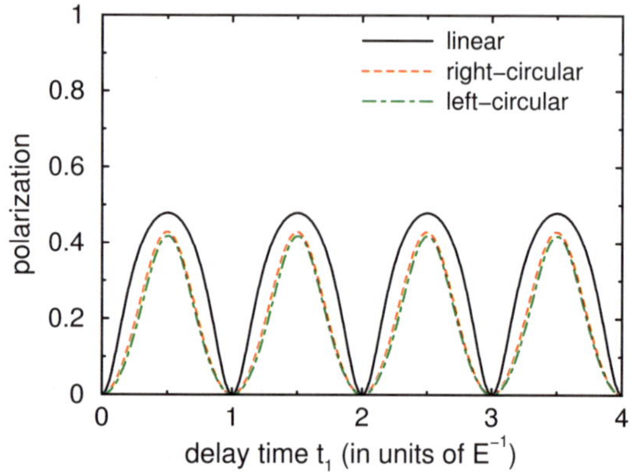

Fig. 2. Nuclear-spin polarization of photoions through the singlet excited state, nsmp $^1P_1$, as a function of time t_1. Results by the probe pulse with linear, right-circular, and left-circular polarization are shown by solid, dashed, and dot-dashed lines, respectively. For illustration the bound-free radial dipole elements are assumed to be $|R_{7p\to ks}| = |R_{7p\to kd}|$, where $7p$ and kl $(l = s, d)$ indicate the electronic configurations of the initial and final states, i.e., $6s7p$ and $6skl$ $(l = s, d)$, respectively.

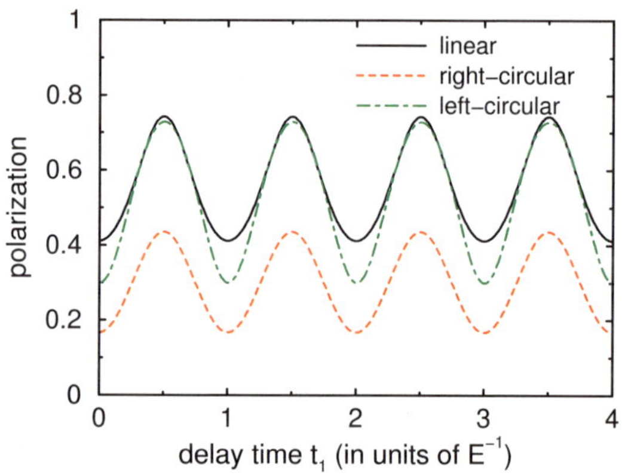

Fig. 3. Same with figure 2 but through the triplet excited state, nsmp $^3P_1$. All other conditions are exactly the same as those for figure 2.

4. On-going experiment

Based on our theoretical analysis, we are currently working on the proof-of-principle experiment. For the experimental convenience we employ Yb atoms, which are not alkaline-earth atoms but have a very similar electronic

Table 1. Natural isotope abundance of Yb.

mass	168	170	171	172	173	174	176
nuclear-spin	0	0	1/2	0	3/2	0	0
abundance (%)	0.14	3.1	14.4	21.9	16.1	31.8	12.7

configuration due to the two valence electrons, $6s^2$, in the outermost shell. The good thing of using the Yb atom is that this element has many stable isotopes with different values of nuclear-spin, including 1/2, with a significant natural abundance. For the natural abundance of Yb, see table 1. In contrast, although some of the alkaline-earth atoms are known to have 1/2 nuclear-spin, they are unstable isotopes.

Figure 4(a) shows the level scheme we are working on now. Yb atoms in the ground state with a natural abundance are resonantly excited to the $6s7p\ ^1P_1$ or $^3P_1$ state by the pump pulse. After some time delay we turn on the probe pulse to induce photoionization. Note that the bandwidth of the pump pulse is 6 GHz and cannot spectrally resolve the specific isotope which is ^{171}Yb in our case. As a result all isotopes are photoionized. To analyze the degree of nuclear-spin polarization of $^{171}Yb^+$, we optically detect the hyperfine transition lines of $^{171}Yb^+$ as shown in figure 4(b). Clearly a

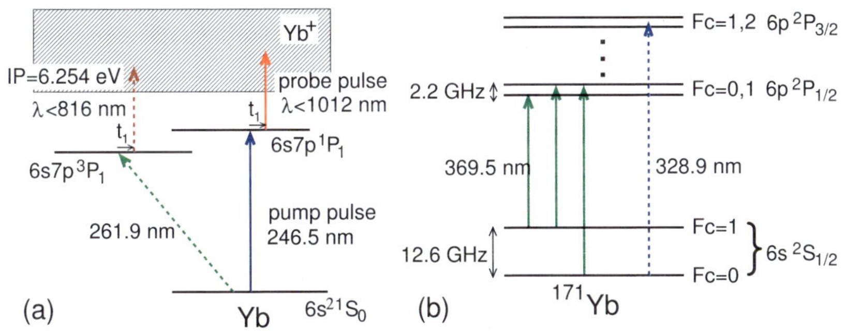

Fig. 4. (a) Level scheme of Yb. A few ns pump-pulse with right-circular polarization resonantly excites Yb atoms *with a natural abundance* in the ground $6s^2\ ^1S_0$ state to $6s7p\ ^1P_1$ or $6s7p\ ^3P_1$. After some time delay, a ns probe pulse is turned on and induces photoionization which is presumably nuclear-spin polarized. Hyperfine structures are not explicitly depicted for simplicity but they are there and play an essential role in our scheme. Note that *all isotopes* are excited and ionized up to this point. (b) Relevant hyperfine transition lines of ^{1}71Yb which we selectively detect for the analysis of nuclear-spin polarization. For this purpose the spin-detection laser must have a bandwidth as narrow as a few hundred MHz.

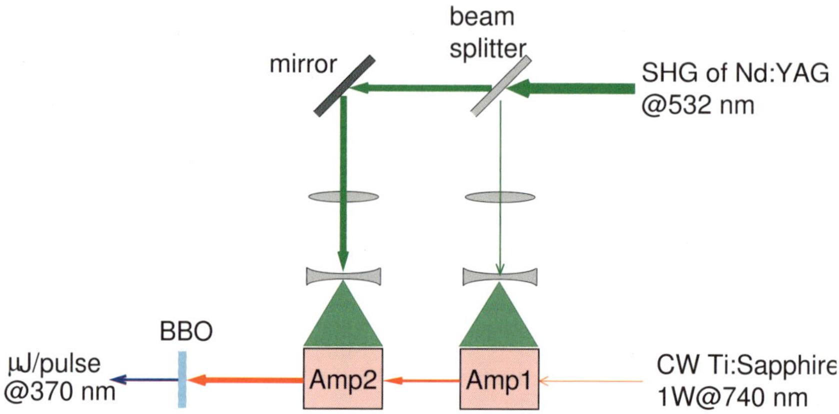

Fig. 5. Optical arrangement to produce tunable narrow-bandwidth ns pulses.

selective detection of $^{171}Yb^+$ through a laser-induced fluorescence requires that the spin-detection laser has a very narrow bandwidth.

To produce a tunable narrow-bandwidth ns pulse, we perform the 2-stage pulse amplification using a tunable continuous-wave (CW) Ti:Sapphire laser operated in a single mode as a seed. The optical arrangement is shown in figure 5. An output from a CW Ti:Sapphire laser is sent into the 2-stage dye-laser pumped by the second harmonic (532 nm) of a Nd:YAG laser with a 2 ns duration operated at 10 Hz. Since we use a single-mode Ti:Sapphire laser as a seed and amplify the seed by the ns dye-laser, the amplified pulse has a < 2 ns duration and a bandwidth as narrow as ~200 MHz, which is significantly narrower than the bandwidth of commercial dye-lasers, a few GHz (6 GHz for our system). The amplified pulse at 740 nm now goes through a nonlinear crystal (BBO) to produce a second harmonic generation (SHG), resulting in a $\sim \mu$ J/pulse with a ~200 MHz bandwidth and a tunability around 370 nm by tuning the CW Ti:Sapphire seed laser. This spectrally-narrowed tunable ns pulse should be able to selectively excite each hyperfine state of $^{171}Yb^+$ while all other Yb^+ ions with a different mass remain unexcited. If the spin-detection laser is chosen to be right/left-circularly polarized, the laser-induced fluorescence corresponding to the three hyperfine transition lines shown in figure 4(b) have different signal intensities depending on the degree of nuclear-spin polarization. Therefore, we should be able to estimate the degree of nuclear-spin polarization from the the laser-induced fluorescence measurement of $^{171}Yb^+$.

5. Conclusions

In conclusion we have theoretically investigated the change of nuclear-spin polarization upon photoionization by the pump and probe pulses with a short pulse duration, and found that the use of the singlet and triplet intermediate state results in the different delay-dependence of nuclear-spin polarization of photoions. The proof-of-principle experiment using Yb atoms with a natural abundance is currently underway in our group.

Acknowledgments

This work was supported by a Grant-in-Aid for Scientific Research from the Ministry of Education and Science of Japan.

References

1. D. Fick *et al.*, *Phys. Rep.* **214**, 1 (1992).
2. H. Reich and H. J. Jänsch, *Nucl. Instr. Meth. A* **288**, 349 (1990).
3. A.N. Zelenskii *et al.*, *Nucl. Instr. Meth. A* **334**, 285 (1993).
4. T. Nakajima *et al.*, *Appl. Phys. Lett.* **83**, 2103 (2002).
5. T. Nakajima, *Appl. Phys. Lett.* **84**, 3786 (2004).
6. T. Nakajima, *Appl. Phys. Lett.* **88**, 111105 (2006).
7. Y. Matsuo *et al.*, *Jpn. J. Appl. Phys.* **46**, 1181 (2007).
8. T. Nakajima, *Phys. Rev. Lett.* **99**, 024801 (2007).
9. T. Nakajima *et al.*, *Phys. Rev. A* **77**, 063404 (2008).
10. T. Nakajima, *J. Opt. Soc. Am. B* **26**, 572 (2009).

RADIATION DAMAGE AND RECOVERY IN POLARIZED $^{14}NH_3$ AMMONIA TARGETS AT JEFFERSON LAB

J. D. Maxwell

Department of Physics, University of Virginia,
Charlottesville, VA 22904, USA
E-mail: jmaxwell@jlab.org

for the University of Virginia polarized target group*

Dynamically nuclear polarized solid ammonia offers an attractive combination of high polarization, comparatively high dilution factor and high radiation damage resistance as a target material in electron scattering experiments. Polarized $^{14}NH_3$, provided by the University of Virginia Polarized Target Group, was used as target material in two simultaneous experiments at Thomas Jefferson National Accelerator Facility in the spring of 2009. Target polarization performance for both experiments is discussed, as is previously unseen behavior in irradiated $^{14}NH_3$.

Keywords: Polarized targets; dynamic nuclear polarization; radiation damage.

1. Introduction

Polarized solid ammonia has proven an effective tool in nuclear and high energy physics experiments requiring control on spin degrees of freedom. After doping with paramagnetic centers, ammonia ($^{14}NH_3$, $^{15}NH_3$ or $^{15}ND_3$) provides high proton (near 100 %) and deuteron (near 50 %) polarizations under Dynamic Nuclear Polarization (DNP).[1] The University of Virginia's Polarized Target Group focuses on target research and creation for use in nuclear physics experiments. In the spring of 2009, all three of Thomas Jefferson National Accelerator Facility's experimental halls were running experiments using polarized targets, and in two of the halls, B and C, our solid ammonia ($^{14}NH_3$) was used. This paper gives a preliminary look at the polarization performance and radiation recovery in the ammonia targets during these experiments.

*UVa Target Group: D. Crabb, D. Day, O. Rondon, H. Baghdasaryan, N. Kalantarians, J. Maxwell, J. Mulholland, K. Kovacs, N. Kvaltine.

1.1. *JLab polarized ammonia experiments in 2009*

The Spin Asymmetries on the Nucleon Experiment (SANE) ran in Jefferson Lab's hall C from January to March, 2009. SANE is an inclusive, double polarization measurement of the proton's spin asymmetry A_1 and spin structure function g_2 using a novel, non-magnetic telescope array of detectors.[2] SANE's target was the "U.Va." $^{14}NH_3$ target, which uses a 5 T superconducting Helmholtz pair magnet allowing target polarizations over a wide range of angles with respect to the incident beam. During SANE, 180° and approximately 90° were used. Beam currents were limited by the target to about 100 nA.

eg1-dvcs (Deeply Virtual Compton Scattering), ran in hall B, and is a generalized parton distribution study, using the CEBAF Large Acceptance Spectrometer (LAS) to make single and double spin asymmetry measurements simultaneously.[3] Hall B beam currents are typically around 7 nA.

2. Dynamic nuclear polarization

During an experiment, these ammonia targets are constantly polarized via Dynamic Nuclear Polarization (DNP), which leverages paramagnetic radicals to provide electron-proton hyperfine splitting in a high magnetic field and at low temperature. Using microwaves, the energy levels are populated via the "forbidden transitions," preferentially filling the desired spin-polarized state. While the proton relaxation time is on the order of tens of minutes at 1 K, the electron relaxation time is on the order of milliseconds, allowing an electron to relax and be used to polarize another proton.

In $^{14}NH_3$ these paramagnetic centers, primarily NH_2, are produced by irradiating the material at a small electron accelerator. The $^{14}NH_3$ used in SANE and eg1-dvcs was irradiated under liquid argon at NIST in Gaithersburg, Maryland, to a total radiative dose of approximately 10^{17} e^-/cm^2 with 19 MeV electrons. Under DNP after this "warm" dose, ammonia material routinely breaks 90 % proton polarization in the SANE and eg1-dvcs refrigerators.[4]

2.1. *Radiation recovery: anneals*

Electron beam incident on the DNP target has both a prompt and cumulative effect on the polarization. The former is due to heating from the beam reducing the efficiency of DNP, and typically results in a polarization loss of around 5 to 7 percent.[5] Over time the radiation dose will damage the material, causing the latter effect by making additional paramagnetic radicals which can allow the aligned spins to relax. Eventually this effect will reduce the polarization beyond practical use.

To restore this cumulative polarization loss, anneals are necessary. During an anneal, the material is removed from the beam and is heated to between 70 and 100 K for between 10 to 60 minutes to allow excess and unwanted radicals to recombine. After cooling and reintroducing microwaves the material can again reach previous maximum polarizations.

However, the material's lifetime is still limited. With more accumulated dose and several anneal cycles, the rate of the polarization decay in the beam will eventually increase beyond practical use. During anneals, not all radicals created by the beam can be recombined; this increase in decay rate is due to the accumulation of these unwanted but unremoved radicals.

3. Polarization during radiation dose deposition

3.1. *SANE polarization*

Hall C's SANE experiment took data for two and a half months, and used 10 samples of ammonia which were annealed 26 times. Shown in figure 1 are two example material lifetimes from SANE, giving the proton polarization of the sample as it accumulated radiation dose. These two samples received among the highest dose in Peta-electrons/cm^2 in the experiment. Also shown in figure 1 is an overview of charge averaged online polarizations for all SANE runs.

The relatively high beam current used, between 80 and 110 mA, meant the ammonia could be used for between six to eight hours of beam time before an anneal was necessary, which corresponds to a dosage between approximately 2 and 4 Pe$^-$/cm^2. Although material exhaustion was avoided by timely replacement of material, a suggestive increase in decay rate can be seen at the end of the lifetime of the material plotted on the top right of figure 1, as it passes 25 Pe$^-$/cm^2.

In both samples, near 90 % in-beam peak proton polarizations can be seen immediately after anneals, which are represented in figure 1 as vertical black lines. Often the polarization can be seen to rise after beam is returned following an anneal; this is due to excessive removal of paramagnetic centers during an anneal (an "over-anneal") and the subsequent replacement of these centers in the electron beam. For later comparison with Hall B data, the material on the left shows a best case SANE decay from 85 % to 75 % in about 1.5 Pe$^-$/cm^2.

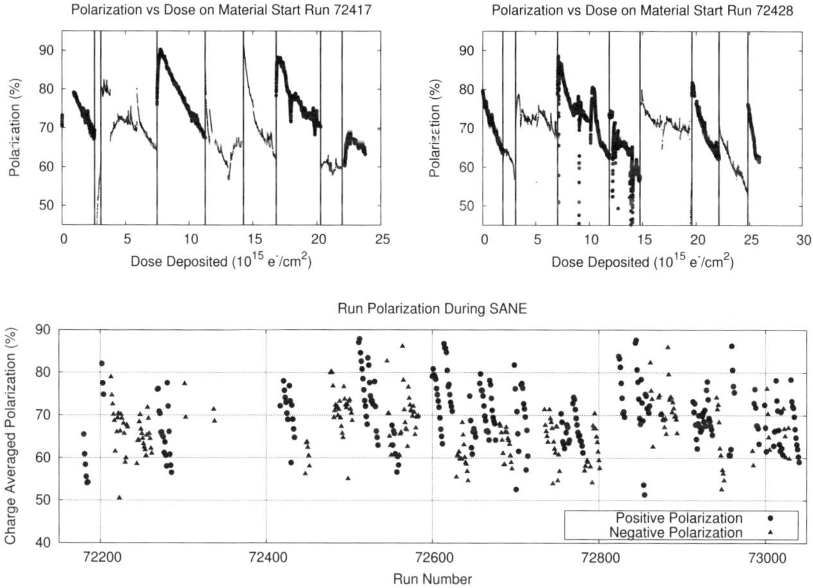

Fig. 1. At top are preliminary SANE polarization vs. dose accumulation for two examples of ammonia material lifetime. Anneals are shown as vertical solid lines and the thicker points represent positive polarization. At bottom are charge averaged online polarizations for all SANE runs.

3.2. *eg1-dvcs polarization*

Shown in figure 2 are the material lifetimes of two $^{14}NH_3$ samples used during eg1-dvcs as top and bottom cup materials. The much lower current allowed by the Hall B detectors makes the lifetime, while similar in Pe^-/cm^2, nearly two months instead of a week.

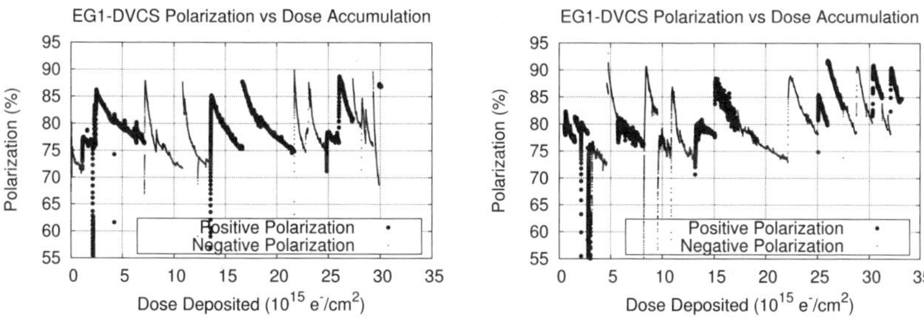

Fig. 2. Preliminary eg1-dvcs top (left) and bottom (right) polarization.

In both samples, in-beam peak polarizations are near or above 90 % after almost every anneal. Using an example decay from 85 % to 75 %, which occurred in approximately 3.5 Pe^-/cm^2, over twice that of the example shown earlier from SANE, points to a decay rate difference between these experiments due to the disparity in beam current. Further investigation of this effect, outlined only qualitatively here, is expected. Material exhaustion is apparent at the end of both materials' lifetimes, with significant increase in decay rate at around 30 Pe^-/cm^2.

4. Optimal microwave frequency

As an ammonia sample accumulates dose after an anneal, the microwave frequency required to maximize the polarization will shift due to accumulation of different radicals. At 5 T, the microwave frequency will begin around 140.20 GHz for the positive polarization, or around 140.40 GHz for the negative, but will then drift apart from these starting frequencies as dose accumulates. Microwave frequencies are chosen by the target operators to maximize the polarization and are thus dependent on their ability and diligence. Shown in figure 3 are microwave frequencies used for all 26 material samples in SANE, and for two material samples in eg1-dvcs. In SANE the horizontal scale is the equivalent of about six to eight hours, while in eg1-dvcs, this is about a week.

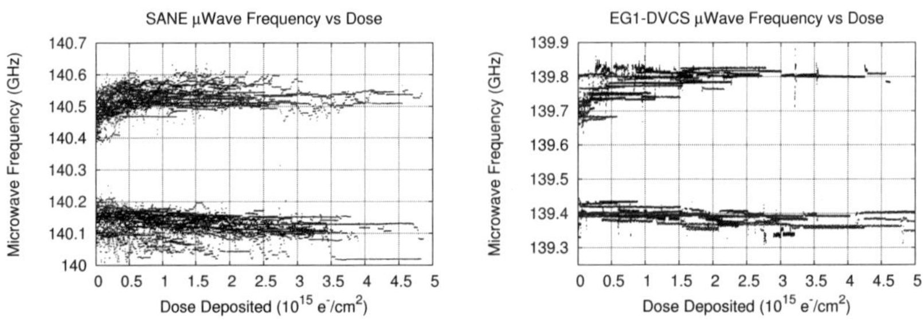

Fig. 3. Preliminary microwave frequency used during DNP in SANE (left) and eg1-dvcs (right). Frequencies for negative polarization are represented in the upper grouping in each case, while the positive polarizations are below.

In the case of DNP for negative polarization, both SANE and eg1-dvcs show a fast increase in the optimum microwave frequency which quickly slows, creating an exponential curve which in SANE goes from 140.40 to around 140.53 GHz at the end of the anneal cycle (close to 4 Pe^-/cm^2). The positive polarization frequencies are more linear as they drift lower, from about 140.20 to near 140.13 GHz in SANE. The microwave frequencies used in eg1-dvcs can be seen at the right of figure 3 to be approximately 0.75 GHz lower than the corresponding points from SANE; this shift is due to the slight magnetic field shift required to adjust the hyperfine transition energy to within frequency range of the hall B microwave source. Despite this shift, and more significantly the great disparity in beam current used, both experiments exhibit a similar pattern and gap between positive and negative polarization frequencies as dose accumulates.

5. Anneal procedures

Shown in figure 4 are the 26 anneals performed during SANE, expressed as the temperature of the top cup and the duration that temperature was maintained. Vertical black lines represent the replacement of ammonia, and points represent the peak proton polarizations in the top and bottom material cups following a given anneal. Over a sample's lifetime, the anneal temperature and time are increased to counter accumulation of excess radicals unaffected by lower temperature and time. As the experiment progressed, the target operators erred on the side of over-annealing a material sample; necessary radicals can be rebuilt in the electron beam, while unwanted, excess radicals will remain until the next anneal.

Difficulties in balancing the temperature in the top and bottom material cups result in disparities in top and bottom cup polarization performance. The top cup, which is further from the heating coil used to control the temperature, could be as much as ten degrees cooler than the bottom cup in some cases.

During eg1-dvcs, a standard anneal procedure was followed for every anneal. Three cernox temperature sensors, below, between and above the two target cups, were used to keep the temperature in the top cup at 85 K for 35 minutes. The temperature disparity between cups in hall B's target are less pronounced due to the "oven" used to heat the material for anneals, but turning back to figure 2 shows slightly higher peak polarizations in the bottom target material which may be a sign of this effect.

6. Unexpected behavior in SANE ammonia

Two previously unseen behaviors were observed in some of the ammonia used in the SANE experiment, one of which was confirmed in eg1-dvcs ammonia as well. The first novel observation was unusual radioactivity in

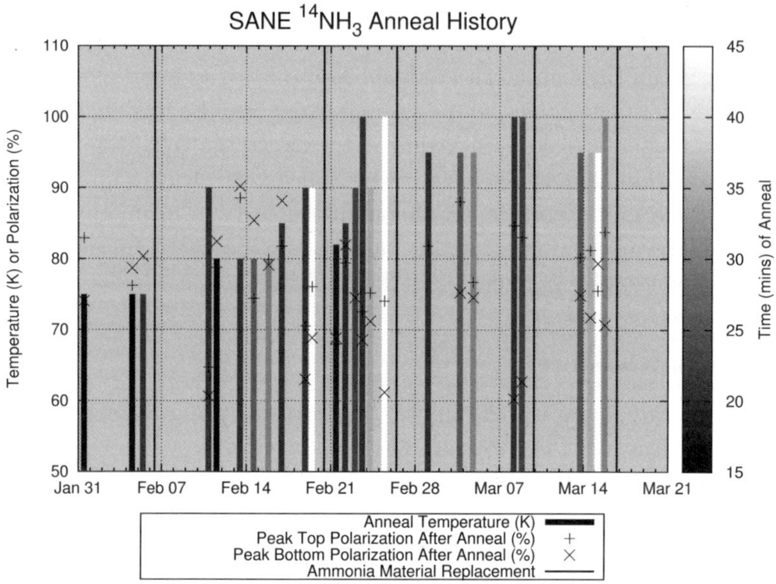

Fig. 4. SANE anneal history. Each vertical bar represents an anneal, with the temperature expressed as the vertical height and the time held at that temperature expressed as the shade. Again, polarizations are preliminary.

two material samples following use; these samples received approximately 15 and 19 Pe^-/cm^2 respectively and were used simultaneously as top and bottom target material. These doses are less than the highest dose for a material during SANE, 26 Pe^-/cm^2, and it is not clear at this time why these two samples exhibited higher persistent activity than other samples.

The high activity in these samples lead to a gamma spectrum analysis, which was performed by JLab's Radiation Control Group. Tests of the two materials with higher persistent activity showed a strong peak at 477.7 KeV, which corresponds to Be^7. The cross section for $^{14}N(\gamma, X)^7Be$ is 0.12 mb.[6] This gamma peak was then confirmed to exist in other ammonia samples, both from SANE and eg1-dvcs.

The second new observation was an abnormal discoloration of the ammonia beads after use in the experiment. Newly frozen ammonia appears white but takes on a dark purple or blue tint after irradiation. This deep hue fades in storage under liquid nitrogen, but will reappear with further dose. During SANE, six of the ten samples exhibited a brownish yellow color in the region under the beam's raster, the pigmentation occurring as anywhere from a scant hue in only a small portion of the material to a deep coloration throughout the material. Other samples exhibited the expected deep purple hue, and some showed a light yellow tint in addition to purple. The coloration differences do not display an obvious correlation to radiation dose and are not fully understood as of this writing.

While previous electron scattering experiments at JLab and SLAC used $^{15}NH_3$, this was our first use of $^{14}NH_3$ at such high doses and is thus our first opportunity to observe these behaviors.

7. Conclusions

The $^{14}NH_3$ used in SANE and eg1-dvcs behaved much like $^{15}NH_3$ used in previous experiments at SLAC and JLab. In-beam peak polarizations exceeded 90 % for both experiments and material exhaustion was observed above 25 Pe$^-$/cm^2. Rate effects due to the beam current disparity between SANE and eg1-dvcs were seen, as the SANE material showed both shorter anneal cycles and shorter overall lifetime in terms of radiation dose. However, some SANE material displayed discoloration and persistent radioactivity that were outside our previous experience.

Acknowledgements

The author would like to thank D. Crabb, D. Day and O. Rondon for helpful discussions and guidance, and also the Jefferson Lab Target Group, especially C. Keith and C. Carlin for access to all Hall B eg1-dvcs data. This work was supported by Department of Energy contract DE-FG05-86ER40261 and by the Institute of Nuclear and Particle Physics of the University of Virginia.

References

1. D. G. Crabb and W. Meyer, *Ann. Rev. Nucl. Part. Sci.* **47**, 67 (1997).
2. O. A. Rondón, The RSS and SANE Experiments at Jefferson Lab, in *Proc. Workshop Spin Structure at Long Distance*, eds. J. P. Chen, K. Slifer and W. Melnitchouk, AIP Conf. Proc., Vol. 1155 (AIP, New York, 2009).
3. A. Biselli, DVCS with longitudinally polarized target using CLAS at 6 GeV, in *Proc. Workshop 18th International Spin Physics Symposium*, eds. D. G. Crabb, Y. Prok, M. Poelker, S. Liuti, D. B. Day and X. Zheng, AIP Conf. Proc., Vol. 1149 (AIP, 2009).
4. P. M. McKee, *Nucl. Instr. Meth. A* **526**, 60 (2004).
5. T. J. Liu *et al.*, *Nucl. Instr. Meth. A* **405**, 1 (1998).
6. J. Liu, S. Roesler, S. Rokni and R. Sit, *Evaluation of Radiological Consequence from Air Activation at NLC BDS Tunnel*, Tech. Rep. RP-00-05, Stanford Linear Accelerator Center (May 2000).

POLARIZED SOLID PROTON TARGET IN LOW MAGNETIC FIELD AND AT HIGH TEMPERATURE

T. Uesaka* and Y. Shimizu

Center for Nuclear Study University Department, University Name,
Bunkyo, Tokyo 113-0033, Japan
**E-mail: uesaka@cns.s.u-tokyo.ac.jp*
http://www.cns.s.u-tokyo.ac.jp/

T. Kawahara

Department of Physics, Toho University,
Chiba 274-8510, Japan

S. Sakaguchi

Riken Nishina Center,
Wako, Saitama 351-0198, Japan

T. Wakui

Cyclotron & Radioisotope Center, Tohoku University,
Miyagi 980-8578, Japan

Proton polarization will find new applications if it is achieved in a low magnetic field and at high temperature. Use as a polarized proton target in studies of neutron-rich unstable nuclei is one of the fascinating examples. A polarized solid proton target based on electron polarization in photo-excited triplet states of aromatic molecule has been constructed at the Center for Nuclear Study (CNS), at the University of Tokyo. The target works in a low magnetic field of 0.1 T and at high temperature of 100 K and has been applied to radioactive isotope beam experiments at RIPS at RIKEN.

Keywords: Polarized target; radioactive nuclear beam experiments.

1. Motivation to high-temperature and low magnetic-field polarization

In the long history of dynamic nuclear polarization, protons in solid materials have been polarized under "extreme" conditions, namely in a high magnetic field of several Tesla and at a low temperature of sub-Kelvin.

Proton polarization under more relaxed conditions of low magnetic field and/or high temperature, would facilitate a variety of applications, such as use in medical diagnosis, quantum computing, and so on. Application to polarization studies of unstable nuclei is one of the fascinating examples.[1]

A polarized proton solid target working in a magnetic field of 0.1 T and at temperature of 100 K has been constructed at the Center for Nuclear Study (CNS), at the University of Tokyo.[2] The target system has been used in radioactive nuclear beam experiments at RIPS, RIKEN.[3–5]

In this report, principles of the proton polarization and an overview of the target system are reviewed.

2. Principles of proton polarization

How can we produce high proton polarization under a low magnetic field and high temperature? It is obvious that we have to discard use of *thermal* electron polarization as a seed of the proton polarization. The alternative method is to use electron "polarization" in photo-excited triplet states of aromatic molecules.[6]

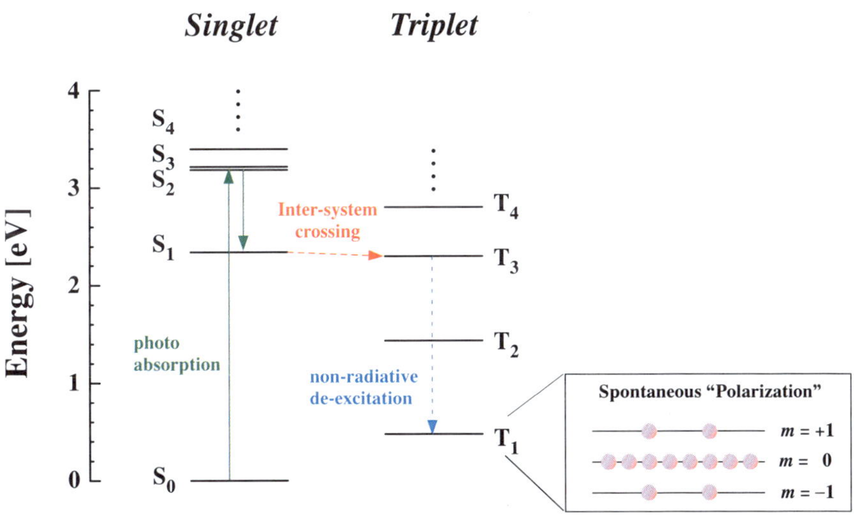

Fig. 1. Spontaneous appearance of electron "polarization" in pentacene.

Energy levels in a pentacene molecule are shown in figure 1. Excited states in an aromatic molecule are categorized into two groups, singlet states and triplet states. When a molecule absorbs a photon with an energy larger than the energy split between the ground (S_0) and the first excited (S_1) states, excited singlet states are populated. In the course of de-excitation, it goes through the lowest singlet state S_1. Although most of the excited molecules de-excite to the ground state directly, a small fraction de-excites to the lowest triplet state T_1 through the third triplet state T_3. The transition between S_1 and T_3, called *"inter-system crossing"*, is caused by a spin-orbit coupling of π electrons. Anisotropy of the spin-orbit coupling brings out population differences among the magnetic sub-levels in T_1. For example, the populations for $m = +1, 0, -1$ states in pentacene molecules doped in a p-terphenyl crystal are reported to be 0.12, 0.76 and 0.12, respectively.[7] This population difference is independent both from strength of external magnetic field and from temperature. The electron population difference between $m = 0$ and $m = -1$ (or $m = +1$ and $m = 0$) states, corresponding to electron polarization of 73 %, can be used to polarize protons even under a low magnetic field and high temperature. In the actual operation, the target material is placed in a magnetic field of 0.1 T and temperature of 100 K in order to suppress proton spin relaxation.

The electron polarization is transfered to protons by a cross polarization method.[8] This method is efficient even in a low magnetic field where the proton Zeeman splitting is not large compared with line-broadening due to an internal field in the crystal.

The proton polarization by the triplet-state method was introduced to the nuclear and particle physics field by the Kyoto group in the 1990s.[9] The CNS group started construction of the target system for use in radioactive nuclear beam experiments.[2] The target system described in the next section has been used in spin asymmetry measurements for the elastic scatterings between a proton and neutron-rich helium isotopes.[3–5]

3. Target system

The target crystal is a single crystal of naphthalene ($C_{10}H_8$) doped with 0.005 mol % of pentacene ($C_{22}H_{14}$). It is formed to the shape of a disk with a thickness of 1 mm and a diameter of 14 mm. The target is attached to a holder and placed in a cryogenic chamber filled with a cold nitrogen atmosphere at 100 K. The target holder and other mechanical parts located around the target are made of hydrogen-free materials, such as polychlorotrifluoroethylene or stainless steal, in order not to produce background protons.

The target crystal is irradiated with a laser light from an Ar-ion laser (Coherent TSM25) with a maximum power of 25 W. The laser light is

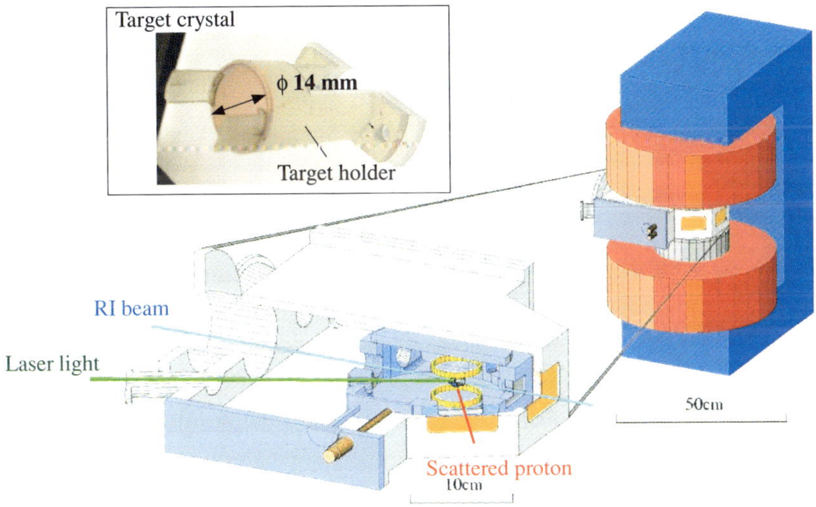

Fig. 2. Overview of the target apparatus.

pulsed with an optical chopper to have a pulse width of several tens of microsecond at a repetition frequency of several kHz. The pulse width dependence of the polarization rate was investigated recently.[10] Immediately after the laser irradiation, the electron polarization is transfered to protons by applying a microwave of $\sim$ 3 GHz with a sweep of external magnetic field. The microwave is fed with a copper-film loop-gap resonator.[11,12] The resonator shown in figure 3 is as thin as 9 mg/cm^2-thick and hardly disturbs the recoiled protons to reach detectors. Figure 4 shows a microwave-power dependence of the polarization rate. The polarization rate takes its maximum at around 2 W (shown with an arrow) where the Zeeman splitting of electrons in the rotating frame coincides with that of protons (the Hartmann-Hahn condition[13]).

The proton polarization is measured with a pulse NMR method. The target material is covered by an NMR coil with a diameter of 19 mm and a length of 3 mm. After addition of a $\sim$ 3 MHz radio-frequency wave, a free induction decay signal is detected with the NMR coil. Figure 5 shows a typical NMR signal. A strong signal at 0–40 μs is due to an added RF wave to tilt the proton polarization. After the strong signal, the free induction decay signal of proton polarization appears and decays with a transverse relaxation time of a few tens of μs. Integrated magnitude of the free induction decay signal is proportional to the proton polarization.

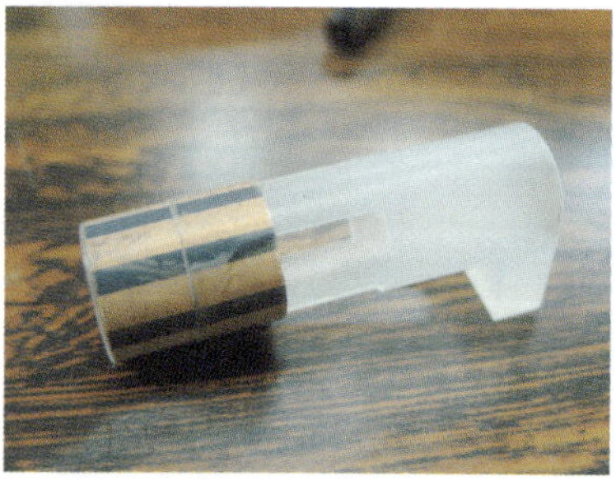

Fig. 3. Photograph of a loop-gap resonator.

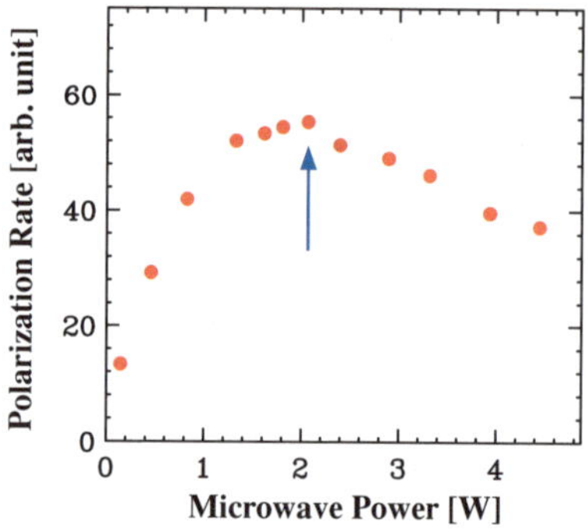

Fig. 4. Microwave-power dependence of the polarization rate. At $\sim$ 2 W (shown with an arrow), the Hartmann-Hahn condition is achieved.

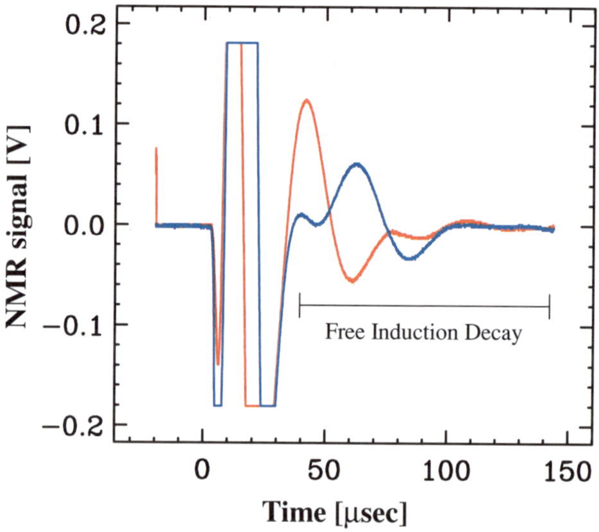

Fig. 5. A typical NMR signal.

The absolute value of the proton polarization can not be determined by the free induction decay signal alone. A calibration to relate the magnitude of the free induction decay signal to the absolute proton polarization was done by measuring the spin asymmetry in the proton-^{4}He elastic scattering at 80 A MeV. The proton polarization was determined by comparing the measured asymmetry and the previously-reported vector analyzing power,[14] as shown in figure 6. The proton polarization during the p-^{4}He measurement was found to be 11 %. The maximum polarization achieved by the system was 20 %.

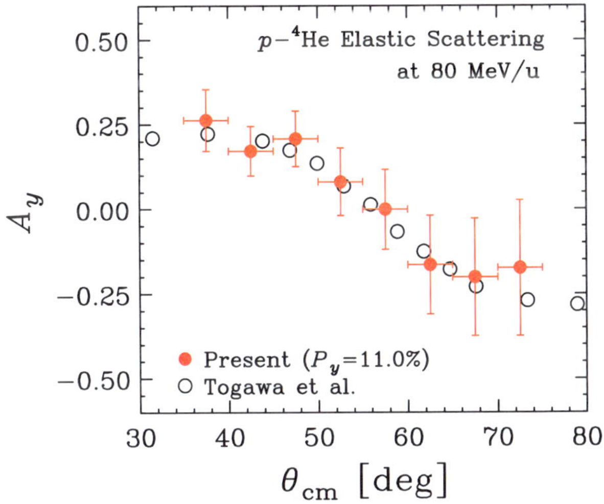

Fig. 6. Vector analyzing power for the p-^{4}He elastic scattering at 80 MeV/u. Filled circles indicate data taken with the present system by assuming the proton polarization of 11 %. Open circles are from reference 10.

4. Current status of the technique

The polarized solid proton target has been applied to vector analyzing power measurements for the p-^{6}He[3,4] and p-^{8}He[5] elastic scatterings. The maximum value of the proton polarization was found to be 20 % for 150 mm^3 (14 mm$\phi\times$ 1 mm) target crystal. This value of the polarization is smaller than that which has been achieved for a smaller crystal of 50 mm^3.[15]

On the other hand, high proton polarization has been achieved for a crystal with a volume of $\sim$ 30 mm^3 by use of a dye laser.[16,17] Compiled data of the proton polarization in naphthalene reported by the Kyoto group and the CNS group are listed in table 1. The data are taken for a variety of experimental setups and conditions: namely, by using different lasers with different output power ($W_{\rm laser}$), and for crystals of different volume ($V_{\rm crystal}$), as shown in the table. Here we introduce a quantity $P_p \cdot V_{\rm crystal}/W_{\rm laser}$ which is a measure of magnetization ($P_p \cdot V_{\rm crystal}$) per unit laser power ($W_{\rm laser}$).

Except for earlier data in references9 and 16, the newly introduced quantity takes similar values of 5.5–7.0. This comparison clearly indicates that the performance of the polarized target is primarily determined by power of the light source. Reinforcement of a light source is needed to achieve higher polarization in a large crystal of $\sim 150\ \mathrm{mm}^3$

Table 1. Compilation of proton polarization reported by the Kyoto and CNS groups.

Laser	λ_{laser} [nm]	W_{laser} [mW]	V_{crystal} [mm^3]	P_p [%]	$P_p \cdot V_{\mathrm{crystal}}/W_{\mathrm{laser}}$ [% mm^3/mW]	Reference
N_2	370	150	30	13	2.6	9
Dye	590	350	30	32	2.7	16
Dye	590	350	35	70	7.0	17
Ar-ion	454–529	200	30	36.8	5.5	15
Ar-ion	454–529	440	150	20	6.8	2

It should be noted that a new activity to develop a neutron spin filter based on the principle has started at PSI[18] recently.

5. Summary

Proton polarization based on the electron polarization in photo-excited triplet states of aromatic molecules will enable new applications of the proton polarization. The CNS polarized proton solid target has been constructed for use in radioactive nuclear beam experiments. The target system works in a low magnetic field of 0.1 T and at a high temperature of 100 K and has been applied to measurements of elastic scatterings between a proton and neutron-rich helium isotopes, conducted at RIPS, RIKEN. The maximum proton polarization was found to be 20 %. This proton polarization is limited by a lack of photo-excitation power. Future reinforcement of a light source will be a promising way to achieve higher proton polarization.

Acknowledgment

This work was supported by the Grant-in-Aid No. 17684005 of the Ministry of Education, Culture, Sports, Science, and Technology of Japan. One of the authors (S.S.) expresses his gratitude for financial support by a Grant-in-Aid from the Japan Society for the Promotion of Science (JSPS) Fellows (No. 18-11398).

References

1. T. Uesaka, T. Wakui, S. Sakaguchi, T. Kawahara and H. Sakai, *Eur. Phys. J. Special Topics* **150**, 71 (2007).
2. T. Wakui, in *Proc. XI Int. Workshop on Polarized Ion Source and Polarized Gas Targets 2005*, eds. T. Uesaka, H. Sakai, A. Yoshimi and K. Asahi (World Scientific, Singapore, 2007).
3. M. Hatano, H. Sakai, T. Wakui, T. Uesaka, N. Aoi, T. Ikeda, K. Itoh, H. Iwasaki, T. Kawabata, H. Kuboki, Y. Maeda, N. Matsui, T. Ohnishi, T. Ohnishi, T. Saito, N. Sakamoto, Y. Satou, K. Sekiguchi, K. Suda, A. Tamii, Y. Yanagisawa and K. Yako, *Eur. Phys. J. A* **25**, 255 (2005).
4. T. Uesaka and S. Sakaguchi, in preparation.
5. S. Sakaguchi, in *Proc. International Symposium on Physics of Unstable Nuclei* (World Scientific, Singapore, 2008).
6. H. W. van Kesteren, W. T. Wenckebach and J. Schmidt, *Physical Review Letters* **55**, 1642 (1985).
7. D. J. Sloop, H.-L. Yu, T.-S. Lin and S. I. Weissman, *J. Chem. Phys.* **75**, 3746 (1981).
8. A. Henstra, P. Dirksen and W. T. Wenckebach, *Phys. Lett. A* **134**, 134 (1988).
9. M. Iinuma, Y. Takahashi, I. Shaké, M. Oda, A. Masaike, T. Yabuzaki and H. M. Shimizu, *Phys. Lett. A* **208**, 251 (1995).
10. T. Kawahara, T. Uesaka, Y. Shimizu, S. Sakaguchi and T. Wakui, these proceedings.
11. B. T. Ghim, G. A. Rinard, R. W. Quine, S. S. Eaton and G. R. Eaton, *J. Magn. Res. A* **120**, 72 (1996).
12. T. Uesaka, M. Hatano, T. Wakui, H. Sakai and A. Tamii, *Nucl. Instr. Meth. A* **526**, 186 (2004).
13. S. R. Hartmann and E. L. Hahn, *Phys. Rev.* **128**, 2042 (1962).
14. H. Togawa and H. Sakaguchi, RCNP Annual Report, 1 (1987).
15. T. Wakui, M. Hatano, T. Uesaka, H. Sakai and A. Tamii, *Nucl. Instr. Meth. A* **550**, 521 (2005).
16. M. Iinuma, Y. Takahashi, I. Shaké, M. Oda, A. Masaike, T. Yabuzaki and H. M. Shimizu, *Phys. Rev. Lett.* **84**, 171 (2000).
17. K. Takeda, K. Takegoshi and T. Terao, *J. Phys. Soc. Jpn* **73**, 2313 (2004).
18. P. Heutle. private communication.

PULSE STRUCTURE DEPENDENCE OF THE PROTON SPIN POLARIZATION RATE

T. Kawahara

Department of Physics, Toho University,
Funabashi, 274-8510, Japan
E-mail: kawahara@cns.s.u-tokyo.ac.jp

T. Uesaka* and Y. Shimizu

Center for Nuclear Study, Graduate School of Science, University of Tokyo,
Bunkyo, 113-0033, Japan
**E-mail: uesaka@cns.s.u-tokyo.ac.jp*

S. Sakaguchi

RIKEN (The Institute of Physical and Chemical Research) Nishina Center,
Wako, 351-0198, Japan
E-mail: satoshi@ribf.riken.jp

T. Wakui

Cyclotron and Radioisotope Center, Tohoku University,
Sendai, 980-8578, Japan
E-mail: wakui@cyric.tohoku.ac.jp

The proton polarization is obtained at high temperature and in low magnetic field by means of photo-excited triplet state of aromatic molecules. For this method, we use a continuous wave Ar-ion laser which is pulsed by an optical chopper. In this system, the photon number is the bottleneck in achieving higher polarization. The proton polarization rate was measured by changing a duty from 5 % to 50 % and a reputation frequency from 0.75 kHz to 10.6 kHz. For a duty factor of 50 % and a repetition frequency of 10.6 kHz, the polarization rate can be a factor five larger than that for the old setting.

Keywords: Proton polarization; polarized target; aromatic molecule.

1. Introduction

The proton polarization is used in many fields such as medical physics, molecular conformation, nuclear and particle physics, and so on. For obtaining high polarization, thermal polarization of electrons is generally used, since electrons can be more easily polarized than protons. Then, the polarization is transferred from electrons to protons. A high proton polarization can be obtained in this way. However, this method requires low temperature and high magnetic field strength for obtaining high polarization. This condition hinders application of proton polarization in various fields. In this situation, a proton polarization system which can produce high polarization at high temperature and in low magnetic field has been developed at the Center for Nuclear Study (CNS), University of Tokyo. The proton polarization can be achieved in a low magnetic field of 0.1 T and at a high temperature of 100 K by making use of the population difference of electrons in Zeeman sublevels in a photo-excited triplet state of aromatic molecules. As a material, we use a naphthalene crystal doped with a small amount (typically 0.005 mol%) of pentacene. The electrons of pentacene molecules are excited to higher singlet states by irradiating a laser light whose wavelength is the same as the absorption wavelength of the pentacene molecule. These excited singlet states decay to the first excited singlet state S_1 by internal conversion. While most of the electrons in the S_1 state decay to S_0 due to short lifetime, a small fraction of them decay into an excited triplet state T_3 by intersystem crossing. The intersystem crossing is a transition process between different spin states caused by the spin-orbit interaction, which mixes the singlet and triplet states. Then, the T_3 state decay to the lowest triplet state T_1. The electron population difference is spontaneously produced among the magnetic sub-states in T_1 state as shown in figure 1. The population rate is 76 % and 12 % for m =0 and $m = \pm 1$ states, respectively. The population difference between m =0 and $m = -1$ states can be regarded as the electron polarization. Finally, the T_1 state decays to the ground state S_0. The most important point for our purpose is that this population is almost independent of the temperature and the magnetic field strength. Therefore, we can polarize proton at high temperature and in the low magnetic field strength with this system. At present, the proton polarization is successfully obtained up to 20 % by using this method at 100 K and in about 0.1 T. Furthermore, this system has been successfully applied to RI beam experiments[1,2] as a polarized proton solid target.

In the method described above, a pulse dye laser is generally used to excite the aromatic molecules. However, we use a continuous wave Ar-ion laser for practical reasons. The laser light is pulsed by an optical chopper.[3] The pulse structure tried before was a pulse width 20 μsec and a repetition rate 2.5 kHz. In this case, the photon number is the bottleneck in

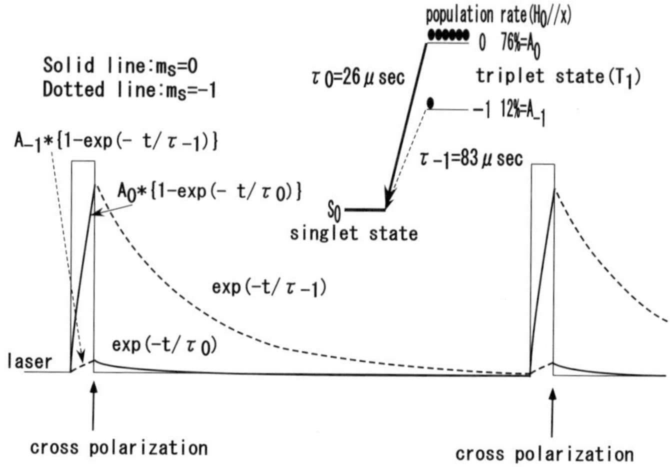

Fig. 1. Time development of electron polarization.

achieving higher polarization. We used only 5 % of the laser power. Simply, for increasing the photon number, the best laser light would be a continuous wave. However, one problem arises that the lifetime of the magnetic sub-level m=0 and that of m=-1 is different. The m=0 state has a larger population, but its lifetime is shorter (τ_0=26 μsec, τ_-=83 μsec[4]) as shown in figure 1. Accordingly, the electron population at m=0 state becomes constant when the pulse width is long. On the other hand, the population in m=-1 state becomes constant more slowly than the m=0 state. Therefore, while the photon number is increased, the electron polarization decreases for long time laser irradiation. Thus, the electron polarization should depend on the pulse structure of the laser light. However, its dependence has not been studied since the pulse structure is fixed in the pulsed laser such as the dye laser. In our target, the pulse structure is determined by slit-opening of the optical chopper and its repetition frequency. Thus, the pulse structure of the laser light can be changed very easily. In this work, we studied the proton polarization by changing the pulse structure.

2. Optical system

In this section, we will describe the optical system for electron polarization. Figure 2 shows the overview of the optical system. The laser light, pulsed by an optical chopper, is coupled into an optical fiber and guided to the target system. The laser power is monitored by sampling the 1 % fraction of the light with a photo diode. A signal from the photo diode is used for a

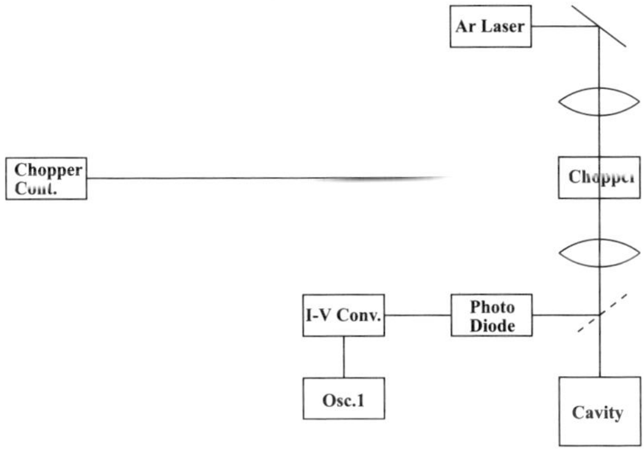

Fig. 2. Schematic overview of optical system.

trigger of the microwave and the magnetic field sweeping for the polarization transfer from electrons to protons. The target crystal is illuminated by the laser light sent through the fiber and optical system consisting of mirrors and lenses. Details of the laser and the optical chopper are described in the following.

For the optical excitation of pentacene molecules, we used an Ar-ion laser (Coherent TSM25) with a wavelength ranging from 454.5 nm to 528.7 nm and a total maximum output power of 25 W in the multiline operation mode. The operation mode of the laser can be selected between the multiline and the single-line modes. Among the available wavelengths of Ar-ion laser, a preferred wavelength for the excitation of pentacene is 514 nm since pentacene has a local absorption maximum near 514 nm. In this measurement, we used the laser light with a power of 8 W in the multiline mode.

To produce pulsed laser light from CW laser light, an optical chopper with two chopper blades is used. The duty factor can be varied easily by shifting the overlap of two chopper blades (fig. 3).
The frequency of laser pulse can be modified by changing the rotating speed of the optical chopper. This optical system enables us to change the duty factor from 5 to 50 % and the repetition frequency from 0.75 to 10.6 KHz. The pulse width is determined by the duty factor and repetition frequency. The relation of the three parameters is expressed as

$$D = ft, \tag{1}$$

where D, f and t are the duty factor, the repetition frequency, and the pulse width respectively.

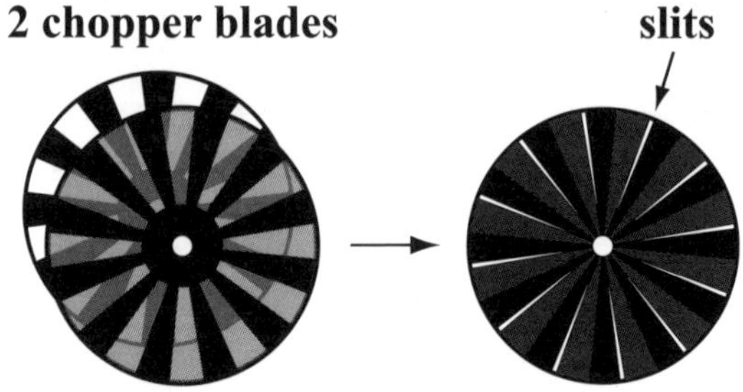

Fig. 3. Optical chopper blades.

3. Measurement

The measurement was carried out under following conditions. A crystal size was 14 mm in diameter and 3 mm in thickness. Protons were polarized at a temperature of 200 K and in a magnetic field of about 60 mT. Figure 4 shows the typical timing chart of polarization process.

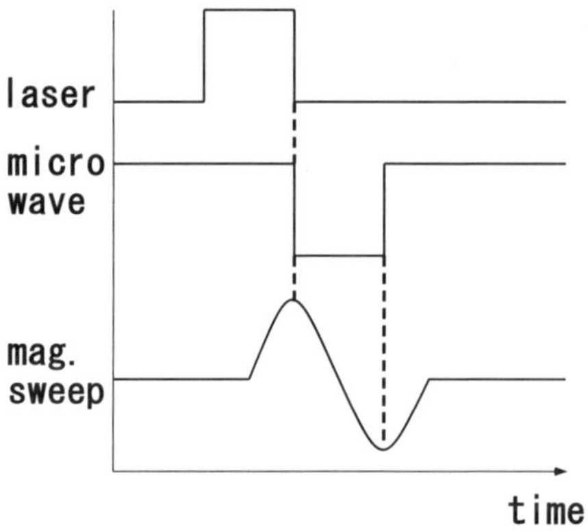

Fig. 4. Timing chart during polarization. When the laser is irradiated onto the crystal, the electron polarization is produced. Just after the irradiation of the laser light, the microwave was irradiated and the magnetic field strength was swept in order to transfer the electron polarization to protons.

In the first step, the laser light was irradiated onto the crystal. Just after the irradiation, the microwave was irradiated and the magnetic field strength was swept in order to transfer the electron polarization to the protons. The steps were repeated at a certain frequency.

The magnitude of proton polarization after five minutes buildup is defined as the proton polarization rate, which is measured with the pulse NMR method. The result is shown in figure 5 where the proton polarization rate is plotted as a function of the repetition frequency.

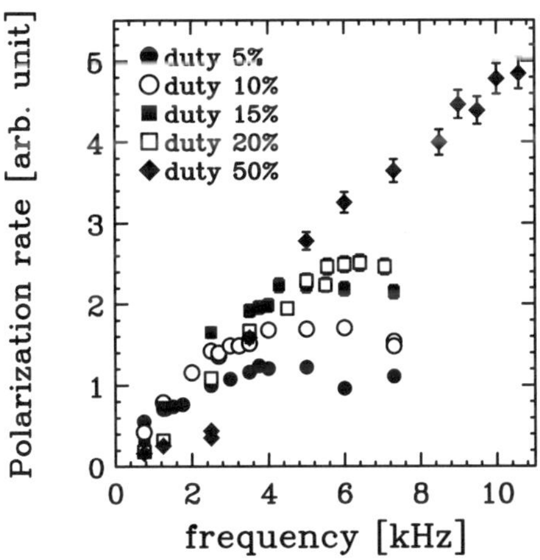

Fig. 5. The polarization rate was measured by changing a duty factor and repetition frequency.

In the previous works,[3] the repetition frequency and the duty factor were fixed at 2.5 kHz and 5 %, respectively. All the measured data are normalized to the data in this condition. At high frequency limits the polarization rate is almost saturated and proportional to the duty factor. In the present work, we found that proton polarization rate takes the maximum value when the repetition frequency is 10.5 KHz and the duty factor is 50 %. The polarization rate was improved by a factor of five compared with the previous works.

4. Theoretical model

We built a simple theoretical model of the electron population in order to estimate the polarization rate. There are three steps in the polarization process. The first one is the electron polarization, the second is the polarization transfer from electrons to protons, and the third is the relaxation of electron polarization. In this model, we assume that the efficiency of the polarization transfer from electrons to protons is 100 %. The electron polarization is the electron population difference between the m=0 and m=-1 states, which are magnetic sub-states of a triplet state of pentacene molecule. Hereafter, the $m=0$ and $m=-1$ states are denoted as m_0 and m_- states, respectively.

In the first step, the electron population increases according to the build up function during laser irradiation. The build up function is given as

$$f_{b,i}(t) = A_i\tau_i\{1 - \exp(-\frac{\mathrm{t}}{\tau_\mathrm{i}})\}, \tag{2}$$

where τ_i is the lifetime of the triplet state, A is the population rate and i is the magnetic sub-state of triplet state. After the laser irradiation, the electron population difference is transfered to protons and the electron population decreases according to a relaxation function. When the electron polarization is transfered, the number of the electrons in two magnetic sub-states turns over by cross polarization.[3] The relaxation function is given as

$$f_{r,i}(t) = \exp(-\frac{\mathrm{t}}{\tau_\mathrm{i}}). \tag{3}$$

Here, the lifetimes of the two sub-states used in the model are taken as free parameters. The solid and dotted lines in figure 1 represent time development of the electron population in m_0 and m_- states, respectively. The ρ_0 and ρ_-, electron population in m_0 and m_- states, are calculated by using the equations (2) and (3). The proton polarization rate is derived as

$$\frac{dP_p}{dt} \propto \rho_0 - \rho_-. \tag{4}$$

The lifetimes τ_0 and τ_- are determined by using the result of measurement and this calculation to be τ_0=26 μsec and τ_-=88 μsec. In reference 4, the lifetimes at 100 K are τ_0=26 μsec and τ_-=83 μsec. Obtained lifetimes at 200 K are almost the same as the values at 100 K. The proton polarization rate calculated by our simple model reproduces the measured data as shown in figure 6.

5. Summary

To pursue possible improvement in proton polarization, we have examined the pulse structure dependence of the proton polarization rate. The proton polarization rate was measured by changing the duty factor from 5 % to 50 % and the repetition frequency from 0.75 kHz to 10.6 kHz. At a duty factor of 50 % and a repetition frequency of 10.6 kHz, the polarization rate was improved by a factor of five compared with the previous works. We have found that the proton polarization rate depends strongly on the pulse structure. We built a simple model of the electron population in order to estimate the polarization rate. The τ_0 and τ_-, which are the lifetimes of magnetic substates in triplet state T_1, are determined by using the result

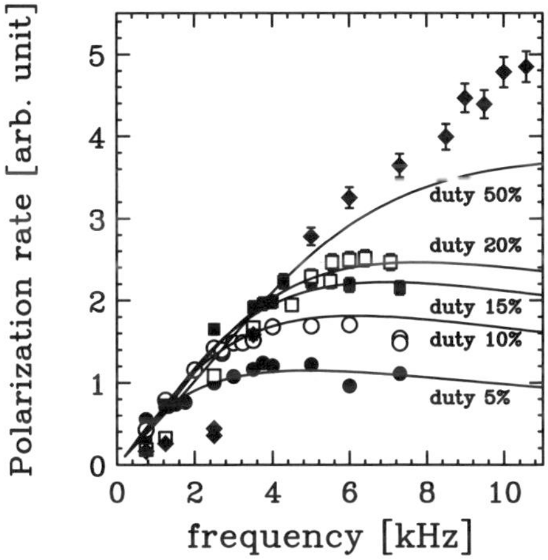

Fig. 6. The polarization rate calculated from our simple model (solid curves) together with the measured polarization rate. The result are calculated for the duty factor of 5 %, 10 %, 15 %, 20 % and 50 %.

of measurement and this calculation to be τ_0=26 μsec and τ_-=88 μsec. Obtained this lifetime at 200 K are almost the same as the value at 100 K.

References

1. M. Hatano *et al.*, *Eur. Phys. J. A* **25**, 255 (2005).
2. S. Sakaguchi *et al.* in *Proc. of ISPUN07*, 245 (World Scientific, Singapore, 2008).
3. T. Wakui *et al.*, *Nucl. Instr. Meth. A* **550**, 521 (2005).
4. M. Iinuma. PhD thesis, Kyoto University (Kyoto, Japan, 1997).

PROTON NMR IN THE LARGE COMPASS $^{14}NH_3$ TARGET

J. Koivuniemi*, F. Gautheron, C. Hess, Y. Kisselev, W. Meyer,
E. Radtke and G. Reicherz

Universität Bochum, Institut für Experimentalphysik I,
44780 Bochum, Germany
**E-mail: Jaakko.Koivuniemi@cern.ch*
wwwcompass.cern.ch

N. Doshita, T. Iwata, K. Kondo and T. Michigami

Yamagata University, Yamagata,
992-8510 Japan

for the COMPASS collaboration

In the large COMPASS polarized proton target the 1508 cm^3 of irradiated granular ammonia is polarized with the dynamic nuclear polarization method using 4 mm microwaves in 2.5 T field. The nuclear polarization up to 90–93 % is determined with CW NMR. The properties of the observed ammonia proton signals are described and spin thermodynamics in high fields is presented. Also the second moment of the NMR line is estimated.

Keywords: polarization; target; proton; ammonia; NMR; DNP.

1. Ammonia target

The polarized solid $^{14}NH_3$ target of the COMPASS experiment at CERN[1–3] is using continuous wave NMR to determine the nuclear polarization. During the physics data taking both longitudinal 2.5 T and 1.0 T fields, and transverse 0.63 T field, are used. The homogeneity of the solenoid field is tuned to be better than 20 ppm with the help of 16 trim coils. The Q-meters[4] are tuned to give the absorption part of the NMR signal. The target proton nuclear polarization is determined with continuous wave NMR. Ten single loop NMR coils 50 mm long and 13 mm wide made from 1.0 mm stainless steel tube are used. Inductance was estimated to be 60–80 nH. The coils were mounted on the outer surface of the target cell and connected with 0.9 mm diameter coaxial cable (Precision tube KA50034) inside the mixing chamber to the Q-meters.[4]

The target material is placed into three cells of 4 cm in diameter. The central cell is 60 cm long and is separated by 5 cm from 30 cm long up- and downstream cells. Each of the cavity volumes can be fed separately with microwaves from outside of the dilution cryostat. Typical ammonia packing factors were 0.46–0.52. The solid $^{14}NH_3$ is polarized through $\dot{N}H_2$ radicals.[5] The natural abundance of deuterons in the protons is 0.0115 % and 0.368 % of nitrogens are nitrogen-15. In addition the volume between the target beads is filled with the mixture of helium-3 and helium-4. The target cells are in the diluted phase of the helium mixture, so this is mostly helium-4.[6] The density of ammonia is about 0.85 g/cm^3.[2]

The target cells were made of very thin and light weight materials.[7] The polyamide mesh had an aramide fiber support structure. Stycast 1266 polymer was used as a glue in the construction. These synthetic fibers and polymers have carbon, hydrogen, nitrogen and oxygen atoms in their molecular structure. A proton background signal can thus be expected in the NMR measurement of the unpolarized signals at temperatures around 1 K. In addition the proton signal could also come from water ice frozen from ambient air during the target filling and loading into the dilution cryostat. The contribution of this proton background was not significant however compared to the cell construction materials.[3]

2. Proton polarization

When the nuclear spin system is in thermal equilibrium with superfluid ^{4}He the polarization can be calculated from the Brillouin function[2]

$$P_I(x) = \frac{2I+1}{2I}\coth\left(\frac{2I+1}{2I}x\right) - \frac{1}{2I}\coth\left(\frac{x}{2I}\right), \tag{1}$$

where the spin number $I = \frac{1}{2}$ for protons and $I = 1$ for ^{14}N. Here $x = Ihf/k_BT_s$ with h the Planck constant, f NMR resonance frequency in the 2.5 T field, k_B the Boltzmann constant and T_s the spin temperature. The polarization of the protons at 1.0 K is about 0.25 %. To estimate the statistical error in thermal equilibrium calibration at temperatures 1.0–1.5 K a similar data set was produced on computer with the same signal-to-noise ratio. This was then analyzed with the same programs that were used for the real data. A histogram of the 800 produced calibration constants was fit to a Gaussian to get an error estimate. For the ammonia protons the error was about 1.7 %, while for the background protons it was 5–7 % due to the smaller signal. Taking into account the size of the signals these gave an overall statistical error of about 4 % to the final calibration constant

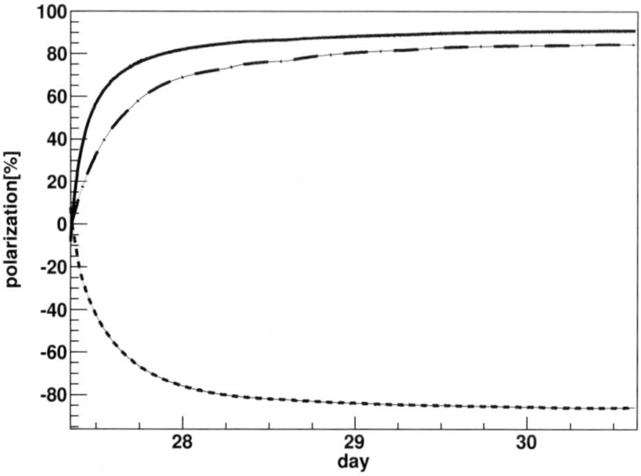

Fig. 1. Typical proton polarization build up. The up- (solid) and downstream (dashed) cells reach +90 % and +85 % polarizations while the central target cell (dotted) is polarized to -87 %. The lower polarization achieved at the downstream cell is probably due to larger microwave guide losses resulting in smaller microwave power compared to the upstream cell. In addition the dilution cryostat cooling is likely not as efficient as at the upstream end of the target, where the helium-3 is circulating more freely.

after subtraction of the background proton signal. In this simulation the uncertainty comes from the noise and from the temperature measurement.

Typical proton polarization build up is shown in figure 1. The helium-3 flow during the polarization was 60–70 mmol/s and the temperature in the mixing chamber was 200–400 mK. The microwave frequency was about 69 940 MHz for positive polarization and 70 270 MHz for negative. The microwave power is reduced a little at the end of the polarization build up. The frequency modulation of the microwaves has only a small effect (about 2 %) on the final polarization. The average cell polarization was calculated from the three coils on the up- and downstream cells and from the four coils on the central cell.

3. Spin Hamiltonian

For the hydrogen in NH_3 at 2.5 T using CW NMR the nuclear spin Hamiltonian is

$$H = H_z + H_{dip} + H_{quad} + H_{hf} + H_{RF}. \tag{2}$$

In frequency units the Zeeman term is about 106 MHz. The dipolar interaction between two protons at distance of 1.653 Å gives about 25 kHz. The quadrupole term vanishes for the spin $\frac{1}{2}$ while the hyperfine interaction is probably small due to the low density of the paramagnetic centers, about 10^{-4}–10^{-3} $\dot{N}H_2$/molecule. The continuous wave NMR is usually done at low RF fields, so the last term can also be neglected. Strictly speaking the magnetic field giving rise to the Zeeman term is the local field, which depends on the sample shape.[8] For a spherical sample the local field is the same as the external field however.

The dipole interaction has the simple form[9,10]

$$D_{zz,jk} = \frac{\mu_0 \hbar \gamma_j \gamma_k}{4\pi r_{j,k}^3}(1 - -3\cos^2\theta_{j,k}), \tag{3}$$

where $r_{j,k}$ is the distance between the two nuclei and $\theta_{j,k}$ the angle between the external magnetic field and the position vector between the nuclei. γ_j and γ_k are the gyromagnetic ratios of the two nuclei.

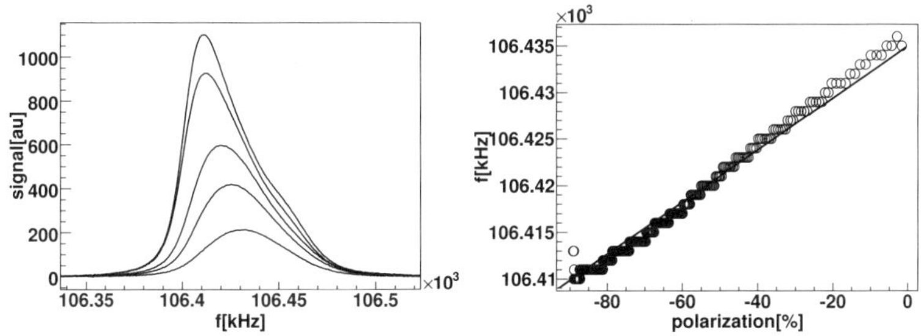

Fig. 2. *Left:* NMR signal line shape for approximate proton polarizations -20 %, -40 %, -60 %, -80 % and -90 %. A memory function fit to the -20 % polarized signal gave $M_2 = 490$ kHz^2 and $\mu = 3.7$. At -40 % polarization the fitting of a single line becomes more difficult due to the asymmetry of the lineshape. In this case the $M_2 = 516$ kHz^2 and $\mu = 3.9$. *Right:* Typical NMR peak frequency shift during polarization build up. The frequency shift of the peak in coil #4 was fit to a line $106435 + 0.279122 \cdot p$.

The second moment is the sum of the squared frequency shifts from the dipole interaction with neighboring nuclei

$$M_2^j = g_I \sum_k D_{zz,jk}^2 + g_S \sum_l D_{zz,jl}^2. \tag{4}$$

Here g_I is $3I(I+1)/4$ for like spins and g_S is $S(S+1)/3$ for unlike spins.[9] I is the spin number of the nucleon in consideration and S the spin number of the neighboring unlike nuclei.

For the calculation of the dipolar interaction and the second moment of the proton NMR line the positions of the nitrogen and hydrogen atoms need to be known. Due to the granular nature of the target material, the crystal direction is random in the magnetic field. Here we assume a cubic lattice with space group $P2_13$.[11] The screw axis operation 2_1 rotates the crystal first 180 degrees and translates it 1/2 of lattice size along the same axis. In the primitive centering P of the Bravais lattice the points are on the cell corners only. From the initial unit cell positions of one nitrogen atom $(x, x, x), x = 0.21032$ and one hydrogen atom $(0.3738, 0.2606, 0.1129)$ a lattice is generated applying the symmetry operations in this space group. The lattice constant a=5.1305 Å agrees well with the measured density of 0.85 g/cm^3. The molecular structure of ammonia is assumed to be preserved with $r_{\mathrm{NH}} = 1.0099$ Å, $r_{\mathrm{NN}} = 3.3769$ Å and H-N-H bond angle of 107.5–109.0°.[11]

After a sufficiently large lattice has been generated, it is cut into a spherical shape. The dipolar interaction and second moments of the central hydrogens are then calculated. Next the sample is rotated with Euler angles (α, β, γ) to a random direction and the calculation of M_2 is repeated. This is done about 1000 times to simulate the number of crystals seen by each NMR coil. The simulation for this static model resulted in an M_2 of about 890 kHz^2. This is of course larger than in a simple one molecule simulation, since the proton spin also sees all the other neighboring molecules.

4. Proton lineshape

In the solids the NMR linewidth is usually determined by the dipole and quadrupole interactions. In the case of spin $\frac{1}{2}$ protons the quadrupole interaction is absent. The indirect coupling gives usually frequency shifts of a fraction of hertz and requires a high resolution NMR to be used.[12] It has been suggested that the covalent bond with the nitrogen could mediate the indirect interaction between the protons.[2] Recently the J-coupling has been observed between two magnetically active nuclei on both sides of a hydrogen bond.[13] Proton tunneling in the hydrogen bond could also play a role in the lineshape.[14] The classic paper of Van Vleck[15] calculated the second moment of the NMR line. In the dipole interaction the odd moments vanish resulting in a symmetric lineshape and the lineshape is independent of polarization or sample shape. Abragam calculated the polarized case of a spherical sample of one species of nuclear spins coupled by pure dipole interaction and found a good agreement to $M_2(p) = M_2(0)[1 - p^2]$ in case of CaF_2.[16] Here at maximum polarization p=1 the second moment goes to zero. While the NMR line can be seen to become narrower due to the polarization build up, it does not seem to follow such a simple law in the

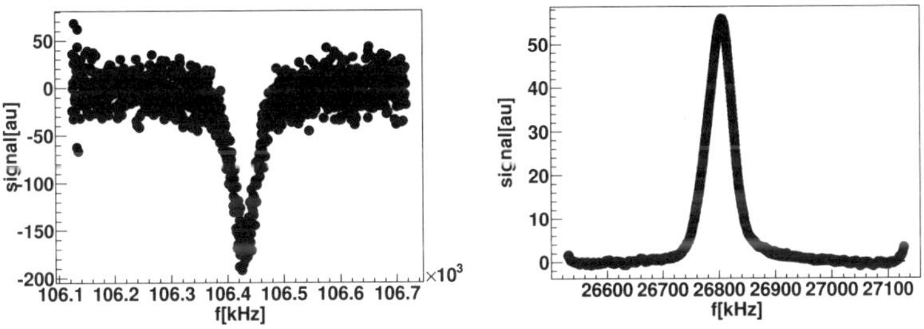

Fig. 3. *Left:* Empty target cell signal at 1.0 K in 2.5 T field from coil #6. From the fit M_2=1071 kHz^2 and μ=5.5. *Right:* Proton NMR signal measured in transverse 0.63 T field with polarization of +82 %. Since the Q-meter is not tuned for this frequency a phase correction of 1.6 rad was done using the Kronig-Kramers dispersion relations. The fit to a memory function gave M_2 = 815 kHz^2 and μ=4.5. The lineshape does not have clear asymmetry like in the much more homogeneous longitudinal 2.5 T field. The reason for this can also be the Q-meter base line subtraction.

case of ammonia. The thermal equilibrium signal at 1.0 K from the background protons is shown in figure 3. It is clearly wider than the signal from ammonia, where $M_2 = 520$–650 kHz^2 and $\mu = M_4/M_2^2 \sim 3.8$–4.0. The integrated intensity of the background protons was about 20 % of the ammonia proton intensity. The +82 % polarized signal in transverse 0.63 T field is also shown. Broader linewidth can be due to a less homogeneous magnetic field. The radio frequency field from the detection coil is not orthogonal to the static field in this case.

5. Spin thermodynamics

Nuclear spin entropy[2,8] in high field, taking into account only the Zeeman term of the spin Hamiltonian in equation 2, is

$$S_n/R = \frac{x}{2I}\coth\frac{x}{2I} - \frac{(2I+1)x}{2I}\coth\frac{(2I+1)x}{2I} + \ln\Bigl(\frac{\sinh\frac{(2I+1)x}{2I}}{\sinh\frac{x}{2I}}\Bigr). \quad (5)$$

Here R is the universal gas constant. The polarization values of 90 % correspond to spin temperatures around 1.7 mK. The entropy density for the protons reduces to 1.65 J/mol K from the initial unpolarized 5.76 J/mol K. If the conditions for dynamic nuclear polarization are assumed to be the same for all of the coils on each target cell, the variations in the final polarization values could reflect small differences in the polarizable spin density along each cell.

Table 1. Typical maximum polarizations probed by the NMR coils. The NMR peak position is determined by the magnetic field homogeneity and the target polarization. Also the calculated spin temperature, entropy and nuclear heat capacity are given.

Coil	p [%]	f [kHz]	T_n [mK]	S_n [J/mol K]	C_B [J/mol K]
1	+90.4	106459	+1.7	1.6	3.4
2	+90.3	106462	+1.7	1.6	3.4
3	+92.5	106463	+1.6	1.3	3.2
4	-89.1	106413	-1.8	1.8	3.5
5	-87.2	106408	-1.9	2.0	3.6
6	-86.8	106406	-1.9	2.0	3.6
7	-83.7	106408	-2.2	2.4	3.7
8	+82.9	106453	+2.2	2.4	3.7
9	+77.2	106456	+2.5	2.9	3.5
10	+93.5	106449	+1.5	1.2	3.0

One of the interesting properties of water ice at low temperatures is its residual entropy.[17] As pointed out by Pauling in his paper the hydrogen bonds do not always lead to residual entropy. It is, however, interesting to estimate upper limit for this in the case of ammonia ice. In a similar way to the Pauling's calculation, the "ammonia rules" require that each nitrogen has three close hydrogen neighbors to preserve the molecule structure. In addition the nitrogen has three distant hydrogens that reside close to the other three nitrogens. For N nitrogens and $3N$ hydrogens the total number of configurations is then 2^{3N}. For the six hydrogens on each of the nitrogen atoms there are $2^6 = 64$ possible combinations. From these only 20 fulfill the "ammonia rules". Thus the upper limit for entropy

$$S_0 = k_B \ln\Big(2^{3N}\Big(\frac{20}{64}\Big)^N\Big) = Nk_B \ln\Big(\frac{5}{2}\Big) \tag{6}$$

gives 7.6 J/mol K compared to the 3.4 J/mol K in the water ice.

The nuclear heat capacity in constant external magnetic field $C_B = T(\partial S_n/\partial T)_B$ becomes

$$C_B/R = \frac{x^2}{(2I)^2}\frac{1}{\sinh^2\frac{x}{2I}} - \frac{(2I+1)^2x^2}{(2I)^2}\frac{1}{\sinh^2\frac{(2I+1)x}{2I}}. \tag{7}$$

Thus the 90 % polarized proton nuclear heat capacity in 2.5 T field is then about 3.4 J/mol K while at zero polarization the heat capacity goes to zero. This is the reason why the relaxation of the polarization in the "frozen spin targets" is very slow at low temperatures. The equation 7 relates the heat flow to the spin system to the polarization relaxation rate. For example, polarization loss in one day from 90 % to 89 % corresponds to a heat flow of about 2 nW/mol in 2.5 T field at temperatures below 60 mK.

References

1. P. Abbon *et al.*, *Nucl. Instr. Meth. A* **577**, 455 (2007).
2. B. Adeva *et al.*, *Nucl. Instr. Meth. A* **419**, 60 (1998).
3. J. Koivuniemi *et al.*, *J. Phys.: Conference Series* **150**, 012023 (2009).
4. K. Kondo *et al.*, *Nucl. Instr. Meth. A* **526**, 70 (2004).
5. W. Meyer, *Nucl. Instr. and Meth. A* **526**, 12 (2004).
6. N. Doshita *et al.*, *Nucl. Instr. Meth. A* **526**, 138 (2004).
7. S. Neliba *et al.*, *Nucl. Instr. Meth. A* **526**, 144 (2004).
8. O. V. Lounasmaa, *Experimental Principles and Methods Below 1 K* (Academic Press, London, 1974).
9. A. Abragam, *The Principles of Nuclear Magnetism*, (Oxford University Press, Hong-Kong, 1989).
10. M. Mehring, *Principles of High Resolution NMR in Solids* (Springer-Verlag, Heidelberg, 1983).
11. R. Boese *et al.*, *J. Phys. Chem.* **101**, 5794 (1997).
12. N. F. Ramsey, *Phys. Rev.* **91**, 303 (1953).
13. A. J. Dingley, F. Cordier, S. Grzesiek, *Concepts in Magnetic Resonance* **13(2)**, 103 (2001).
14. A. J. Horsewill and W. Wu, *J. Magn. Reson.* **179**, 169 (2006).
15. J. H. Van Vleck, *Phys. Rev.* **74**, 1168 (1948).
16. A. Abragam *et al.*, *J. Magn. Reson.* **10**, 322 (1973).
17. L. Pauling, *J. Am. Chem. Soc.* **57**, 2680 (1935).

DNP WITH TEMPO AND TRITYL RADICALS IN DEUTERATED POLYSTYRENE

L. Wang*
Physics Department, School of Science, Donghua University, Shanghai, 200051, China
**E-mail: wang_li@dhu.edu.cn*

W. Meyer, C. Hess, E. Radtke and G. Reicherz
Institut of Experimental Physics AG I, Ruhr-University Bochum, Bochum, D-44780, Gemany

N. Doshita, K. Kondo and T. Iwata
Physics Department, Faculty of Science, Yamagata University, Yamagata, 990-8560, Japan

N. Horikawa
College of Engineering Chubu University, Kasugai, 487-850, Japan

Chemically doping with TEMPO and trityl radicals was performed in fully deuterated polystyrene samples. The deuteron polarizations and the behavior of paramagnetic centers have been investigated. 7.3 % deuteron polarization with TEMPO has been obtained at 2.5 T and 1 K and a deuteron polarization of 12.3 % with a trityl radical.

Keywords: Spin; polarization; target; trityl radical.

1. Introduction

The Dynamic Nuclear Polarization (DNP) process is a very useful method to enhance the polarization degree of protons or deuterons. For this, paramagnetic radicals are necessary and can be introduced into the solid target materials by chemical or radiation methods.

In various target materials, protons can be almost completely polarized even at temperatures around 1 K. For deuterons, with much lower magnetic moments compared to the protons, it is more difficult to achieve a high polarization degree. For several decades, the standard solid-state deuteron target materials such as deuterated butanol,[1,2] deuterated ammonia ND_3[3–5] and lithium deuteride ^6LiD[6] have been polarized to more than 50 % at magnetic fields between 2.5 T and 6.0 T in dilution refrigerators. For deuterated polymers target materials the situation is presently as follows:

- In deuterated polyethylene (CD_2) prepared for the DNP process by irradiation a deuteron polarization of 30 % has been obtained at 2.5 T and in a dilution refrigerator[7] and a deuteron polarization about 35 % at 6.5 T in a 1 K evaporator.[2]
- In deuterated polystyrene C_8D_8-chemically doped with TEMPO-deuterons have been polarized up to 40 % at 2.5 T in a dilution refrigerator.[8]

Trityl radicals as new radicals in the polarized target business have been used in recent years. Doping deuterated butanol or deuterated propanediol with such radicals even allowed a polarization of 80 % at magnetic fields as low as 2.5 T.[1]

In this paper, the deuteron polarization results obtained in TEMPO doped deuterated-polystyrene (d-PS) are compared with that of trityl-doped samples at the same DNP conditions, i.e. at 2.5 T and 1 K.

2. Samples with TEMPO doping: preparation and polarization results

Deuterated polystyrene samples with TEMPO doping were prepared by dissolving powder or small pieces of polystyrene in toluene at a temperature of 65–70 °C. As TEMPO dissolves in toluene, TEMPO was added as soon as deuterated polystyrene was getting transparent inside toluene. Subsequently, the mixture was stirred at 65–70 °C until toluene had evaporated, poured into Petri glasses and dried at room temperatures for 2–4 hours. The obtained foils with a diameter of 10 cm were not pressed. Their thickness was less than 1 mm. A typical EPR-line is shown in figure 1 with a width (FWHM) of 3.35 mT. The polarization results at 2.5 T and 1 K of a sample with a TEMPO concentration of $2.3 \cdot 10^{19}$ spins/g are given in table 1 with the relaxation time $T_{l,d}$ and the polarization build-up time.

3. Samples with the trityl doping: preparation and polarization results

The trityl molecule, much heavier than the TEMPO molecule, is very stable at room temperature. As solvent of the “Finland D36” radical, iso-butanol

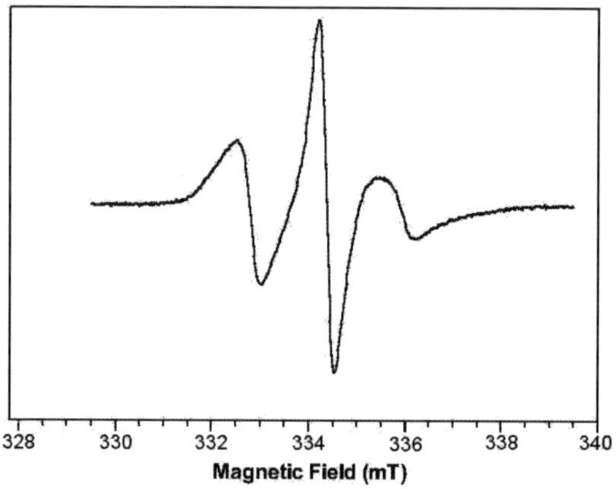

Fig. 1. EPR lines of TEMPO in deuterated polystyrene

Table 1. Polarization results in TEMPO doped deuterated polystyrene.

Sample	Microwave frequency (GHz)	d-polarization (%)	$T_{l,d}$ (min)	$T_{build-up}$ (min)	$FWHM_{bolometric}$ (mT)
d-PS(98 %-d) +TEMPO	69.850 70.050	+7.3 -7.7	12	33	6.73

Note: $f_{d,NMR}$=16.4MHz

was used. As iso-butanol has an solubility with toluene, iso-butanol with the Finland radical was poured together with the toluene-polystyrene mixture (made as described in section 2). The entire mixture was stirred at 65–70 °C and in four hours toluene and iso-butanol evaporated. The remaining mixture (polystyrene plus the radicals) was poured into Petri glasses and dried at room temperature to become a thin foil.

A typical EPR-line with a width of 0.16 mT is shown in figure 2. The polarization results at 2.5 T and 1 K of a sample with a trityl radical concentration of $1.16 \cdot 10^{19}$ spins/g are given in table 2.

The NMR signal of deuterated polystyrene with a polarization of -12.3 % is shown in figure 3.

4. Comments and conclusions

The polarization results listed in table 1 and 2 clearly indicate that deuterated polystyrene doped by the trityl radical "Finland D36" leads to much

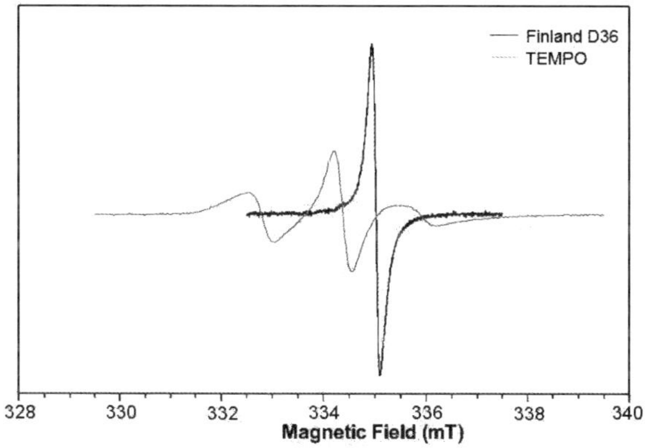

Fig. 2. EPR lines of "Finland D36" in deuterated polystyrene and as comparison of TEMPO.

Table 2. Polarization results in "Finland D36" doped deuterated polystyrene.

Sample	Microwave frequency (GHz)	d-polarization (%)	$T_{l,d}$ (min)	$T_{build-up}$ (min)	$FWHM_{bolometric}$ (mT)
d-PS(98 %-d)	69.976	+11.8	24	23	1.86
+Finland D36	70.020	-12.3			

Note: $f_{d,NMR}$ = 16.4 MHz.

higher polarization values compared to those obtained by TEMPO doping. From the quite different EPR-linewidths (see fig. 2) it is expected that the distance of the optimum microwave frequencies for both polarization directions is different: about 200 MHz in TEMPO-doped samples compared to 44 MHz in trityl-doped polystyrene samples. In the future, the optimum trityl radical concentration will be investigated and measurements at higher magnetic fields and at lower temperatures are planned. Finally, it should be mentioned that the reproducibility of the sample preparation has to be improved.

Acknowledgments

We thank Dr. F. Piegsa for discussions concerning the target material. We are also grateful to Dr. J.Heckmann, Dr. St. Goertz and Dr. Taishan Hu for open and fruitful discussions.

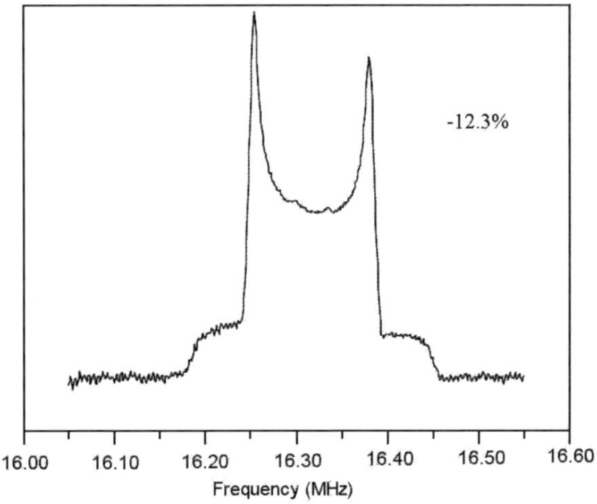

Fig. 3. NMR signals of deuterated polystyrene with a polarization of -12.3 %.

References

1. St. Goertz, *et al.*, *Nucl. Instr. Meth. A* **526**, 43 (2004).
2. D. G. Crabb, *Nucl. Instr. Meth. A* **526**, 56 (2004).
3. W. Meyer, *et al.*, *Nucl. Instr. Meth. A* **227**, 35 (1984).
4. W. Meyer, *et al.*, *Nucl. Instr. Meth. A* **244**, 574 (1986).
5. W. Meyer, *et al.*, *Nucl. Instr. Meth. A* **526**, 12 (2004).
6. N. Takabayashi, PhD thesis, Nagoya University (Nagoya, Japan, 2003).
7. L. Wang *et al.*, in *Proc. Int. Conf. on Symmetries and Spin, Praha, 2009.*
8. B. van den Brandt *et al.*, *Nucl. Instr. Meth. A* **526**, 53 (2004).

THE CLIC ELECTRON AND POSITRON POLARIZED SOURCES

L. Rinolfi*

for the COMPASS collaboration

*BE Department - CERN, 1211 Geneva Switzerland
E-mail: Louis.Rinolfi@cern.ch

The CLIC polarized electron source is based on a DC gun where the photocathode is illuminated by a laser beam. Each micro-bunch has a charge of $6 \cdot 10^9$ e^-, a width of 100 ps and a repetition rate of 2 GHz. A peak current of 10 A in the micro-bunch is a challenge for the surface charge limit of the photo-cathode. Two options are feasible to generate the 2 GHz e^- bunch train: 100 ps micro-bunches can be extracted from the photo-cathode either by a 2 GHz laser system or by generating a macro-bunch using a ~200 ns laser pulse and a subsequent RF bunching system to produce the appropriate micro-bunch structure. Recent results obtained by SLAC, for the latter case, are presented. The polarized positron source is based on a positron production scheme in which polarized photons are produced by a laser Compton scattering process. The resulting circularly-polarized gamma photons are sent onto a target, producing pairs of longitudinally polarized electrons and positrons. The Compton backscattering process occurs either in a Compton ring, where a 1 GeV electron beam interacts with circularly-polarized photons in an optical resonator or in a 1.8 GeV Compton Energy Recovery Linac (ERL) or in a 6 GeV Linac with several optical cavities. The undulator scheme is also studied. The nominal CLIC e^+ bunch population is $6.7 \cdot 10^9$ particles per bunch at 200 MeV. The tradeoff between e^+ yield and level of polarization is an important topic. The overall scheme for both polarized electron and positron beams is described.

1. Introduction

The general CLIC (Compact Linear Collider) parameters, as defined in 2008, are given in reference 1.

The CLIC parameters at 3 TeV are derived from an optimised CLIC structure with a new RF frequency (12 GHz) and a new accelerating field (100 MV/m). The CLIC baseline study at 3 TeV assumes polarized electron and unpolarized positron but beam parameters at 500 GeV are also considered where the bunch charge is doubled. The layout of the CLIC Main Beam Injector Complex is shown in figure 1 and a detailed description is given in reference 2.

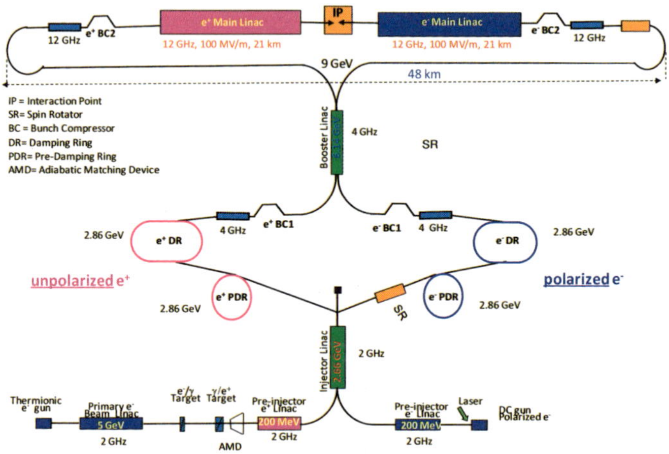

Fig. 1. CLIC main beam injector complex for 3 TeV: Baseline configuration.

2. Generation of unpolarized positron

The positron generation is based on a thermionic gun followed by a primary electron beam linac accelerating the beam up to 5 GeV onto the e^+ target. The positron source itself is composed of hybrid targets, i.e. one thin W crystal target, followed by one W amorphous target (see fig. 2) and an Adiabatic Matching Device (AMD). Then a pre-injector linac accelerates e^+ up to 200 MeV.

The yield is 0.9 e^+/e^- at 200 MeV, corresponding to a normalized yield of 4.5 (e^+ / (e^- x GeV)) at the exit of the pre-injector linac.

Table 1 gives a summary for the CLIC hybrid targets. The PST2009 workshop is focused on polarized particles therefore we give here only the basic information on the system. More details on the unpolarized e^+ source are found in various publications. Experiments using crystal targets were performed at CERN[3] and at KEK.[4] The positron production efficiency was measured with an electron beam of 4 and 8 GeV. Hybrid positron source based on channelling study is described in 5. Radiation damages, using a SLAC beam for a W crystal are reported in 6. A conventional positron

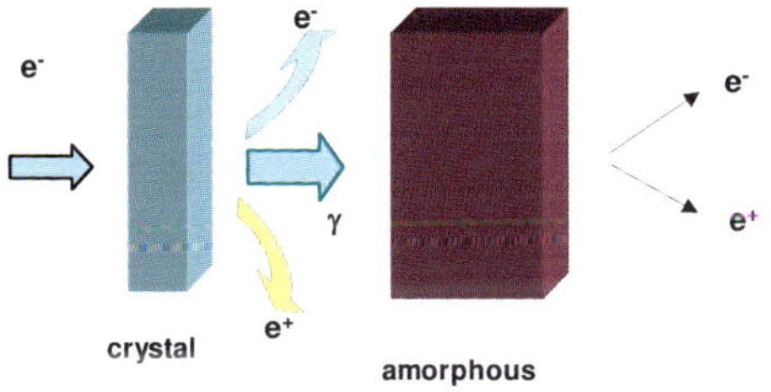

Fig. 2. CLIC hybrid targets for e^+ source.

Table 1. CLIC e^+ targets.

Parameter		CLIC	3 TeV
Target		Crystal	Amorph.
Material		W	W
Length	mm	1.4	10
Radiation lengths	0	0.4	2.9
Beam power deposited	kW	0.2	7.5
Deposited P / Beam Power	%	0.2	8
Energy lost per volume	10^9 GeV/mm3	0.8	1.9
Peak Energy Deposition Density (PEDD)	J/g	7	15

source based on EGS4 simulations and a complete tracking for the e^+ capture and the acceleration up to 200 MeV is described in 7. Simulations from the e^+ target up to the Pre-Damping Ring (PDR) are reported in 8 with the tracking results.

3. Generation of polarized electron

The DC gun should produce 1 nC/bunch, i.e. a charge of $6{\cdot}10^9$ e^-/bunch for 3 TeV configuration. Table 2 gives the beam parameters for the electron source at 3 TeV and at 0.5 TeV. A laser producing 100 ps pulses at a repetition rate of 2 GHz seems rather challenging.

A proposal was made by SLAC to use a CW laser instead and bunch the electron beam at 2 GHz downstream.[9] The formula to calculate the laser energy EL, in order to get the charge Q, at the exit of the photo-cathode, for a cw optical pulse is:

$$E_L = \frac{hc}{q} \cdot \frac{Q}{\lambda \cdot QE}, \tag{1}$$

where $h = 6.62 \cdot 10^{-34}$ J s, $c = 3 \cdot 10^8$ m/s, $q = 1.6 \cdot 10^{-19}$ C, λ is the laser wavelength and QE the quantum efficiency of the photocathode.

Table 2. CLIC electron source parameters.

Parameter	Symbol	0.5 TeV	3 TeV
Number Electrons per microbunch	Ne	$10 \cdot 10^9$	6 x 109
Number of microbunches	nb	354	312
Width of microbunch	tb	100 ps	100 ps
Time between microbunches	tb	0.5002 ns	0.5002 ns
Microbunch rep rate	fb	1999 MHz	1999 MHz
Width of macropulse	TB	177 ns	156 ns
Macropulse repetition rate	FB	50 Hz	50 Hz
Charge per micropulse (e x Ne)	Cb	1.6 nC	0.96 nC
Charge per macropulse (Cb x nb)	CB	566 nC	300 nC
Average current from gun (CB x FB)	Iave	28 A	15 A
Average current in macropulse (CB / TB)	IB	3.2 A	1.9 A
Duty Factor w/in macropulse (tb/tb)	DF	0.2	0.2
Peak current of micropulse (IB / DF)	Ipeak	16 A	9.6 A
Current density (Ipeak/) [spot size radius 1 cm]	D	5 A/cm2	3 A/cm2
Polarization		> 80 %	> 80 %

Important first results were obtained by Nagoya University and KEK,[10] followed by SLAC[11] and JLAB.[12] From table 2, Q (0.5 TeV) = 566 nC and Q (3 TeV) = 300 nC. The wavelength for the GaAs photocathode is 780 nm. The quantum efficiency measured on the SLAC photocathode is 0.7 %. Therefore, the requested laser pulse energy is:

$$E_L(3\text{ TeV}) = 68\ \mu\text{J} \tag{2}$$

$$E_L(0.5\text{ TeV}) = 128\ \mu\text{J} \tag{3}$$

The current density of 3 to 5 A/cm^2 is a challenge for the photocathode regarding the surface charge limit. Nevertheless a total charge of 600 nC has been produced by SLAC from a DC gun. The photocathode was illuminated by a cw laser (156 ns pulse length) and the extracted charge was measured.

Figure 3 shows the experimental results obtained by SLAC.[13] The laser energies used for the experiment are consistent with the theoretical ones. The emittance was not measured but the polarization was measured and found to be around 82 %. Simulations were performed downstream of the DC gun, assuming a bunching system at 2 GHz. The latter is composed of 2 pre-bunchers cavities followed by a buncher and an accelerating cavity.[9] Table 3 gives the simulation results. Such performance would satisfy the CLIC requirements for the polarized electron source.

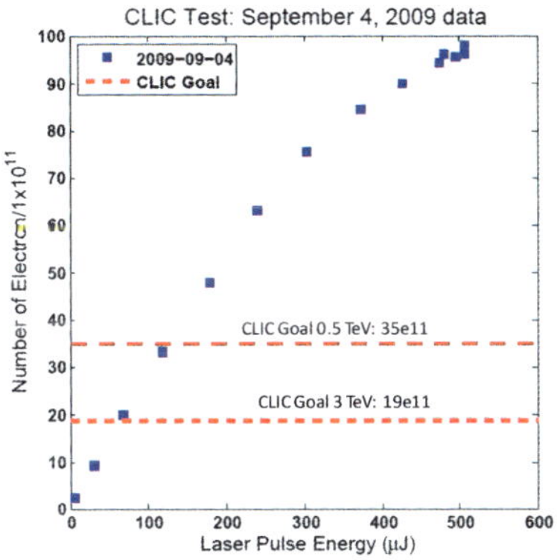

Fig. 3. Production of polarized e^- at SLAC.

Table 3. Electron parameters simulations up to 20 MeV.

Parameters	Units	CLIC 3 TeV
Gun voltage	kV	140
Injector energy	MeV	20
Initial charge at the gun	nC	1
Capture efficiency	%	88
Initial bunch length at the cathode	ns	156
Final bunch length (FWHM)	ps	14
Energy spread (FWHM)	keV	100
Normalized RMS emittance	mm mrad	22

4. Generation of polarized positron

The generation of polarized positrons for CLIC is an enormous challenge. There are mainly two possible approaches. One is based on the undulator scheme where an electron beam, with an energy in the range of 100 GeV or more, is sent through a short-period undulator[14].[15] The other one is based on laser Compton back scattering. Here, three variations of this latter concept are used. The Compton LINAC scheme[16],[17] the Compton ring scheme[18][19] and the Compton ERL scheme[20].[21] In each of these, an electron beam interacts with a powerful circularly polarized laser beam.

The CLIC undulator scheme assumes the electron beam passing through an helical undulator with energy of 250 GeV. The undulator is 100 m long with $K = 0.7$ and $\lambda_u = 1.5$ cm. The Ti target is 0.4 radiation length and it is not immersed in the magnetic field of the adiabatic matching device.

The capture sections are working at 2 GHz with a gradient of 25 MV/m. Simulations results[22] show that a yield of 1.4 e^+/e^- is obtained after the capture at 200 MeV with a peak magnetic field of 4 T. A polarization of 60 % for e^+ is achieved with a collimator radius below 1 mm also reducing the yield down to 0.8 e^+/e^-. The scheme has very strong impacts on the CLIC main beam and needs more detailed studies.

The CLIC Compton ring assumes a double chicane where the energy spread is different inside and outside of the chicane.[23] The present study is based on a regime of laser cooling, with continuous generation of photons allowing a yield enhancement. The interaction between the unpolarized electron beam, with an energy of 1.06 GeV, and the laser occurs at the IP with a collision angle as small as possible. In the present design it is 8 °. Inside the chicanes, the square of energy spread of the electron beam remains constant. In the proposed CLIC Compton ring scheme,[24] the electron beam is composed of 312 bunches with a charge of $6.2 \cdot 10^9$ e^-/bunch (1 nC) corresponding to 2 A circulating beam. The ring circumference ($\sim$ 47 m) corresponds to the pulse length of 156 ns. The RF system is composed of 2 cavities working at 2 GHz and 200 MV each. The YAG laser produces circularly-polarized photons at 1.164 eV and the energy stored in the optical cavity is 600 mJ. The laser spot size at the collision point has a radius of 0.005 mm and a length of 0.9 mm. Figure 4 gives a simplified layout of the CLIC injector based on Compton ring. Simulations results give a yield of 0.063 photon per electron and per turn.[23] This corresponds to a flux of $3.3 \cdot 10^{16}$ photons/s. The polarized photons are collimated, reducing the photon flux down to $1.33 \cdot 10^{16}$ photons/s but increasing the polarization level. They are sent onto a target to produce polarized e^+. Simulations give a level of polarization about 84 % which is the present tradeoff between the yield and the polarization. However the photons flux is not enough to get the requested e^+ bunch charge after the capture. Therefore, stacking into the PDR is necessary.

Simulations have been performed for longitudinal stacking into the PDR.[25] An optimization of parameters increases the stacking efficiency. The simulations show that efficiency of 90 % could be obtained with specific PDR parameters. However, outstanding questions remain open and further improvements will be necessary. The energy spread of injected positron beam should be as small as possible and the PDR momentum acceptance as large as possible. Today this scheme is the one which would be preferable for the CLIC upgrade when polarized positrons will be produced for the physics.

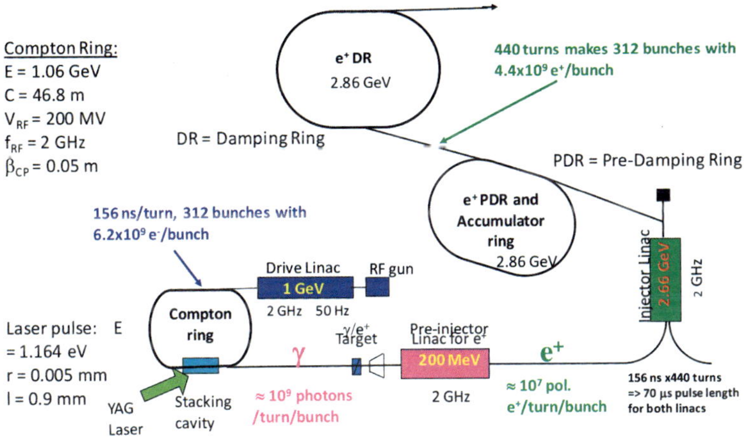

Fig. 4. CLIC injector based on a Compton ring for polarized e^+ source.

The CLIC ERL is a continuous low-charge high-repetition frequency electron linac. The beam energy is 1.8 GeV. For this scheme the requested bunch spacing into the ERL is 32 ns.[26] With the bunch charge of $3 \cdot 10^9$ e^-/bunch (0.48 nC), the beam current is 15 mA and the repetition frequency is 31.25 MHz. Preliminary simulations give a yield of $5 \cdot 10^8$ photons/bunch assuming that laser energy of 0.6 J could be stored in one optical cavity installed in the ERL ring. Based on a conservative e^+ yield of 0.005, $2.5 \cdot 10^6$ e^+/bunch would be produced.

The scheme provides a good solution to avoid stacking into the PDR. For CLIC, the repetition rate is 20 ms and the idea is to separate the functions of stacking and damping to use this repetition rate efficiently. For that purpose, 2 small storage rings (SR1 and SR2) between the ERL and the PDR are implemented: 20 ms are used for stacking in the SR1, followed by 20 ms of damping in SR1 (25 Hz). During the same 20 ms of damping in SR1, 20 ms of stacking are performed, in parallel into SR2 followed by 20 ms of damping in SR2 (25 Hz) and so on.

Assuming that 2000 bunches could be stacked into the same bucket of SR1 and SR2, then $5 \cdot 10^9$ e^+/bunch could be obtained. The two rings SR1 and SR2, with a circumference of $\sim$ 47 m, are designed for energies around 1 GeV. The e^+ beams are extracted from SR1 and SR2 and 312 bunches are accelerated up to the PDR energy (2.86 GeV). A 2 GHz linac working at 50 Hz repetition rate needs to be implemented between the SR1/SR2 and the PDR. No more stacking would be required into the PDR. With these parameters the CLIC requirements are fulfilled.

The CLIC Compton LINAC scheme uses a 6 GeV LINAC where the electron beam is sent through several CO_2 laser amplifier cavities. It requires powerful lasers but does not require e^+ stacking into the PDR. The needed number of e^+ per bunch is produced in every laser shot at 50 Hz repetition rate. The main features of this scheme are based on the use of mid-IR CO_2 laser (1 J, $\lambda = 10\ \mu m$) and the most energy-efficient back-scattering geometry. The RMS bunch length for the electron bunch and for the laser pulse is 3 ps. The production of 1 photon per electron has been demonstrated.[17] With a conservative conversion efficiency of the polarized photons into polarized e^+, 50 photons are necessary for each e^+. Assuming that ten consecutive optical Compton cavities could be implemented with ten IPs to accumulate the photons flux, 5 nC per electron bunch would produce 1 nC per positron bunch, which is the CLIC requested charge. The LINAC's electron beam is formatted into a train of 312 bunches at 50 Hz repetition rate. The 1 nC positron bunches, produced on a target by the Compton-scattered photons, will be injected into the PDR without stacking.[27]

5. Summary

Based on the SLAC experiment, the polarized electron source for CLIC is feasible without major issues.

At present all proposed schemes for polarized positrons need substantial R&D to fulfil the requested CLIC performance. The present requirements from physics do not stress the need for a polarized positron source, therefore, the CLIC study group assumes unpolarized positrons as the baseline for the CDR (Conceptual Design Report). The latter is expected in 2010, with polarized positrons as a possible upgrade. Nevertheless, due to the clear potential advantages for physics, studies and R&D, regarding the various issues, are ongoing, in close collaboration with many institutes around the world.

6. Acknowledgement

Corsini read carefully the report and made very useful comments.

References

1. F. Tecker (ed.) *et al.*, CLIC 2008 parameters, CLIC Note 764 (2008).
2. L. Rinolfi, The CLIC Main Beam Injector Complex. A review in 2009, CLIC Note 750 (2009).

3. R. Chehab *et al.*, Experimental study of a crystal positron source, *Phys. Lett. B* **525**, 41 (2002).
4. T. Suwada *et al.*, Measurement of positron efficiency from a tungsten monocrystalline target using 4 and 8 GeV electrons, *Phys. Rev. E* **67**, 016502 (2003).
5. O. Dadoun *et al.*, Study of a hybrid positron source using channelling for CLIC, CLIC Note 808.
6. R. Chehab *et al.*, Radiation damage study of a monocrystalline tungsten positron converter, CLIC Note 369, LAL/RT 98-02, EPAC98, Stockholm, CERN/PS 98-17(LP) (1998).
7. T. Kamitani and L. Rinolfi, The CLIC positron production scheme, CLIC Note 537, in *Proc. XXI Linear Accelerator Conference, Gyeongju, Korea, 2002.*
8. A. Vivoli *et al.*, The CLIC positron capture and acceleration in the Injector linac, (waiting for a CLIC number).
9. F. Zhou *et al.*, Preliminary design of a bunching system for the CLIC polarized electron source, CLIC Note 813.
10. T. Omori *et al.*, *Phys. Rev. Lett.* **67**, 3294 (1991).
11. T. Maruyama *et al.*, Systematic study of polarized electron emission from strained GaAs/GaAs superlattice photocathodes *Appl. Phys. Lett.* **85**, 13 (2004).
12. J. Grames *et al.*, Lifetime Measurements of High Polarization Strained Superlattice Gallium Arsenide at Beam Current > 1 mA Using a new 100 kV Load Lock Photogun, in *Proc. Particle Accelerator Conference, Albuquerque, NM, June 25-29, 2007.*
13. J. Sheppard, CLIC electron beam experiment, in *Proc. 2009 Linear Collider workshop of the America, Albuquerque, 2009.*
14. N. Phinney, N. Toge, N. Walker, ILC Reference Design Report, ILC-Report 2007-001 (2007).
15. J. Clarke, Sources, in *Proc. Workshop TILC09, Tsukuba, 2009.*
16. T. Omori *et al.*, Design of a polarized positron source for linear colliders, *Nucl. Instr. Meth. A* **500**, 232 (2002).
17. V. Yakimenko and I. Pogorelski, Polarized gamma source based on Compton backscattering in a laser cavity, *Phys. Rev. ST Accel. Beams* **9**, 091001 (2006).
18. T. Omori *et al.*, Efficient propagation of polarization from laser photons to positrons through Compton scattering and electron-positron pair creation, *Phys. Rev. Lett.* **96**, 114801, (2006).
19. F. Zimmermann *et al.*, CLIC polarized positron source based on laser Compton scattering, CLIC Note 674, in *Proc. EPAC06, Edinburgh, Scotland, UK, 2006.*
20. A. Variola *et al.*, Proposal for a unique lepton source ERL based, ILC Positron source meeting, RAL (2006).
21. T. Omori, ERL based Compton scheme, in *Proc. POSIPOL 2007, Orsay 2007.*
22. W. Gai *et al.*, Update on Undulator based positron source for CLIC, in *Proc. CLIC09 workshop, CERN, 2009.*

23. E. Bulyak *et al.*, Analytic study on Compton rings, in *Proc. POSIPOL09, Lyon, 2009.*
24. L. Rinolfi *et al.*, The CLIC positron source based on Compton schemes, in *Proc. PAC09, Vancouver*, CLIC Note 788 (2009).
25. F. Zimmermann *et al.*, Stacking simulations for Compton positron sources of future linear colliders, in *Proc. PAC09, Vancouver*, CLIC Note 814 (2009).
26. T. Omori, L. Rinolfi, ERL Compton scheme for CLIC, in *Proc. CLIC09, CERN, 2009.*
27. V. Yakimenko *et al.*, Compton linac for polarized positrons in *Proc. CLIC09, CERN, 2009.*

STATUS OF HIGH INTENSITY POLARIZED ELECTRON GUN AT MIT-BATES

E. Tsentalovich* and J. Bessuille

MIT-Bates,
Middleton, MA, 01949, USA
**E-mail: evgeni@mit.edu*

M. Tiunov

Budker Institute for Nuclear Physics,
Novosibirsk, Russia, 630090

MIT-Bates, in collaboration with BNL, has developed a high intensity polarized electron gun for the eRHIC project. The gun implements large area cathode, ring-shaped beam and active cathode cooling. The paper describes the current status of the project.

Keywords: Polarized electron gun; high intensity.

1. Introduction

The development of highly polarized electron beams has led to many new advances in nuclear and particle physics in recent decades. Polarized electron beams evolved from the development of the laser and semiconducting materials, when research in electron spin-polarization from III-V based photoemitters made it possible to produce electron beams with polarization using bulk GaAs photocathodes. Since that time, polarized electron sources have been established at numerous facilities worldwide.[1–7]

Modern polarized electron sources routinely produce an average current of hundreds of μ A with a polarization approaching 90 %. This intensity satisfies the requirements of the existing accelerator facilities. New advances in nuclear physics are expected with the development of the high luminosity electron-ion collider (EIC). The concept of such a collider has been discussed in the nuclear physics communities around the world for the past decade. One of the most advanced projects of EIC is eRHIC, based on the existing Relativistic Heavy Ion Collider (RHIC) complex located at Brookhaven National Laboratory (BNL).[8]

Two alternative versions of eRHIC have been developed. The ring-ring version is based on the construction of the electron storage ring which will intersect the RHIC ion ring in one of the existing interaction regions. The linac-ring version of eRHIC provides the possibility to achieve a higher luminosity. This version is based on the construction of the very high intensity energy recovery linac (ERL). The linac version excludes the possibility of stacking. Therefore, the polarized electron source must be able to provide a very high average current. In order to achieve a luminosity of $1 \cdot 10^{33}$ $cm^{-2}s^{-1}$ an average current of at least $I_{av} \approx 50$ mA is required. Meanwhile, the highest average current produced in existing polarized electron guns on the test benches is in the mA region, but with rather low lifetime.[9,10]

MIT-Bates in collaboration with BNL investigates the possibility of building a very high intensity polarized electron gun.[11] This paper reports the results of phase I of the project.

2. Approach

The major limitation in achieving high average current is produced by ion back bombardment.[12] It is difficult to expect a significant improvement of the vacuum conditions over present state-of-the-art installations. However, the ion induced damage could be mitigated by using a large area cathode and thus reducing the density of the ion current.

It was demonstrated[13,14] that ions originated in the vicinity of the anode (those ions have the largest energy and are presumably more harmful) tend to strike the central area of the cathode. It could be beneficial to use a ring-shaped laser beam for photoemission and not waste laser intensity on the most damaged center of the cathode.

Using a ring-shaped electron beam with large diameter could lead to beam losses. At this intensity, losses should not exceed 10^{-6} in the gun vicinity. Accurate simulations including space charge effects are required to ensure that the beam losses are acceptable.

Tens of Watts of average laser power will be deposited on the GaAs crystal. Active cathode cooling is necessary to avoid the crystal overheating.

The project is divided in two phases. Phase I includes full simulations of the gun and the beam line, design and manufacturing of the test chamber with active cathode cooling, and conducting tests with this chamber validating sufficient efficiency of the cathode cooler, high voltage (120 kV) compatibility and viability of the vacuum manipulating design. In phase II the gun equipped with a cooled cathode and load lock; the beam line will be designed and manufactured and the measurements of the life time at different current will be conducted. Currently phase I is completed and its results are reported here.

3. Gun and beam line simulations

A ring-shaped beam will be used in this project. However, the exact shape of the beam cannot be guaranteed due to possible optical misalignments and non-uniformity of the quantum efficiency across the crystal. To ensure that changes in beam shape do not increase losses, three different shapes have been used for simulations: ring-shaped, Gaussian and flat distributions.

The 120 kV DC gun was designed to be operated with a current from 0 to 100 mA. The anode-cathode gap is 100 mm and the maximum electric field on the surface of the cathode is about 39 kV/cm. The gun features a biased (1 kV) anode in order to repulse ions produced in the beam line.

The main purpose of the beam line is separation of the UHV conditions in the gun from the inferior vacuum conditions of the beam dump. The beam line consists of two 90° dipole magnets and a doublet of solenoidal lenses between these dipoles. A third solenoidal lens is used to increase the size of the beam in the dump in order to reduce the power density. The dipole magnets have the same focusing properties in both directions to maintain the axial symmetry of the beam. The lenses have a large internal diameter (90 mm) and produce very linear focusing.

SAM code[15] was used for the simulations. Beam propagation through the beam line has been simulated for all three distributions. Since it would be too computationally intensive to simulate losses of the order of 10^{-6} directly, the following approach was used. For the given beam configurations, the electrical and magnetic fields were calculated, including the fields produced by the beam itself. In the next step, only electrons emitted from the very edge of the cathode were considered (only these electrons could contribute to beam losses). Since the emitting current density is rather low at the edge of the cathode (except for the flat distribution), a very significant gain in statistical accuracy was achieved.

Simulations demonstrated that everywhere in the beam line, losses of the order of 10^{-6} happen at apertures of less than 20 mm. The only exception is the entrance into the first dipole for the gaussian beam, where this critical aperture is about 23 mm. Since the actual apertures of the beam line are about 30 mm, no losses of the order of 10^{-6} are expected.

Calculations of the trajectories of ions produced in the cathode-anode gap demonstrated that ions produced close to the anode are indeed focused into the center of the cathode, and their footprint on the cathode has a rather small overlap with the emitting pattern of the ring-shaped electron beam.

4. Test chamber design and fabrication

The conceptual design of the chamber is shown in figure 1. The test chamber was designed as a prototype of the actual gun. It utilizes the same mechanism for vacuum manipulation as the real gun will utilize. The GaAs crystal installed in the molybdenum puck is delivered into the test chamber with a magnetic-coupled manipulator.

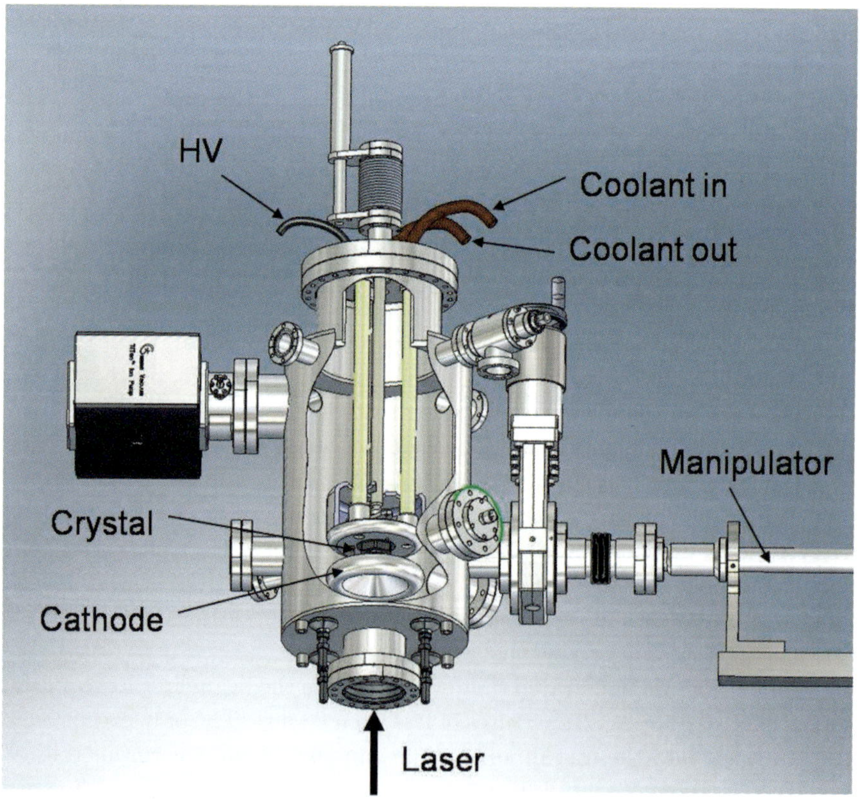

Fig. 1. Conceptual design of the test chamber.

The cathode assembly consists of the cathode electrode, heat exchanger and field shield. It is suspended on three ceramic pipes that insulate the assembly from the ground. One of the pipes is used to deliver HV to the cathode, the other two serve as conduits for the cooling agent. An additional ceramic pushing rod installed on the Linear Transfer Mechanism (LTM) allows the cathode electrode to be lowered, separating it from the heat exchanger. The cathode puck is inserted into the gap with the manipulator. The cathode assembly is then raised, pressing the puck to the heat exchanger. Cone-to-cone surfaces on the puck and the heat exchanger center the puck and provide good thermal contact. The field shield protects the unpolished inside parts of the assembly from an electric field.

The heat exchanger consists of two copper plates brazed together; a spiral channel machined in the plates conducts the cooling agent, providing an effective heat transfer. Fluorinert has been chosen as the cooling agent. This liquid has an extremely low electrical conductivity, a high electrical strength and acceptable viscosity and thermal conductivity. The cooling agent circulates through a chiller with adjustable flow and temperature.

The test chamber is equipped with several view ports to monitor vacuum manipulations. Illumination is provided with halogen bulbs installed inside the vacuum chamber.

5. The results of the tests

The first tests demonstrated very reliable vacuum manipulation. A transfer of the puck with a crystal into and out of the test chamber has been performed a dozen times. The manipulator engages and disengages the puck at the cathode assembly reliably. The cone-to-cone centering system works very well. For test purposes the puck was disengaged from the manipulator several millimeters off the center, and yet it was centered perfectly when the cathode electrode was raised. The viewports provide good observation of the manipulation. Halogen lights provide excellent illumination. They give more than enough light, even with the voltage significantly lower than nominal voltage, so one may expect a very long lifetime of these bulbs. Since four bulbs have been installed inside the chamber, we never expect to have a need to open the vacuum chamber to replace a burned bulb.

The gun was successfully processed to 120 kV. However, multiple electrical breakdowns have been observed during the HV processing. These breakdowns are not dangerous for the GaAs crystal since processing takes place without the crystal puck in the gun. The source of the breakdowns has been identified. Some electrons originated by field emission during processing accumulate on the ceramic pipes and finally produce a potential so great that a breakdown occurs. Potentially, such breakdowns could punch through the ceramic, resulting in vacuum failure, so it is very desirable to avoid them. Several sources of field emission capable of producing such electrons have been identified. The field shield will be modified to reduce the electrical field in these locations significantly.

The cathode cooling tests were conducted with a thermocouple attached to the outer edge of the molybdenum puck. This part of the puck is the farthest from the cooling surfaces and it is expected that the temperature of the thermocouple is close to the temperature of the crystal. The tests were conducted in vacuum. The ring-shaped laser beam was directed to the crystal through a viewport at the bottom of the test chamber. The laser was able to produce up to 38 W of laser power. Taking into account losses in the optics and the viewport, the maximum laser power delivered to the crystal was about 34.2 W.

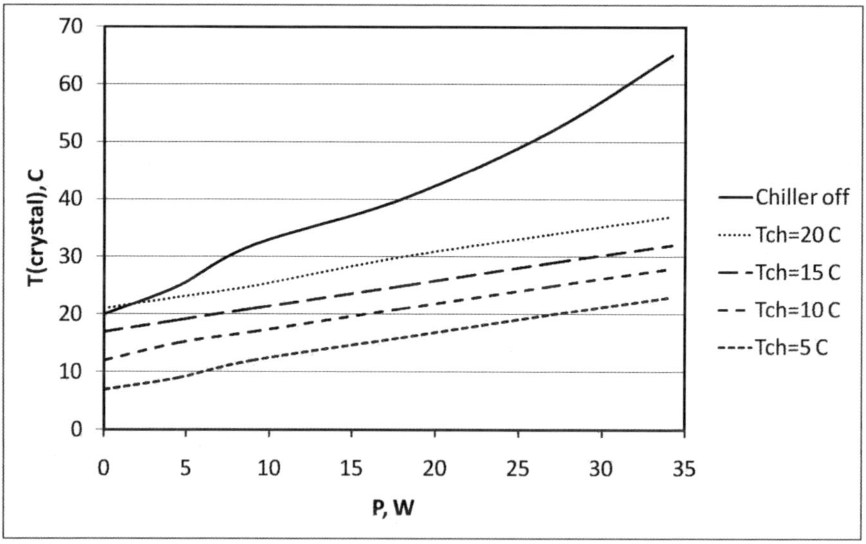

Fig. 2. Cathode cooling. Temperature of the cathode as a function of laser power at different set points of the chiller (Tch).

Results of the tests are presented in figure 2. The temperature of the cooling agent (Tch) varied from 5 °C to 20 °C. The temperature difference between the cooling agent and the thermocouple was about 17 °C at the maximum laser power. Therefore, by setting the chiller set point to 5 °C, we were able to keep the crystal temperature at about 22 °C at this laser intensity.

6. Conclusion

Design of the vacuum manipulation has proved to be successful. The cooling power is sufficient to keep the crystal at room temperature with a laser power of at least 35 W. Modifications of the field shield will be implemented in order to reduce the probability of breakdowns during processing.

References

1. R. Alley *et al.*, *Nucl. Instr. Meth. A* **365**, 1 (1995).
2. K. Aulenbacher *et al.*, *Nucl. Instr. Meth. A* **391**, 498 (1997).
3. M. J. J. van den Putte *et al.*, *Nucl. Instr. Meth. A* **406**, 50 (1998).
4. J. Grames *et al.*, in *Proc. 15th Spin Phys. Symp.*, AIP Conf. Proc. **675**, 1047 (AIP, New York, 2003).
5. Wolther von Drachenfels *et al.*, in *Proc. 15th Spin Phys. Symp.*, AIP Conf. Proc. **675**, 1053 (AIP, New York, 2003).
6. E. Tsentalovich *et al.*, *Nucl. Instr. Meth. A* **582**, 413 (2007).
7. Y. Poltoratska *et al.*, in *Proc. 18th Spin Phys. Symp.*, AIP Conf. Proc. **1149**, 983 (AIP, New York, 2009).
8. V. Ptitsyn, in *Proc. Particle Accelerator Conference*, 1927 (2007).
9. M. Poelker *et al.*, in *Proc. 12th Int. Workshop on PST*, AIP Conf. Proc. **980**, 73 (AIP, New York, 2008).
10. R. Barday and K. Aulenbacher, in *Proc. 17th Spin Phys. Symp.*, Kyoto, AIP Conf. Proc. **915**, 1019 (AIP, New York, 2007).
11. E. Tsentalovich, in *Proc. 12th Intern. PST Workshop*, AIP Conf. Proc. **980**, 79 (AIP, New York, 2008).
12. M. Poelker and J. Grames, in *Proc. 11th Int. Workshop on PST*, 127 (World Scientific, 2007).
13. J. Grames *et al.*, in *Proc. 12th Int. Workshop on PST*, AIP Conf. Proc. **980**, 110 (AIP, New York, 2008).
14. E. Tsentalovich, in *Proc. 18th Spin Ph. Symp.*, AIP Conf. Proc. **1149**, 997 (AIP, New York, 2009).
15. M. A. Tiunov *et al.*, in *Proc. of 18th Int. Symp. on Electron Beam Ion Sources*, AIP Conf. Proc. **572**, 155 (AIP, New York, 2001).

TARGET SECTION FOR SPIN FILTERING STUDIES AT COSY AND CERN/AD

C. Barschel*, F. Rathmann, J. Sarkadi and H. Ströher

for the PAX-collaboration

Institute for Nuclear Physics, Jülich Center of Hadron Physics,
Forschungszentrum Jülich, Leo-Brandt-Str. 1, 52425 Jülich, Germany
**c.barschel@fz-juelich.de*

G. Ciullo, P. Lenisa and M. Statera

Istituto Nazionale di Fisica Nucleare and Università,
44100 Ferrara, Italy

K. Grigoryev

High Energy Physics Department, St. Petersburg Nuclear Physics Institute,
188300 Gatchina, Russia

A. Nass and E. Steffens

Physikalisches Institut, Universität Erlangen-Nürnberg,
91058 Erlangen, Germany

G. Tagliente

Istituto Nazionale di Fisica Nucleare Bari,
70126 Bari, Italy

The PAX (Polarized Antiproton eXperiment) collaboration aims to polarize a stored antiproton beam by means of spin filtering. The setup requires a polarized internal gas target (PIT) surrounded by silicon detectors.[1] An overview of the target configuration necessary for spin filtering is presented. The setup of the PIT is discussed with an emphasis on the working principle of the Breit-Rabi Polarimeter (BRP) including the calibration procedure. Furthermore, some preliminary results of the BRP signal and transitions tuning are presented.

Keywords: PAX; Breit-Rabi polarimeter.

1. Setup of the polarized target section

The purpose of the target section is to provide a polarized hydrogen or deuterium gas as a target for the PAX spin-filtering. The high areal densities of up to 10^{14} atoms/cm^2 required for spin-filtering are achieved by containing the injected gas in a storage cell. A schematic view of the target section is shown in figure 1. The ABS consist first of a set of sextupole magnets,

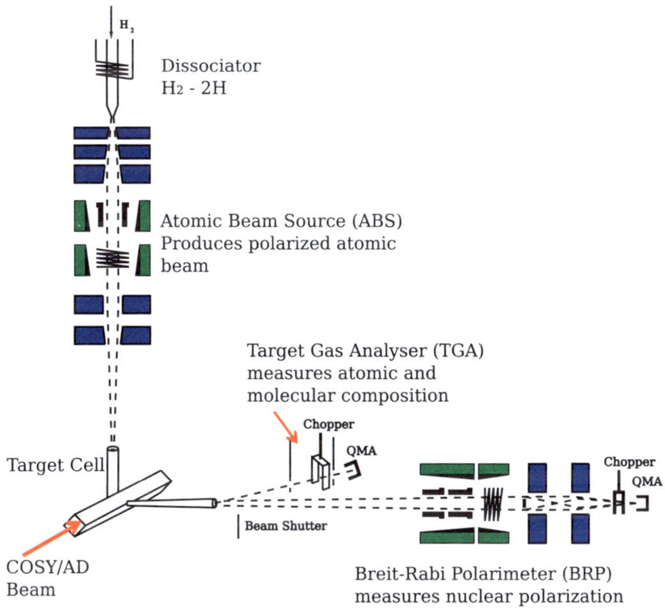

Fig. 1. Schematic overview of the PAX target section showing the Atomic Beam Source (ABS), the target cell and the Breit-Rabi Polarimeter (BRP). Following the path of atoms, first the dissociator breaks H_2 molecules into atoms in a microwave induced plasma. The atoms then enter the vacuum through a nozzle cooled down to 100 K and are focused by the sextupole system into the target cell. A sample extracted from the cell is continuously analyzed by the BRP.

followed by two adiabatic RF-transition units[2] and again two sextupole magnets. The combination of sextupole magnets and RF-transitions is arranged in the ideal case to let only the hyperfine state $|1\rangle$ through and thus to inject a polarized atomic gas into the storage cell. See reference 3 for more details. Finally a gas sample is extracted from the cell and analyzed by the so-called Breit-Rabi Polarimeter (BRP).

2. Polarization of hydrogen atomic gas

The nuclear polarization is defined by the relative number of atoms with nuclear spins parallel or anti-parallel to the holding magnetic field. The expectation values for the nuclear (and electron) polarization depend on the individual hyperfine state population of the hydrogen and deuterium atoms.

The orbital angular momentum being absent in the hydrogen ground state $1S_{1/2}$, the nuclear magnetic dipole moment couples directly to the magnetic field generated by the electron. Since both proton and electron are particles with spin 1/2, the resulting total angular momentum is either $F = 0$ or $F = 1$.

In an external magnetic field the triple degeneracy of $F = 1$ is split. This effect is shown in the left part of figure 2 using the basis $|F, m_F\rangle$. The mixing of the two states $|2\rangle$ and $|4\rangle$ depends on the mixing angle θ which itself depends on the magnetic field:

$$\cos 2\theta = \frac{x}{\sqrt{1+x^2}}, \quad \text{with} \quad x = \frac{B}{B_C^H} \quad \text{and} \quad B_C^H \approx 50.7\ \text{mT}. \tag{1}$$

The dimensionless parameter x is the ratio of the external field to the critical field.[2] The polarization of an atomic gas therefore dependents on the external magnetic field and is plotted in figure 2 (right). It is defined as

$$P_z = n_1 - n_3 - (n_2 - n_4)\cos 2\theta; \quad \text{with} \quad \sum_i n_i = 1. \tag{2}$$

In a weak holding field the mixed states $|2\rangle$ and $|4\rangle$ will not contribute to the polarization compared to the pure states $|1\rangle$ and $|3\rangle$. The PAX experiment will operate in weak holding fields.

3. Working principle of the Breit-Rabi polarimeter

The BRP measures the polarization by determining the relative intensities of all hyperfine states of a sample beam from the storage cell. The main components are shown in the top part of figure 3. The Target Gas Analyzer (TGA) is a Quadrupole Mass Analyzer (QMA) with a chopper located at an offset angle relative to the extracted effusive beam. The TGA measures the relative amount of atoms and recombined molecules coming from the cell. At a low holding field $B \ll B_C^H$, the recombination reduces the polarization in the cell and thus needs to be taken into account.

The BRP itself is composed of two adiabatic high frequency transitions (HFT),[2] followed by two sextupole magnets. A QMA with chopper

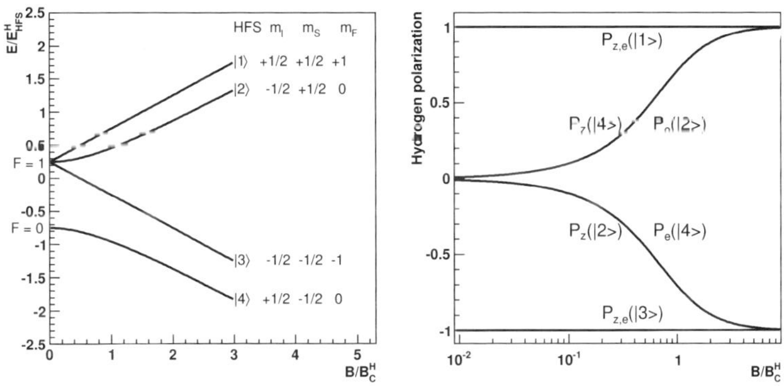

Fig. 2. Left: The energy eingenvalues of the hydrogen hyperfine structure in an external magnetic field (Breit-Rabi diagram). Right: the contribution to the atomic polarization of each hyperfine state depends on the holding magnetic field. The states $|1\rangle$ and $|3\rangle$ do not depend on the external field.

measures the beam intensity. The HFTs produce a gradient magnetic field together with a high frequency electrical field. The transitions effectively exchange two hyperfine populations by transferring the polarization of the electron to the nucleon. Transitions exchanging the populations within the $F = 1$ triplet with $\Delta F = 0$ are called weak- and medium- field transitions (WFT/MFT), while transitions exchanging the populations with $\Delta F = 1$ are called strong field transitions (SFT).[4]

The sextupole magnets induce a Stern-Gerlach force to the atoms and separate them based on the electron spin. The atoms with spin $m_S = -\frac{1}{2}$, effectively the states $|3\rangle$ and $|4\rangle$, are defocused while the atoms with spin $m_S = +\frac{1}{2}$ ($|1\rangle$ and $|2\rangle$) are focused into the QMA. A circular beam blocker placed in front of the first sextupole magnet blocks atoms with a trajectory close to the axis. This ensures that atoms with negative electron spin are effectively deflected by the Stern-Gerlach force and will not reach the QMA.

The combination of the HFTs exchange hyperfine populations and the sextupoles magnets exclude the final states $|3\rangle$ and $|4\rangle$. Those two effects always result in a signal composed of two different initial states. The transitions are sequenced to measure all relative occupation numbers n_i by measuring different intensity combinations. The polarization is then determined by equation 2.

The BRP calculates the polarization by measuring different combinations of transitions. For the given example, the transitions SFT_{14} and MFT_{13} are switched on, this signal mode is thus called $s14m13$, the effect of the transitions and sextupoles is shown in figure 3. The transition SFT_{14} exchanges the populations $N|1\rangle \leftrightarrow N|4\rangle$, while the MFT_{13} is a double transition. First the populations $N|1\rangle \leftrightarrow N|2\rangle$ are exchanged, then followed by $N|2\rangle \leftrightarrow N|3\rangle$. This distinction is important for the modeling.

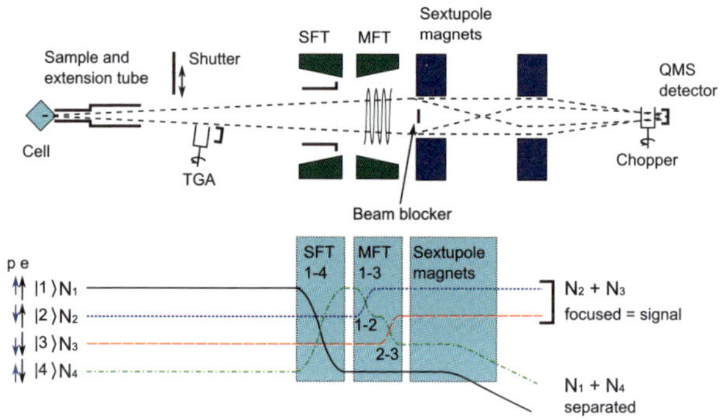

Fig. 3. Example of an acquisition mode. The signal $s14m13$ is generated by activating the transitions SFT_{14} and MFT_{13}. The resulting intensity is ideally composed of the populations $N|2\rangle + N|3\rangle$.

The SFT transition exchanges the population numbers with $\Delta F = 1$. This transition occurs with a single photon exchange between the states $|1\rangle$ or $|2\rangle$ in the triplet and the singlet state $|4\rangle$. The frequency necessary for the transition depends on the magnetic field seen by the atoms. The gradient magnetic field used in the setup ensures that the transition occurs only once. The transition conditions are shown in figure 4.

As mentioned before the signals are a linear combination of the four hyperfine state intensities I_i. For hydrogen there are 11 different signal combinations S_i, the signals are modeled as[5]

$$S_i = \sum_a M_{ia} I_a; \quad M^{ia} = \sum_b \sigma_b T^i_{ba}. \tag{3}$$

The so-called measurement matrix M_{ia} describes the physical effect of the transitions T^i_{ba} and the sextupole magnets σ_b. For example, the signal $S_{s14m13} = SFT_{14} + MFT_{13}$ is described by one row of the measurement matrix. In an ideal case this leads to:

$$M^{s14m13} = \sigma \cdot T_{23} T_{12} T_{14} = \begin{pmatrix} 0 & 1 & 1 & 0 \end{pmatrix}. \tag{4}$$

This overdetermined linear system of 11 signals and 4 unknown intensities can be solved with a least square fit algorithm. The intensities represent

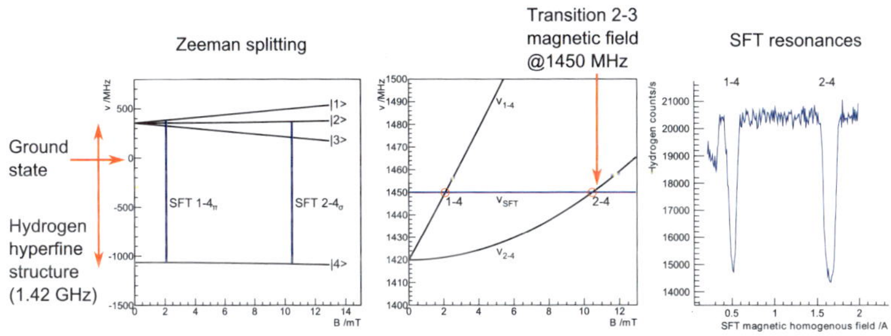

Fig. 4. Transition conditions for the SFT. Left: the Breit-Rabi diagram shows the transition conditions for an operating frequency of 1450 MHz. Right: the frequency difference of states 1-4 and 2-4, the intersection points with the operating frequency are the necessary magnetic field for the transition to occur. Figure c is an example of a magnetic field scan for the SFT transition showing the resonances 1-4 and 2-4. The peaks are shifted to the right due the additional presence of a gradient field.

the absolute population numbers of each hyperfine state. With hydrogen, a minimum of four signals are necessary to calculate the individual hyperfine states.

4. Calibration procedure

The transition efficiencies must be taken into account in the previous modeling. The efficiencies are indirectly measured by a calibration procedure. The principle of the calibration is to generate enough signal combinations to fit the system of equations describing the BRP with the acquired data.

For the calibrations, both the efficiencies (including the sextupole and the intensities) are unknown, leading to an underdetermined system of equations. This problem is solved by injecting different modes with the ABS. Each new injection mode adds four new unknown intensities and 11 more signals, however the efficiencies remain unchanged regardless of the cell content. Each new injected mode thus effectively adds seven additional signals to the model. Table 1 gives a comparison of signals and unknowns for 1 and 3 ABS injection modes. The minimum required to solve the system is 3 ABS modes, this is also the maximum possible with the installed MFT transition. The injection modes are $|1\rangle + |2\rangle$, $|2\rangle$ and $|1\rangle$ by selecting the MFT modes "off", "m12+m23" and "m23".

The ABS and BRP are modeled as a succession of transitions and sextupole magnets which can be switched on or off leading to $N_{ABS} \cdot N_{BRP}$ different signals composed of a combination four hyperfine state intensity (I_k) for hydrogen. The transitions are described with ϵ_i efficiencies and the

Table 1. Hydrogen unknown variables for the calibration.

Using 1 ABS mode		Using 3 ABS modes	
H signals	11	H signals	33
H unknown	$10+1+4=15$	H unknown	$11+3\cdot4=23$
D signals	29	D signals	87
D unknown	$40+2+6=48$	D unknown	$42+3\cdot6=60$

BRP sextupole magnets are described with σ_j transition probabilities. The signals S_i are thus modeled with $g_i(x;\beta_1,\beta_2,...,\beta_n)$ functions, with $\boldsymbol{\beta}$ being the vector of unknown parameters $\boldsymbol{\beta}=(\epsilon_i,\sigma_j,I_k)$. For example, the transition s14 is now modeled with

$$T_{s14}=\begin{pmatrix} 1-\epsilon_{s14} & 0 & 0 & \epsilon_{s14} \\ 0 & 1 & 0 & 0 \\ 0 & 0 & 1 & 0 \\ \epsilon_{s14} & 0 & 0 & 1-\epsilon_{s14} \end{pmatrix}.$$

This indirect and nonlinear problem can be linearized and solved with a Gauss-Newton iteration algorithm. The least-square-fit method minimizes the sum S of weighted residuals and converges to the best values for the unknown vector $\boldsymbol{\beta}$:

$$S=\sum_{i=1}^{m}\left[\frac{y_i-g(x_i,\boldsymbol{\beta})}{\sigma_i}\right]^2=\sum_{i=1}^{m}\left[\frac{r_i}{\sigma_i}\right]^2. \quad (5)$$

The overdetermined system makes it possible to fit the BRP model to the data acquired for the calibration.

5. Signal and transition example

An advantage of the BRP design is the capability to measure the polarization of the storage cell with only a small extracted intensity. This is important in order to maximize the target density. The sensitivity of the BRP is achieved by using a chopped quadrupole mass spectrometer. The spectrometer is tuned to detect only atomic hydrogen (mass 1) and the incoming flux focused by the sextupoles is interrupted by a chopper rotating in front of the ionization volume. This technique of single ion counting ensures not only a reliable background subtraction, but also provides a direct evaluation of the error by measuring the variance of each ion count per time bin. The schematic view of the chopper and QMA setting and a measured signal is shown in figure 5.

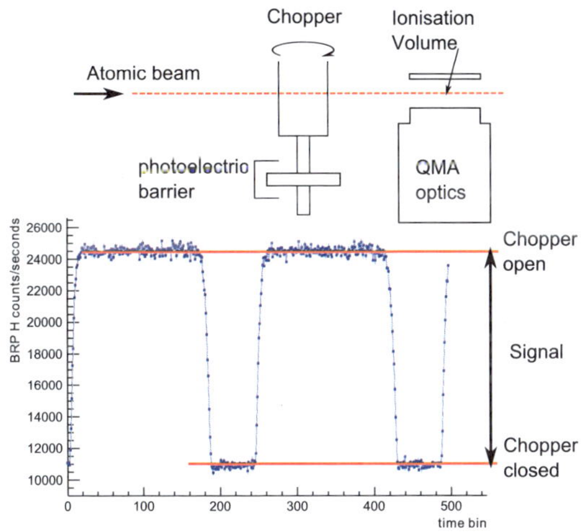

Fig. 5. Example of a chopped QMA signal from the BRP. The signal is averaged over 1000 chopper revolutions and clearly shows the background level on the bottom and the background + signal level on top. The flux intensity is directly proportional to the difference of both levels.

The SFT is the first transition to be tuned in the BRP as the resonances are well separated and can be identified unambiguously due to the single photon exchange. Figure 4(c) shows the result of a magnetic field scan. The SFT operates at a fixed frequency; the resonance is thus tuned with the magnetic field. At the transition point the populations 1-4 or 2-4 are exchanged and the atoms transfered into the state $|4\rangle$ after the SFT are then removed by the sextupole magnets and the beam blocker. If the storage cell contains only atoms in the states $|1\rangle$ and $|2\rangle$, the signal is thus reduced by a factor of two.

6. Timeline and conclusion

The BRP transitions and the sextupole magnets have been shown to work with a 300 K effusive beam from the storage cell. Additionally, the data acquisition and slow control system are being prepared for the hydrogen calibration.

The BRP will be connected to the PAX target chamber in 2010 and commissioned for the spin-filtering experiment at COSY. The first experimental spin-filtering studies using the new PIT and the BRP will be carried out at COSY by the end of 2010. In the second phase of the PAX experiment, the polarization build-up with antiprotons will be measured at AD/CERN.

References

1. A. Nass, Experimental Setup for Spin-Filtering Studies at COSY and AD, these proceedings.
2. W. Haeberli, *Ann. Rev. of Nucl. Sci.* **17**, 373 (1967).
3. A. Nass, L. Barion, M. Capiluppi, H. Kleines, P. Lenisa, F. Rathmann, J. Sarkadi, M. Stancari and E. Steffens, The Polarized Target for Spin Filtering Studies at COSY and AD, in *17th Int. Spin Phys. Symp. (SPIN06)*, eds. K. Imai and *et al.*, AIP Conf. Proc., Vol. 915 (New York, 2007).
4. A. Abragam and J. M. Winter, *Phys. Rev. Lett.* **1**, 374 (1958).
5. C. Baumgarten, Studies of spin relaxation and recombination at the hermes hydrogen/deuterium gas target, PhD thesis, Ludwig-Maximilians Universität München, (Munich, Germany, 2000).

FIRST EXPERIMENTS WITH THE POLARIZED INTERNAL GAS TARGET AT ANKE/COSY

M. Mikirtychyants[*,†], K. Grigoryev[†], R. Engels, F. Rathmann, D. Chiladze, A. Kacharava, B. Lorentz, D. Prasuhn, J. Sarkadi, R. Schleichert, H. Seyfarth and H. Ströher

Institut für Kernphysik, Forschungszentrum Jülich, Leo-Brandt-Str. 1, 52425 Jülich, Germany
**E-mail: m.mikirtychyants@fz-juelich.de*
† delegated from Petersburg Nuclear Physics Institute

F. Klehr

Zentralabteilung Technologie, Forschungszentrum Jülich, Leo-Brandt-Str. 1, 52425 Jülich, Germany

S. Barsov, S. Mikirtychyants and A. Vasilyev

Petersburg Nuclear Physics Institute Orlova Rosha, 188300 Gatchina, Russia

S. Dymov

Laboratory of Nuclear Problems, Joint Institute for Nuclear Research, 141980 Dubna, Russia

H. Paetz gen. Schieck

Universität zu Köln, Zülpicher Str. 77, 50937 Köln, Germany

E. Steffens

Physikalisches Institut II, Universität Erlangen-Nürnberg, 91058 Erlangen, Germany

A Polarized Internal gas Target (PIT) has been developed for the ANKE spectrometer at COSY. After its first installation at the ANKE target position in summer 2005, commissioning studies were carried out. In March 2006, the first single polarization measurements with the polarized hydrogen beam from an Atomic Beam Source (ABS) were performed. The beam was injected into a storage cell made from aluminum foil. The data analysis showed that the events from the extended gas target can be clearly identified in the ANKE forward detection system. Using unpolarized nitrogen, the background from the cell walls could be determined as well. The thickness of the gas in the storage cell with the hydrogen atoms in hyperfinestate $|1\rangle$ was measured as 2×10^{13} atoms/cm^2. The ABS jet target thickness was $(1.5 \pm 0.1) \times 10^{11}$ atoms/cm^2. In November 2006, the commissioning of a Silicon Tracking Telescope (STT) was successfully finished. In the following beam time in January 2007, a new storage cell made from aluminum coated with teflon was used together with the STT. The Lamb-shift polarimeter (LSP) was mounted below the target chamber to allow online tuning of the transition units and monitoring of the ABS jet polarization. A first double-polarized experiment was performed in January 2007. The results will be presented.

Keywords: Atomic beam source; ANKE; polarized target; internal target; polarimeter; detection of atomic beams; spin polarized hydrogen; deuterium.

1. Introduction

In 2004, the Atomic Beam Source (ABS)[1] and the Lamb-shift polarimeter (LSP)[2] were transferred from the laboratory to the COSY building. After all necessary tests, where the parameters listed in table 1 were determined, in the summer of 2005 the source was installed at the spectrometer ANKE for further commissioning. Measurements to determine the COSY-beam dimensions at the ANKE-target position and first tests with storage-cell prototypes were carried out parallel to these studies.

Table 1. Main parameters of the polarized atomic beam source of ANKE/COSY.

Gas Type	Intensity, at/s	P_z	P_{zz}
Hydrogen	$(7.8 \pm 0.2) \times 10^{16}$	$+0.89 \pm 0.01$	-
		-0.96 ± 0.01	-
Deuterium	$(3.9 \pm 0.1) \times 10^{16}$	$+0.73$	$+0.77$
		-0.82	-1.17

2. PIT at ANKE

In order to achieve the maximum luminosity in the experiments with the internal gas target, it is important to minimize the dimensions of the storage-cell tube. During the first test in February 2004, the diameter of the COSY beam at injection and after acceleration at the ANKE target position was measured. For this, a frame carrying various diaphragms was constructed. The diaphragm, which was mainly used, had dimensions of $50_{hor.}\text{x}25_{vert.}$ mm^2, i.e. larger than the expected beam size. During the tests, the supporting frame was moved across the beam by stepper motors. First, the center of the diaphragm was placed at the expected center of the COSY beam. By moving the diaphragm, the COSY beam current was gradually decreased and the beam's full size could be measured[a]. At injection, the beam had elliptical shape and its full size was $38_{hor.}$x$17_{vert.}$ mm^2. The accelerated beam without target had a size of $9_{hor.}$x$14_{vert.}$ mm^2. With the cluster-target beam (density: 10^{12} atoms/cm^2) it increased to $17_{hor.}$x$17_{vert.}$ mm^2 due to beam heating by the target.

In addition to the determination of the cell dimensions, special care was taken to shield various ABS components (e.g. vacuum pumps) against strong stray field of the central ANKE dipole D2. This issue is especially important for the weak-field transition unit (WFT), which is located about 400 mm away from the D2 gap.

3. Polarized internal target commissioning

Based on the measured results, two storage-cell prototypes were built from a 25 μm thick aluminum foil (99.95 Al). With acceleration of an unpolarized deuteron beam through the large cell ($30_{hor.}$x$20_{vert.}$ mm^2) to an energy of about 2.1 GeV, it was possible to store and accelerate more than 2/3 of the injected deuterons ($\sim 9 \times 10^9$) in the COSY ring. Using beam scrapers in the opposite section of the accelerator ring, the dimensions of the stored beam in the cell were decreased to $13_{hor.}$x$11_{vert.}$ mm^2. With a small cell of $15_{hor.}\times 15_{vert.}$ mm^2, 1.7×10^9 deuterons, i.e. about 15 % of the injected deuterons, were successfully stored in the COSY ring. The length of both cells was 220 mm.

For the first beamtime an aluminum foil, covered with PTFE to minimize depolarization on the surface, was used for the new prototype of the storage cell. The beam tube of this prototype was 400 mm long and had a cross section of $20_{hor.}\times 20_{vert.}$ mm^2. During the run, stacking injection[3] and electron cooling were employed to increase the number of stored and accelerated protons with the storage cell in place (fig. 1, left-hand side). As a last step, the ANKE spectrometer magnet D2 was moved to the position which corresponds to a deflecting angle of 9.2° for the first beam bending magnet D1. In this configuration, 6.4×10^9 protons could be stored and accelerated in the ring. This corresponds to about 50 % of the number of particles which can be accelerated without cell and stacking at injection.

[a]These measurements were done with a 2.1 GeV proton beam without using any cooling procedures (electron cooling at injection or stochastic cooling after acceleration) and without stacking procedure at injection.

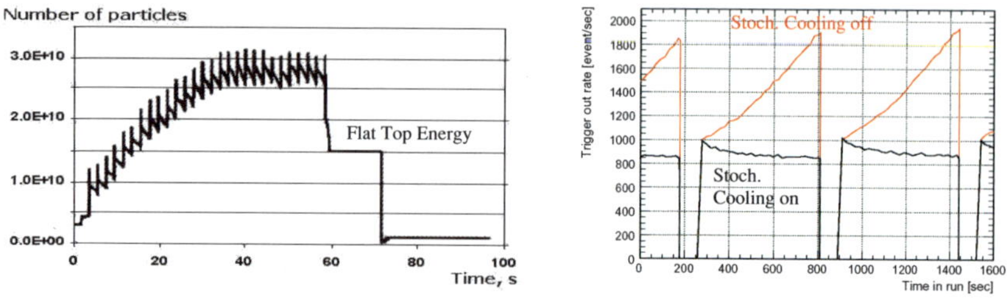

Fig. 1. Left: The beam-current transformer signal and the number of the stored protons in the COSY ring during stacking (28 stacks) through a storage cell followed by 2 s of electron cooling and accelerating to flat top energy. Right: The trigger rate during data taking with stochastic cooling switched on and off during a set of cycles. The strong increase of the trigger rate at the flat top energy after acceleration with stochastic cooling off occurs due to the increase of beam-cell walls' interactions, caused by beam heating.

This number yields an appreciable luminosity of about 10^{29} $cm^{-2}s^{-1}$ for double polarization experiments.

For beam energies higher than 831 MeV stochastic cooling can be used at COSY. This will compensate for the beam heating by the target. On the right-hand side of figure 1 the total trigger rate during data acquisition is shown as a function of time during different beam cycles. Without stochastic cooling being employed, beam heating leads to an intensive interaction of beam halo and thus extensive background growth.

In addition to the storage cell tests, the ABS beam was used as a jet target. In a first experiment the target position along the COSY beam direction could not be identified by the ANKE detector system due to very high rest gas pressure. For a second experiment, an ABS beam cryo catcher was built and installed below the interaction point with the COSY beam. This allowed improvement of the pressure in the target chamber by one order of magnitude to 3.7×10^{-8} mbar. With use of vertex reconstruction, the jet-target position could be clearly identified. The measured integral jet-target thickness of 1.5×10^{11} cm^{-2} perfectly matches the predicted value.

4. Results of the commissioning

In early 2007, the LSP was used to tune and to control the polarization of the ABS beam. However, its use in the strong magnetic stray field of the spectrometer magnet D2 revealed a number of problems. Firstly, the reduced sensitivity of the LSP due to the deflection of the slow protons behind the Glavish-type ionizer. Secondly, the deflection of the quantization

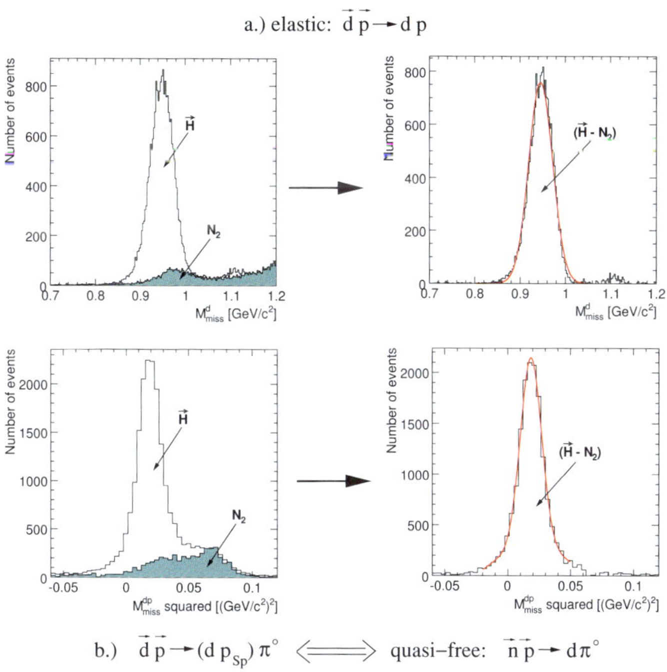

Fig. 2. Missing-mass spectra for the proton from elastic $\vec{d}\vec{p}$ scattering (a) and the π^0 from the quasi-free $np \to d\pi^0$ reaction (b) before (on the left-hand side) and after background substraction (on the right-hand side). The background substraction is based on the additional measurement with N_2 gas in the target cell.

axis (longitudinal solenoid field of the ionizer) due to superpositioning with the stray field leads to a "magnetic misalignment" of the LSP, which could not be compensated by a Wien-filter. This resulted in an underestimation of the measured polarization[b] and, furthermore, in a wrong sign of the vector polarization. Nevertheless, the transition units could be tuned and the polarization, which was measured once per day, could be controlled and was found to be stable within 5 % during one week of operation.

During this beam time, a storage cell ($15_{ver.} \times 20_{hor.} \times 380$ mm) was used. In addition, H_2 and N_2 could be injected into the cell by two separate gas feeding tubes. A first silicon tracking telescope (STT) was mounted around the cell. Polarized or unpolarized deuterons were accelerated to the flat-top energy of $T_d = 1.2$ GeV through the storage-cell tube, filled with

[b]The measured polarization was about 22 % of the expected value.

polarized hydrogen from the ABS or with unpolarized H_2 or N_2 gas from the calibrated gas supply system.

Figure 2 shows sample spectra from the elastic scattering $\vec{d}\vec{p}$ (upper panels) and the $\vec{d}\vec{p} \to (dp_{sp})\pi^0$ reaction (lower panels). Both missing-mass spectra could be corrected for events, which are produced at the cell walls. For this reason, data were taken with N_2 in the storage cell. The N_2-target density was adjusted in a way to provide the same COSY beam heating as for the $\vec{H}$-target.[4] The event distributions of these runs were subtracted from the original measured spectra with hydrogen in the cell, and the results are shown on the right-hand side of figure 2.

The analyzing powers for the reaction $d\vec{p} \to (dp_{sp})\pi^0$ for different scattering angles are known with good precision (see ref. 5). Therefore, the polarization of the storage-cell target was determined with an unpolarized deuteron beam at COSY as $\sim 0.79 \pm 0.07$ based on the measured asymmetries. Vice versa, the polarization of the COSY beam can be observed with unpolarized hydrogen gas in the storage cell as well.

5. Outlook

In the fall of 2009 a long beam time at ANKE on double-polarized $\vec{p}\vec{d}$ breakup is planned at flat top energies $T_d = 1.2$ or $T_d = 2.23$ GeV.[6] At this time, a modified LSP with a rotatable Wien filter will be available to compensate for the deflection of the quantization axis by the magnetic stray fields of the ANKE spectrometer magnet.

6. Acknowledgments

The authors want to thank the members of the ANKE collaboration and the COSY crew, who helped us during the installation and the commissioning studies of the polarized internal target.

References

1. F. Rathmann *et al.*, in *Proc. 15th Int. Spin Physics Symposium*, AIP Conf. Proc. **675**, 553 (AIP, New York, 2003).
2. R. Engels *et al.*, *Rev. Sci. Instrum.* **74**, 4607, (2003).
3. H. J. Stein *et al.*, in *Proc. 18th Conf. Charged Particle Accelerators (RUPAC 2002)*, ed. I. N. Meshkov (NRCRF, Obninsk, 2004).
4. F. Rathmann *et al.*, *Phys. Rev. C* **58**, 658 (1998).
5. D. Chiladze, *Phys. Rev. ST Acc. Beams* **9**, 050101 (2006).
6. COSY Proposal No. 172, Spokespesons: A. Kacharava, F. Rathmann and C. Wilkin, `http://www.fz-juelich.de/ikp/anke/en/proposals.shtml` .

EXTRA PHYSICS WITH AN ABS AND A LAMB-SHIFT POLARIMETER

R. Engels*, K. Grigoryev, M. Mikirtychyants, F. Rathmann, G. Schug, H. Seyfarth and H. Ströher

for the ANKE collaboration

Institute for Nuclear Physics, Jülich Center for Hadron Physics, Forschungszentrum Jülich, Wilhelm-Johnen-Str., 52428 Jülich, Germany
**r.w.engels@fz-juelich.de*

L. Kochenda, P. Kravtsov, V. Trofimov and A. Vasilyev

High Energy Physics Department, St. Petersburg Nuclear Physics Institute, Orlova Rosha 1, 188300 Gatchina, Russia

H. Paetz gen. Schieck

Institut für Kernphysik, Universität zu Köln, Zülpicher Str. 77, 50937 Köln, Germany

R. Emmerich, S. Paul and W. Schott

Physik-Department E18, Technische Universität München, James-Franck Str., 85748 Garching, Germany

M. Westig

I. Physikaliches Institut, Universität zu Köln, Zülpicher Str. 77, 50937 Köln, Germany

The polarized internal gas target of the ANKE experiment is only used for a few months per year for hadron physics at the cooler synchrotron COSY. In the meantime, the whole setup or components like the ABS or the Lamb-shift polarimeter can be used for other experiments. We present various projects, including nuclear fusion, atomic and molecular physics and a neutrino experiment, for which the existing hardware can be used.

Keywords: Polarized source; Lamb-shift polarimeter; polarized fusion; hyperfine spectroscopy; polarized molecules; bound beta decay.

1. Polarized fusion

For more than 40 years it is has been known that polarizing fuel particles will change the total cross section of the nuclear fusion reactions. For the d+^{3}He and the d+t reaction it is expected and has been shown, that aligned spins will increase the fusion rate by a factor of up to 1.5[1] because both reactions have a $J^\pi = 3/2^+$ resonance at low energies. For the d+d reactions no valid theoretical guidance exists. They require consideration of s, p, and d waves in 16 transition matrix elements, which would allow a neutron-lean fusion reactor via the ^{3}He+d reaction if the d+d neutrons could be suppressed. This had been postulated in the d+d spin-quintet state for which quite different predictions[2–9] exist (fig. 1). To determine the degree of quintet-state suppression, a direct spin-correlation cross section experiment at low energies is in preparation.

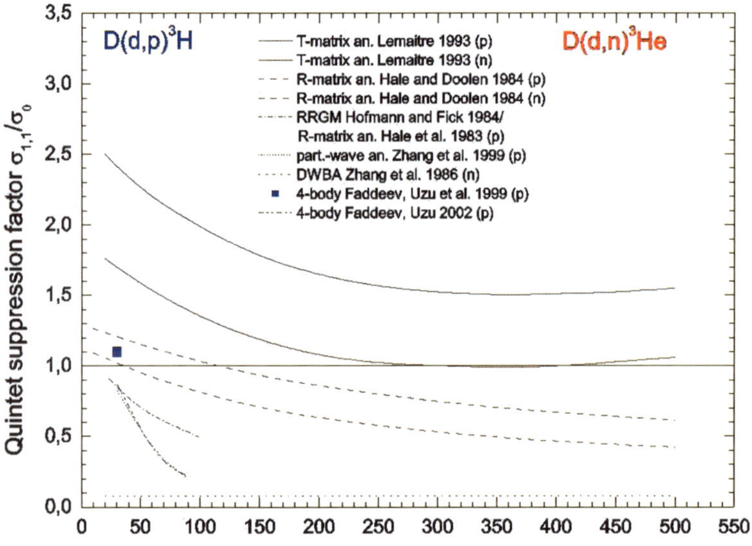

Fig. 1. Different predictions for the ratio of the double polarized total cross section $\sigma_{1,1}$ and the unpolarized cross section σ_0 for both dd-fusion reactions.

In an earlier setup it was planned to use a polarized atomic beam source (ABS), a donation from the University of Cologne,[10] to produce the polarized deuterium jet target and, by ionizing these atoms and deflecting them back, the polarized deuteron beam at the same time. Just a few days before this conference the KVI, Groningen, the Netherlands, generously agreed to provide their polarized ion source POLIS[11] for the present project. This source consists of an ABS, an ECR ionizer and a Lamb-shift polarimeter. At deuteron beam energies up to 32 keV, intensities up to 20 μA can be provided. Together with the expected jet-target density of 2×10^{11} atoms/cm^2 a luminosity of 4×10^{25} cm^{-2}s^{-1} will be possible. This means that at 30 keV, which will be a reasonable energy for coming fusion reactors, a count rate of 50 counts/s is possible. Therefore, the quintet suppression factor can be measured within two months of beam time with a statistical error of 1 %. For an ion-beam energy of 20 keV it will take eight months due to the lower total cross section. In addition, several spin-correlation coefficients can be measured, which will help to understand the reaction mechanism.

2. Hydrogen spectroscopy

With the spin-filter,[12] the central component of the Lamb-shift polarimeter,[13] it is possible to produce a beam of metastable hydrogen (deuterium) atoms in just one Zeeman state. With induced single transitions between the different Zeeman states of the $2S_{1/2}$ and the $2P_{1/2}$ hyperfine states, the Breit-Rabi diagrams, including the hyperfine energy splittings, the Lamb shift and the Lande factors can be measured very precisely.

2.1. *The Breit-Rabi diagram of the 2S state of hydrogen and deuterium*

With a setup similar to that of the atomic beam resonance method[14] (fig. 2), the complete Breit-Rabi diagram of the 2S state of hydrogen and deuterium can be measured. In our setup, the analyzing magnets of the Rabi apparatus are replaced by spin-filters.

In an electron-impact ionizer H_2 (D_2) molecules are dissociated and ionized. With acceleration of the ions to energies between 300 and 2000 eV, beam intensities up to 10 μA can be achieved. After deflection to a horizontal beam direction, a Wien filter is used to separate the protons from the other ions, which originate from the residual gas. By charge-exchange with cesium vapour,[15] metastable hydrogen atoms in the state $2S_{1/2}$ are produced from about 15 % of the protons. In the first spin-filter all metastable atoms except those in one Zeeman state are quenched into the ground state. Only metastable atoms in the Zeeman states $\alpha 1$, $\alpha 2$ or, using a subsequent Sona transition, $\beta 3$ remain in the beam. In a homogeneous magnetic field, magnetic dipole transitions are induced with the magnetic field vector parallel to the velocity direction of the hydrogen atoms in order to avoid longitudinal Doppler shift and broadening. The direction of a static magnetic field, produced by two Helmholtz coils, can be aligned either parallel or perpendicular to the velocity direction of the atoms. Thus, for a transverse magnetic field, the transitions $\alpha 1 \rightarrow \alpha 2$, $\alpha 1 \rightarrow \beta 4$, $\alpha 2 \rightarrow \alpha 1$ and $\alpha 2 \rightarrow \beta 3$ can be measured. The second spin-filter is used to verify the induced transitions. As an example, in figure 2 atoms in the metastable state $\alpha 1$ are leaving the first spin-filter and transfered into $\alpha 2$ in the transition unit. The second spin-filter is set to transmit the state $\alpha 2$ only. Therefore, metastable atoms can produce light in the quenching chamber only if the transition from state $\alpha 1$ into state $\alpha 2$ has occured in the transition unit. The large number of 10^5 photons/s detected in the photomultiplier

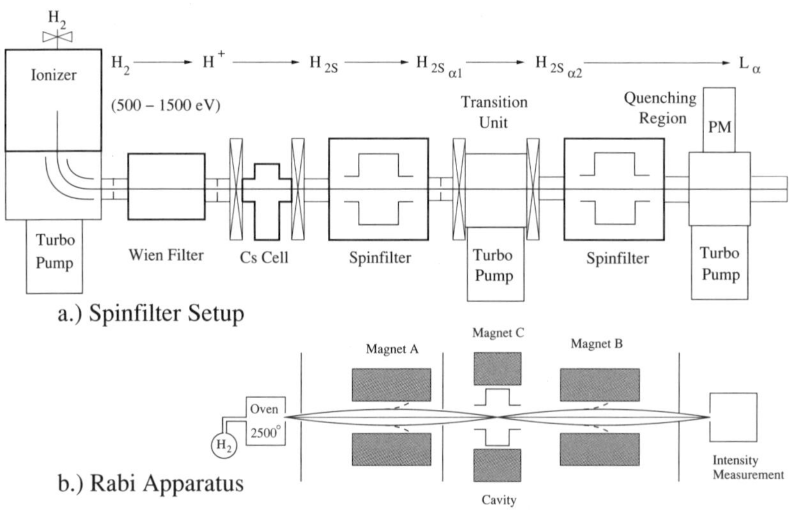

Fig. 2. (a) The setup for the measurement of the complete Breit-Rabi diagram of the 2S states of hydrogen or deuterium. (b) The schematic of the classical Rabi apparatus.

allows it to reach a statistical error comparable to that of the best measurements[16] in 20 minutes. In addition, the hyperfine-splitting energy can be measured independently of the magnetic field as a combination of the transition frequencies $(\alpha 1 \rightarrow \beta 4)$ - $(\alpha 2 \rightarrow \beta 3)$ or $(\alpha 1 \rightarrow \alpha 2) + (\beta 3 \rightarrow \beta 4)$. Therefore, a huge number of individual and direct measurements is possible. To increase the precision of the experiment, the halfwidth of the measured resonances can be decreased with use of the so called *separated oscillatory field method*[17] and with much slower metastable hydrogen beams produced, e.g. with the ABS and excited by electron bombardment. By that, it will be possible to reach the precision of current laser-spectroscopy experiments and to measure the g-factor of the 2S state with a precision of 10^{-8}.

2.2. *The Breit-Rabi diagram of the 2P state of hydrogen and deuterium*

With induced electric-dipole transitions from the single 2S-Zeeman states into single 2P states behind the spin-filter, the classical Lamb shift[18] and the complete Breit-Rabi diagram of the 2P state can be measured. In contrast to the original Lamb measurements today it is possible to change the RF frequency without changing of the RF power in a constant magnetic field by a Lecher TEM waveguide.

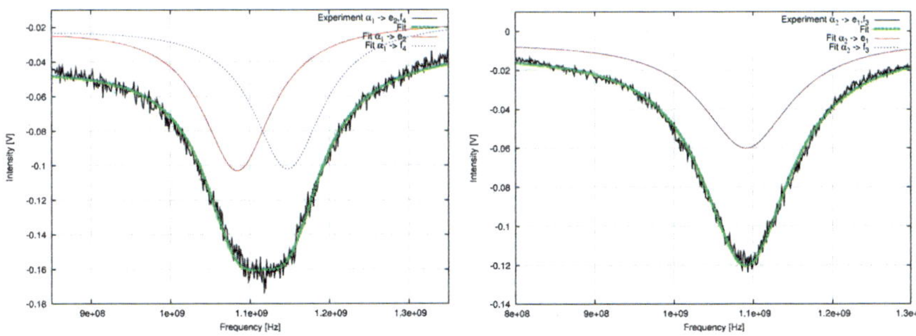

Fig. 3. The observed transitions ($\alpha 1 \rightarrow f4$) and ($\alpha 1 \rightarrow e2$) (left) or ($\alpha 2 \rightarrow f3$) and ($\alpha 2 \rightarrow e1$) (right) at a small vertical magnetic field in the transition region.

In a proof-of-principle measurement metastable hydrogen atoms in the HFS $\alpha 1$ or $\alpha 2$ are selected in the spin-filter and reach the TEM waveguide. For a small transverse magnetic field close to 0 G the transitions ($\alpha 1 \rightarrow f4$) and ($\alpha 1 \rightarrow e2$) (fig. 3, left) or ($\alpha 2 \rightarrow f3$) and ($\alpha 2 \rightarrow e1$) (fig. 3, right) are observed. The difference between the two resonances of the transitions ($\alpha 1 \rightarrow f4$) and ($\alpha 1 \rightarrow e2$) corresponds to the hyperfine splitting energy of the $2P_{1/2}$ state. The result from a fit of 60 (2) MHz agrees with the earlier result of 59.22 (14) MHz.[17] The transition frequencies from the $2S_{1/2}$ into the $2P_{1/2}$ state together with the HFS of these states allows to obtain the classical Lamb-shift. As a first result, a value of 1057(1) MHz could be determined. The dominant error, which is up to three orders of magnitude larger compared to the best values, is produced by the inhomogeneity and the uncertainty of the magnetic field. An error of $\Delta B = 0.5$ G, for example, corresponds to an error of $\Delta f = 1$ MHz.

3. Polarized molecules

When polarized hydrogen atoms recombine in a storage cell, the residual H_2 molecules may still be nuclear polarized.[19] In a collaboration between PNPI, University of Cologne and FZ Jülich a device was built (fig. 4) to measure the polarization of hydrogen atoms and hydrogen molecules after recombination of polarized atoms depending on different materials, temperatures and magnetic fields. In a superconducting solenoid, polarized atoms

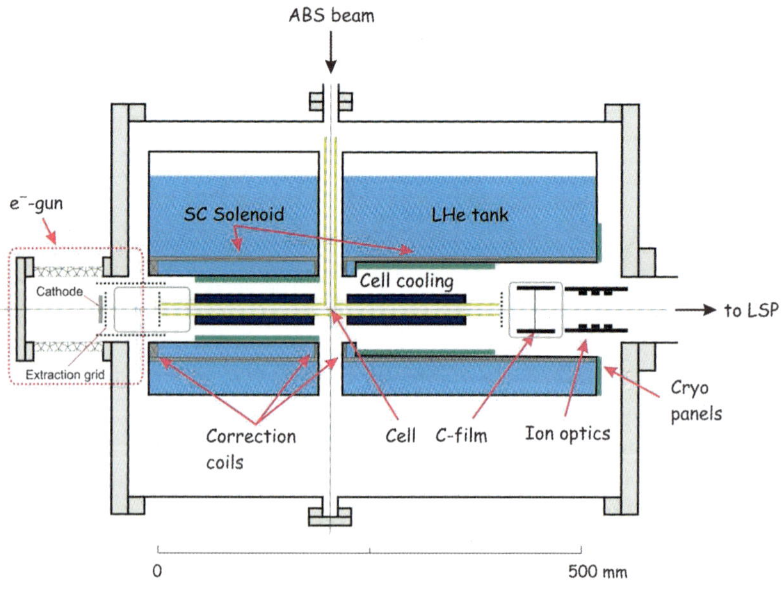

Fig. 4. Setup of the experiment to measure the polarization of hydrogen (deuterium) molecules after recombination of polarized atoms.

from the ABS partly recombine in a T-shape storage cell, where the inner surface can be covered with different materials. Both atoms and molecules, are ionized afterwards by electron bombardment and the protons and H_2^+ ions produced are accelerated to an energy of a few keV. Inside the solenoid both ions have to pass a thin carbon foil, where the last electron of the H_2^+ ions is stripped off and, therefore, two protons are produced. These protons share the kinetic energy of the H_2^+ ion and can be separated by the Wien-filter of the LSP from the protons, which originate from the initial atoms. In this way, the nuclear polarization of the atoms and the molecules can be measured under various conditions.

After solving a long list of minor problems, in the summer of 2009 the first experiments were performed on a gold surface. They showed the surprising result shown in figure 5. The polarization for a fixed magnetic field was stable even at temperatures as low as 47 K. Until now, no material has been found which could preserve the polarization at temperatures below 80 K.[19]

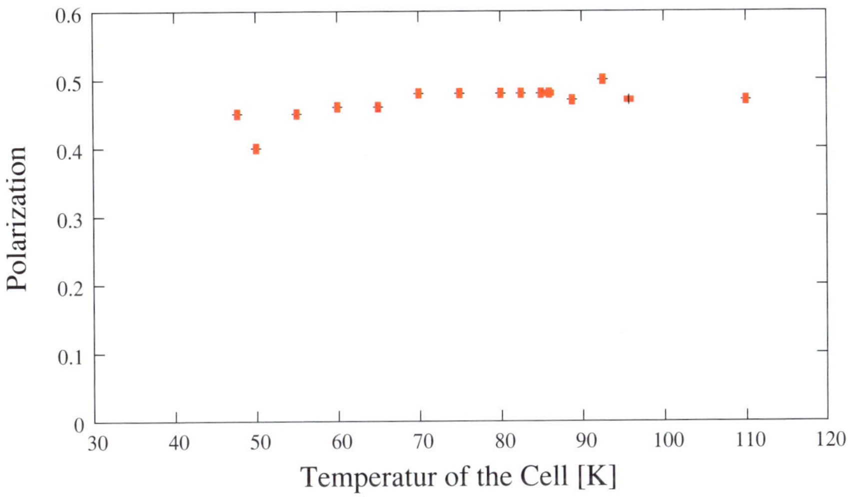

Fig. 5. One of the first results with polarized hydrogen atoms in a storage cell with a gold surface. The polarization for atoms in HFS 1 is stable and independent of the temperature. (Magnetic field: 0.28 T, ion beam energy: 4 keV)

4. Rare neutron decay

In the neutron decay n→p+e+$\bar{\nu}_e$, the proton and electron can be found in different bound S states of the hydrogen atom.[20] The kinetic energy of 326.5 eV for this 2-body decay fits to the energy range of the Lamb-shift polarimeter. Therefore, the polarization of the metastable hydrogen atoms can be measured, i.e., the proton and electron spin after the decay. If the helicity of the antineutrino is completely positive (right-handed), then the probabilities to find the different combinations of electron and proton spin, i.e. the different Zeeman states of the hydrogen atom, can be calculated and tested with this experiment. Therefore, left-handed admixtures to the helicity or scalar and tensor contributions to the weak force can be measured with high precision.[21,22] The challenge of the experiment is the low count rate of metastable hydrogen atoms behind the spin-filter. For the through-going FRM II beam tube, less than 1 count per second is expected. For the detection of the outgoing atoms 4 different methods are suggested:

- Ionization of the metastable atoms only by two different lasers and collection of the outcoming protons (efficiency: ~ 50 %).
- Detection of the Lyman-α photons after quenching the metastable atoms with a photomultiplier and a setup of optimized mirrors (efficiency: $\sim$ 5 %).
- Detection of the Lyman-α photons by photoeffect on a CsI or a tungsten surface and collection of the electrons with a channeltron (efficiency: > 50 %).
- Selective charge-exchange of metastable hydrogen atoms with argon and separation of the H^- ions produced by a velocity filter (efficiency: $\sim$ 10 %).

Which method will give the best signal-to-noise ratio and a reasonable efficiency
will be tested at the Technical University of Munich, Physik Department E18.

References

1. Ch. Leemann *et al.*, *Helv. Phys. Acta* **44**, 141 (1971).
2. E. Uzu *et al.*, *Progr. Theor. Phys.* **90**, 937 (1993).
3. S. Lemaitre and H. Paetz gen. Schieck, *Ann. Phys. (Leipzig)* **2**, 503 (1993).
4. G. Hale and G. Doolen, *LA-9971-MS*, Los Alamos (1984).
5. K. A. Fletcher *et al.*, *Phys. Rev. C* **49**, 2305 (1994).
6. J. S. Zhang *et al.*, *Phys. Rev. Lett.* **57**, 1410 (1986).
7. H. M. Hofmann *et al.*, *Phys. Rev. Lett.* **57**, 2038 (1984).
8. E. Uzu, nucl-th/0210026 (2002).
9. J. S. Zhang *et al.*, *Phys. Rev. C* **60**, 054614 (1999).
10. R. Emmerich and H. Paetz gen. Schieck, *Nucl. Instr. Meth. A* **586**, 387 (2008).
11. H. R. Kremers *et al.*, *Nucl. Instr. Meth. A* **536**, 329 (2005).
12. J. L. McKibben *et al.*, *Phys. Lett. B* **28**, 594 (1969).
13. R. Engels *et al.*, *Rev. Sci. Instr.* **74**, 11 4607 (2003).
14. I. I. Rabi *et al.*, *Phys. Rev.* **55**, 526 (1939).
15. P. Pradel *et al.*, *Phys. Rev. A* **10**, 797 (1974).

16. N. Kolachevsky *et al.*, *Phys. Rev. Lett.* **102**, 213002 (2009).
17. S. R. Lundeen *et al.*, *Phys. Rev. Lett.* **34**, 377 (1975).
18. W. E. Lamb and R.C. Rutherford, *Phys. Rev.* **81**, 222 (1951).
19. T. Wise *et al.*, *Phys. Rev. Lett.* **87**, 042701 (2001).
20. L. L. Nemenov, *Sov. J. Nucl. Phys.* **31**, (1980).
21. W. Schott *et al.*, *Hyp. Int.* **193**, 269 (2009).
22. W. Schott *et al.*, *Eur. Phys. J. A* **30**, 603 (2006).

SYSTEMATIC STUDIES FOR THE DEVELOPMENT OF HIGH-INTENSITY ABS

L. Barion*, G. Ciullo, M. Contalbrigo, P. F. Dalpiaz, P. Lenisa and M. Statera

for the PAX-collaboration

INFN - Sezione di Ferrara , Polo Scientifico e Tecnologico Building C,
via Saragat 1 - 44122 Ferrara, Italy
**email: barion@fe.infn.it*

The effect of the dissociator cooling temperature has been tested in order to explain the unexpected RHIC atomic beam intensity. Studies on *trumpet* nozzle geometry, compared to standard *sonic* nozzle have been performed, both with simulation methods and test bench measurements on molecular beams, obtaining promising results.

Keywords: Atomic beam source; ABS intensity; PAX; polarized; antiprotons; Spinlab; nozzle; trumpet; sonic.

1. Introduction

Atomic Beam Sources (ABS) are widely used in nuclear and particle physics. They are used as polarimeters (for example at the RHIC facility[1]) or in conjunction with a storage cell[2] to produce an intense internal gaseous target, as for example in HERMES.[3,4]

Recently the PAX collaboration[5] has proposed to use an internal gas target as a filter[6] to polarize a stored antiproton beam. For this purpose, however, the target thickness achieved so far is quite low and an increase of one order of magnitude is highly desirable. Since the intensity achieved by the storage cell is proportional to the intensity of the ABS jet, systematic studies on the atomic beam sources are necessary. In an ABS, molecular hydrogen is fed to a dissociator that produces atomic hydrogen, which expands in a vacuum chamber through a nozzle; then a skimmer selects the central part of the divergent beam. In this way a collimated beam of atomic hydrogen is produced. A set of sextupolar magnets and radio-frequency transitions polarize the beam by focusing the atoms of the selected spin state, with the Stern-Gerlach effect.

The ABS are complex apparatuses, whose performance is still not completely understood; for example the beam intensity produced by the RHIC ABS[7] is surprisingly higher (~ 50 %) than what was expected from the design calculations.

The Spinlab[8] laboratory of *Università di Ferrara* focuses on the combination of test bench measurements and beam simulations to both understand the limits of present ABS and to find ways to overcome them.

Among the facilities available there is an unpolarized ABS (originally located at CERN) connected with a movable diagnostic system (see fig. 1). The diagnostic system can measure beam intensity (with a compression

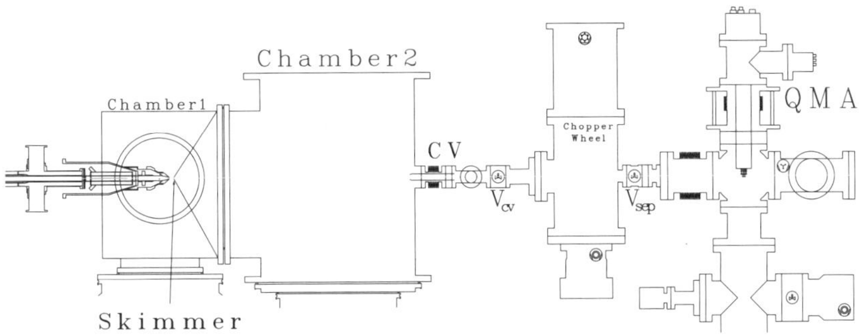

Fig. 1. Atomic beam source connected to the diagnostic system: CV Compression Volume, QMA Quadrupole Mass Analyzer

volume, CV), beam density (with a Quadrupole Mass Analyzer, QMA), and beam velocity distribution with a Time Of Flight (TOF) device.

In this paper two tests will be presented: the effect, on beam intensity, of different dissociator cooling temperatures and the effect of a different nozzle geometry.

2. Dissociator cooling effect

As already mentioned, the RHIC ABS intensity is surprisingly higher than expectations. An analysis carried out by our group, using among others the program SCAN, demonstrates that the increase in the intensity of RHIC ABS is not due to the accurate design of the magnetic system,[9] but that the relevant difference with the other ABS seems to be the cooling system of the dissociator.[10] The cooling of the RHIC dissociator produces a smooth gradient of temperature along the dissociator tube, from the hot discharge region to the cold nozzle.

At Spinlab we have addressed this problem using the test bench shown in figure 1. The unpolarized atomic beam either enters a Compression Volume CV, or it reaches the QMA, that can measure the atomic and molecular components of the beam. A copper collar has been added to the dissociator to create a temperature gradient similar to RHIC. The detailed layout of the dissociator is shown in figure 2. Measurements have been performed, varying

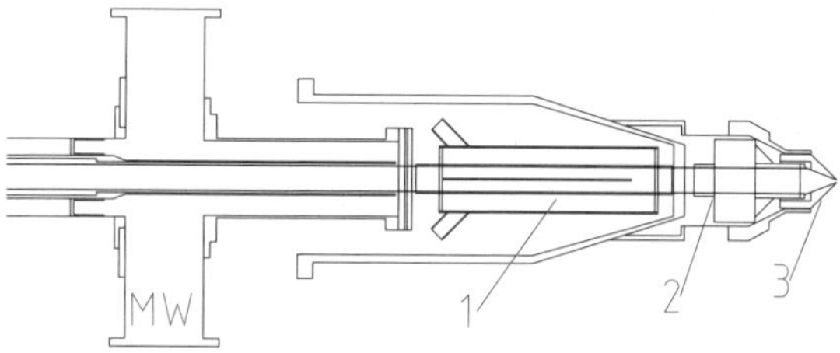

Fig. 2. Dissociator cooling system. (1) Water cooling, (2) Collar, (3) Nozzle.

the collar temperature. Two different inner diameter nozzles have been tested: 2 mm (the most commonly used, similar to RHIC) and 4 mm. The detailed measurements will be presented below. Both nozzles measurements show a clear temperature effect on the beam intensity, that seems connected with the dissociation in the collar region. TOF measurements show that the effect is not related to differences in the beam velocity distribution after the nozzle. While the gradient of temperature of the RHIC ABS is fixed, the dissociator of the Ferrara ABS is cooled with four separate stages, whose temperature can be varied independently from the others. The temperature of each stage is also precisely measured with thermocouples and CLTS sensors. The details of the four cooling stages, shown in figure 2, are listed below:

- Air cooling: forced flow of air at room temperature (fixed) outside the discharge tube, just around the plasma region.
- Water cooling: flow of water inside a copper jacket; the temperature is stabilized and can be set in the range -20 °C to +10 °C.
- Collar: copper collar connected to the first stage of a cold head and two resistive heaters; the temperature is stabilized and can be set in the range 70–290 K.
- Nozzle: aluminum sonic nozzle connected to the second stage of the same cold head and a resistive heater; the temperature is stabilized and can be set in the range 70–220 K.

During the measurements, the ABS was operated at the following parameters:

- Water cooling: 5 °C.
- Nozzle: 100 K.
- H_2/O_2 flux: 75/0.375 sccm (0.5 % O_2).
- Microwave power: 600 W.

In the first test, the temperature of the collar was varied from ~70 K to ~220 K and the total beam intensity (atoms+molecules) was measured with the CV; the result is presented in figure 3. It is clearly visible that there

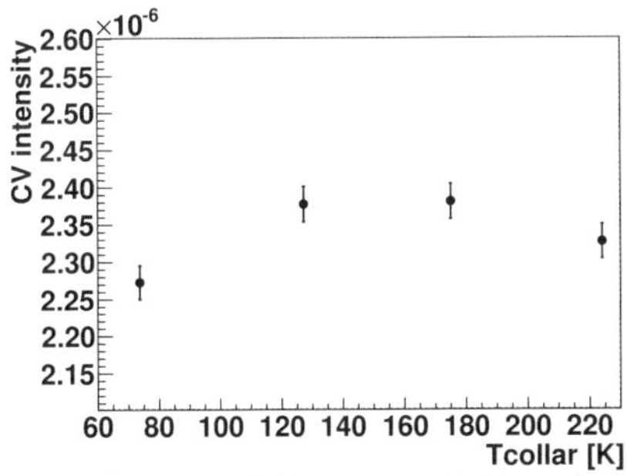

Fig. 3. Compression volume intensity as a function of collar temperature.

is a range of temperature of the collar that produces higher beam intensity. In the tested temperature range the change in intensity is about 5 %. In

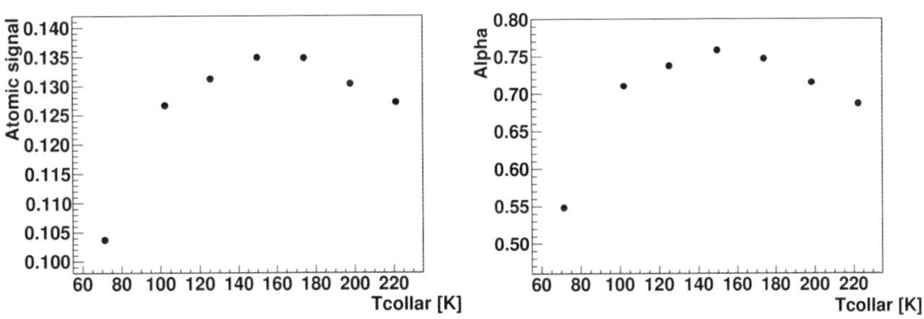

Fig. 4. Quadrupole mass analyzer measurements on atomic component of the beam (left panel) and calculated dissociation ratio (right panel); 2 mm nozzle, 75 sccm H_2, 0.5% O_2.

order to check what influence the collar temperature has on the beam in detail, the QMA was used to measure the atomic and molecular part of the beam separately. The results of these measurements are presented in figure 4, where, in the left panel, the measured atomic beam density is plotted as a function of the collar temperature and in the right panel the dissociation ratio (α) is plotted as a function of the collar temperature. It is evident that the range of temperature that gives higher total beam intensity, matches the one that gives also a higher dissociation.In order to test if this effect is related to the particular set of operating parameters listed above, different conditions were used and the measurements were repeated. The next test consisted in replacing the standard 2 mm diameter sonic nozzle with a 4 mm diameter one. The result of the measurements with QMA is presented in the left panel of figure 5 and shows that the same effect is still present.

An additional test was carried out increasing the oxygen percentage, from the standard 0.5 % to 2 %. The resulting graph is shown in the right panel of figure 5. In this case a different behaviour is clearly visible. In order to check the possible influence of the collar temperature on the beam expansion after the nozzle, a beam velocity distribution was recorded for all the measurements above. As a reference in figure 6 the velocity distribution corresponding to the second measurement (4 mm nozzle, 75 sccm H_2, 0.5 % O_2) are reported for atoms (left panel) and molecules (right panel). The temperature of the collar does not influence the velocity distribution of the beam and confirms that hydrogen thermalizes to the nozzle temperature while passing through it.

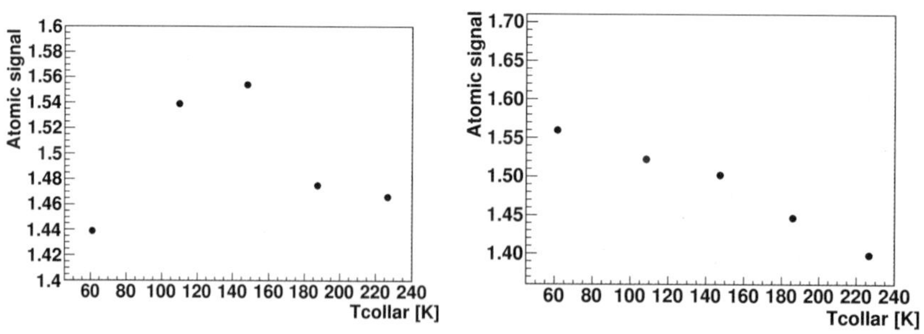

Fig. 5. Quadrupole mass analyzer measurements on atomic component of the beam, 4 mm nozzle, 75 sccm H_2; 0.5 % O_2 (left panel) and 2 % O_2 (right panel).

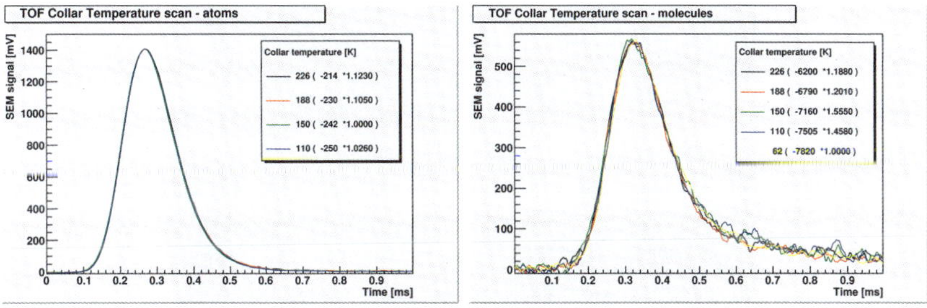

Fig. 6. TOF distribution of atomic beam (left panel) and molecular beam (right panel) for different collar temperatures; 4 mm nozzle, 75 sccm H_2, 0.5 % O_2.

3. Trumpet nozzle

Another series of tests, still dedicated to increasing the ABS intensity, was carried out, modifying the nozzle shape. Using the simulation software for molecular beams, written by G. A. Bird,[11] W. Kubischta tried different geometries and optimized the most promising one, that is the so called *trumpet* nozzle. Simulations were carried out by W. Kubischta and independently at SpinLab, for two different nozzle shapes: the standard *sonic* nozzle and the *trumpet* nozzle, as shown in figure 7. In both cases it was

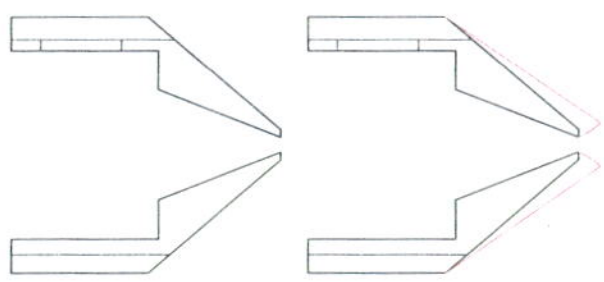

Fig. 7. Sonic nozzle (left) and trumpet nozzle (right).

found that the calculated increase in beam intensity produced by the *trumpet* nozzle, respect to the *sonic* one, is about 60 % for the skimmer flux and 40 % for the collimator flux. Figure 8 shows the simulation geometry and the molecular number density, obtained with the simulation program, for the *sonic* nozzle.

Two nozzles have been produced in aluminum and the corresponding molecular beam intensities measured with the ABS of our testing bench, to compare the Monte Carlo previsions with experimental data. The molecular beam flux has then been measured at the skimmer and CV positions (see fig. 1). Three different nozzle–skimmer distances were tested and compared with the simulations. The results of the measurements on the skimmer flux and CV intensity are presented in figure 9, in the left and right panel

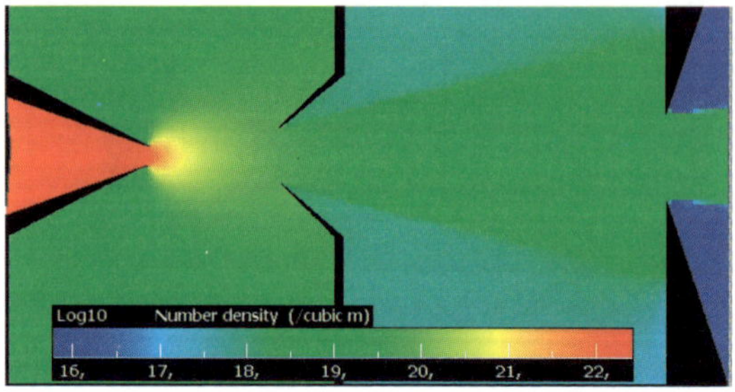

Fig. 8. Simulation geometry for sonic nozzle standard position; (A) nozzle (trumpet), (B) skimmer, (C) collimator.

respectively. Each series of data (sonic and trumpet) has been repeated twice: each nozzle was installed twice, in order to check the reproducibility of the measurements. The data on the left part of each panel correspond to the standard sonic nozzle, while the one on the right correspond to the trumpet nozzle; the different colors correspond to different nozzle-skimmer distances. The reproducibility is clearly confirmed and the measurements confirm the results of the simulations.

The different performance of the two nozzles have also been measured with the QMA; since it is quite far from the nozzle (more than 1 m) and the simulation program is presently limited to a simulation region in the orders of centimeters, only the experimental results have been considered. The experimental data are presented in figure 10: the plot shows that a trumpet nozzle produces a beam density that is about 6 % higher than the one produced by a sonic nozzle.

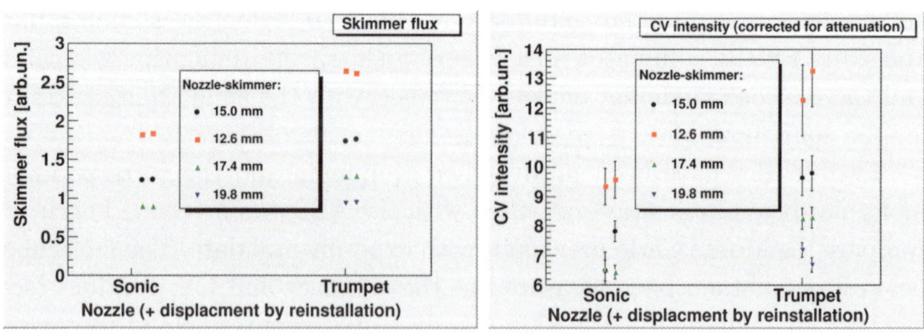

Fig. 9. Flux through the skimmer (left panel) and CV intensity (right panel).

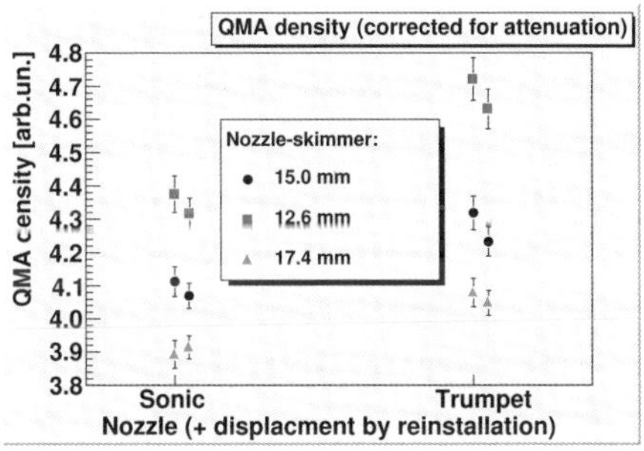

Fig. 10. Beam densities measured by QMA.

4. Conclusions

The results of the measurements perfromed changing the collar temperature, show that the observed enhancement in the RHIC ABS intensity (the "RHIC effect") could be connected with the cooling of the dissociator, in the region between the plasma and the nozzle.

The results of the tests with the *trumpet* nozzle are promising, although a definitive answer on the conveninence of a trumpet nozzle will be available only after testing the same technique on a polarized atomic beam.

References

1. RHIC web site – `http://www.bnl.gov/rhic/` .
2. W. Haeberli, in *Proc. Nuclear Physics with Cooled Stored Beams*, AIP Conf. Proc. **128**, 251 (AIP, New York, 1984).
3. C. Baumgarten *et al.*, The storage cell of the polarized internal H/D gas target of the HERMES experiment at HERA, *Nucl. Instr. Meth. A* **496**, 277 (2003).
4. HERMES web site – `http://www-hermes.desy.de/` .
5. PAX web site – `http://www.fz-juelich.de/ikp/pax/` .
6. F. Rathmann *et al.*, A Method to Polarize Stored Antiprotons to a High Degree, *Phys. Rev. Lett.* **94**, 014801 (2005).
7. A. Zelenski *et al.*, Absolute polarized H-jet polarimeter development for RHIC, *Nucl. Instr. Meth. A* **536**, 248 (2005).
8. L. Barion, "Internal polarized gas targets: systematic studies on intensity and correlated effects", PhD thesis Università di Ferrara (Ferrara, Italy, 2008).
9. M. Stancari, *private communication.*
10. A. Zelenski, *private communication.*
11. DS2G simulation program (Direct Simulation MonteCarlo) by G. A. Bird – `http://www.gab.com.au/` .

UPGRADE OF THE 50 KEV GaAs SOURCE OF POLARIZED ELECTRONS AT ELSA

D. Heiliger*, W. Hillert and B. Neff

University of Bonn, Physics Institute,
Nussallee 12, 53115 Bonn, Germany
**E-mail: heiliger@physik.uni-bonn.de*

Since 2000, an inverted source of polarized electrons has been operated routinely at the electron stretcher accelerator ELSA, providing a pulsed beam with a current of 100 mA and a polarization of about 80 %. One micro-second long pulses with 100 nC charge are produced in space-charge limitation by irradiating a strained-layer superlattice photocathode (8 mm in diameter) with laser light from a flash lamp pumped Ti:Sapphire laser.

Part of the future hadron physics programme requires significantly higher beam intensity, which can be supplied by enlarging the emission area or by improving the quantum efficiency (QE). Both will significantly influence the beam parameters and the optics of the transfer line. Numerical simulations of the space-charge dominated beam transport show that a quasi lossless transport to the linear accelerator is achievable using the existing setup of magnets. Dedicated beam diagnostics like wire scanners and fluorescence monitors will allow an *in situ* optimization of the optics and the transfer efficiency.

Keywords: Polarized electrons; electron sources.

1. Introduction

Since 2006, experiments on baryon spectroscopy have been performed at the University of Bonn, requiring circularly polarized photons which are generated by bremsstrahlung of longitudinally polarized electrons.[1] The polarized electrons required for the irradiation of the bremsstrahlungs-target cannot be produced via self-polarization according to the Sokolov-Ternov mechanism[2] due to the long polarization time. Thus in Bonn, polarized electrons are generated in a dedicated source[3] and are transported to the experiment while aiming at the highest possible conservation of polarization.

The actual setup of the source of polarized electrons and its load-lock system is shown in figure 1. The main parameters of the source are determined by the properties of the injector chain of the ELSA stretcher ring. A beam energy of 48 keV is required for the buncher section of the pulsed injector linac, the pulse length of 1 μs and repetition rate of 50 Hz are determined by the booster synchrotron.

Fig. 1. Picture of the 50 keV source at the University of Bonn.

Polarized electrons are generated by irradiating a strained-layer superlattice photocathode with circularly polarized laser light from a flash lamp pumped pulsed titanium sapphire laser. The generated laser pulse with a pulse length of 10 μs shows a spiking behaviour and is chopped into a 1 μs long pulse (see fig. 2).[4]

The emitted current (by default 100 mA) is limited by space-charge limitation. In order to vary the beam intensity the perveance[a] can be adjusted by changing the distance between the anode and the cathode. Using this space-charge limited operation mode a rectangular current pulse is generated even though the laser pulse is not rectangular.

For future hadron physics experiments a significantly higher beam intensity of approximately 200 mA is required. Such intensities will have an

[a]The perveance is the constant of proportionality between the emitted current and the applied voltage and is only dependent on the geometry.

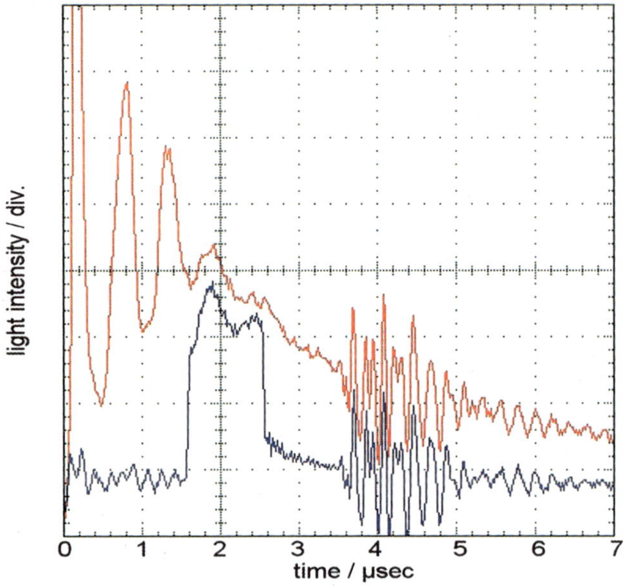

Fig. 2. The laser pulse generated by the Ti:Sa-laser (in red) and the chopped pulse (in blue). The oscillations after 3.6 μs are caused by interfering electromagnetic signals.

impact on the beam parameters and the optics of the transfer line. In this paper, measurements of the charge attainable with the existing setup are shown and the results of numerical simulations of the beam transport will be presented.

2. Attainable charge

Using the currently installed photocathode, the emitted charge was measured for different laser light intensities and perveances provided by different settings of the distance between the electrodes. The charge was determined by integrating over the pulse profile measured inductively with a ferrite-based current transformer installed at the high voltage cable. The energy of the light pulse was derived from integrating over the light intensity profile measured with a photodiode. The measurement is presented in figure 3, where the theoretically expected result is indicated by solid lines. For light pulse energies above 0.3 mJ one can clearly observe an increasing discrepancy between the measurement and the expectation, which can be attributed to the following reasons:

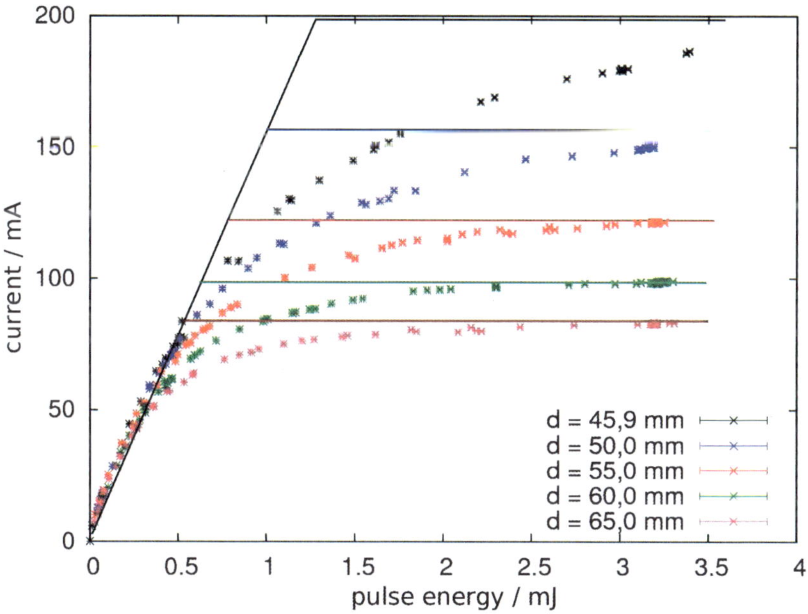

Fig. 3. Emitted charge against laser energy per pulse for different distances d between the electrodes.

(1) The emission current is limited by the surface charge limitation, resulting in a dependence of the quantum efficiency on the laser light energy.
(2) As mentioned before, the energy per pulse is determined by integrating over the light pulse profile. Due to the spiking (see fig. 2), the light power is not high enough to assure a space-charge limited emission over the whole pulse duration. The integrated emitted current then reflects a superposition of emission both within and outside the space-charge limit.

Both effects superimpose and cause the decrease of the emitted charge below the expected values.

With the present setup currents from 80 mA up to 190 mA can be generated. In order to guarantee a long term stability, when operating with higher intensities, the photoemission area has to be increased in order to produce a current of 200 mA safely. This will significantly influence the beam parameters and the optics of the transfer line due to the change of the emittance and the space-charge. In the following, the transfer line, its beam diagnostics and the simulated beam transport will be presented.

3. Intensity upgrade

3.1. *Transfer line*

Figure 4 shows a schematic drawing of the transfer line. In order to avoid a degradation of the ultra high vacuum in the operating chamber by the pressure in the linear accelerator, a 6 m long differential pumping section and a beam pipe with a small diameter (35 mm) is essential. In figure 4 the decrease in total pressure along the transfer line is indicated. The folded beam line with two α-magnets has two symmetry planes, one between the α-magnets and one in the Mott polarimeter. The symmetric setup reduces the required beam diagnostics to three wire scanners and three luminescence monitors.

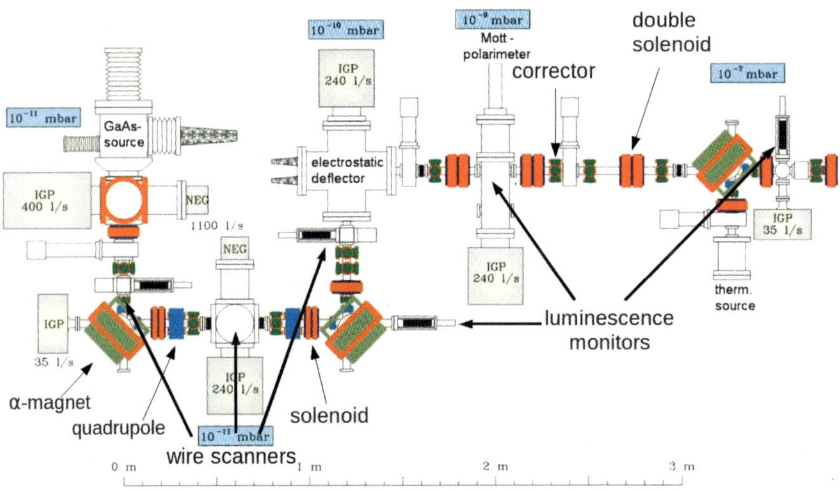

Fig. 4. The transfer line between the operating chamber and the linear accelerator.

The electrostatic deflector rotates the longitudinal spin transverse to the momentum, which is necessary to conserve the spin in the following accelerators. For polarization measurements the Mott polarimeter is used.

3.2. *Beam transport*

The transversal beam dynamics of a low-energy electron beam with a homogeneous, elliptical charge distribution in the presence of external electromagnetic fields are described by the so-called paraxial differential equations:[5,6]

$$\frac{\mathrm{d}^2 x}{\mathrm{d}s^2} + [k_x(s) + S(s) + T(s)] \cdot x - \frac{\varepsilon^2}{x^3} - \frac{2K}{x+z} = 0\,, \tag{1}$$

$$\frac{\mathrm{d}^2 z}{\mathrm{d}s^2} + [k_z(s) + S(s) + T(s)] \cdot z - \frac{\varepsilon^2}{z^3} - \frac{2K}{x+z} = 0\,. \tag{2}$$

The linear term describes the restoring forces of quadrupoles ($k_{x,z}(s)$), solenoids ($S(s)$) and the electrostatic deflector ($T(s)$). The expansion of the beam due the emittance ε and the space-charge[b] is included in the third and fourth term.

In order not to degrade the vacuum, the beam must be transported quasi lossless to the linear accelerator. For the optimization of the magnetic optics of the transfer line the differential equations were solved numerically for a current of 100 and 200 mA. The emitting surface of the photocathode is a full circle which results in a homogeneous, cylinder symmetric charge distribution of the beam when operating in space-charge limitation. Due to the geometry of the electrodes the beam is focused leading to a beam waist downstream and close by the anode. The waist position was chosen as the initial point of the simulation whereas its position varies for different currents and diameters of the emitting surface. The initial parameters, like the position of the waist, the beam radius at the waist and the emittance were taken from numerical simulations using the software EGUN.

Figure 5 shows the optimized results of simulations performed for different settings of the optics for currents of 100 and 200 mA. The optimization criteria were a minimal beam radius along the whole transfer line and a local minimum of the beam radius in the symmetry planes.

In figure 5 the evolution of the beam radius is presented by solid lines, above the abscissa for the horizontal and below the abscissa for the vertical beam plane. The shaded areas represent the aperture of the transfer line. As mentioned before, the origin of the diagram was set to the position of the beam waist.

---

[b] K is called the generalized perveance.

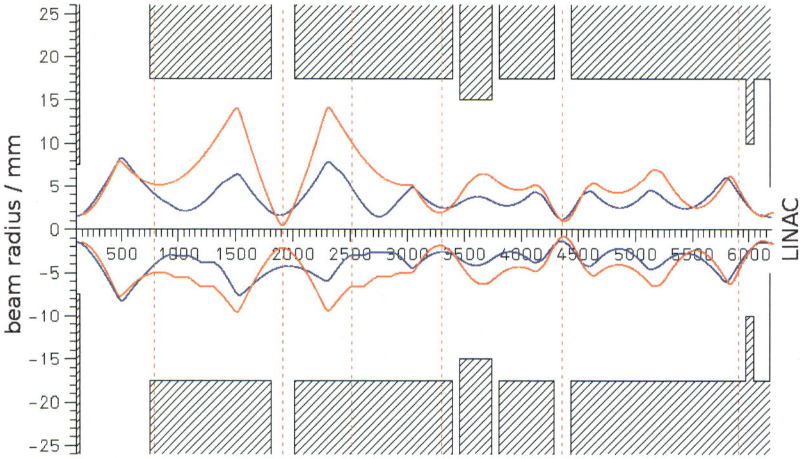

Fig. 5. Optimized results of the numerical simulations for a beam current of 100 mA (blue lines, diameter of the photocathode Ø = 8 mm) and 200 mA (red lines, diameter of the photocathode Ø = 10 mm).

For a current of 100 mA the beam radius is always smaller than one third of the aperture, which implies that a quasi-lossless beam transport should be possible. In the symmetry points, the beam radius has a minimum. The operational experience with a default current of 100 mA shows that an overall transfer efficiency close to 100 % could be obtained routinely and verifies the simulation.

The beam radius for a current of 200 mA is larger than for 100 mA due to the higher intensity and space-charge. Except near the alpha magnets, the radius is always smaller than one half of the aperture, so that a quasi-lossless transport appears to be feasible with 200 mA.

For the practical adjustment of the magnetic optics, dedicated beam diagnostics are needed. Furthermore, the assumption of a cylinder-symmetric beam profile as used for the solution of the paraxial differential equations has to be verified. The available beam diagnostics are wire scanners and luminescence monitors (see fig. 4). Near the operation chamber only wire scanners are installed in order not to degrade the vacuum in the chamber. A wire scanner consists of two wires with a diameter of 50 μm mounted on a frame. The wire scanner is able to scan both planes of the beam by moving the wires through the beam and collecting the charge. The collected charge is converted into a voltage signal, amplified, integrated (every 20 ms for a duration of 1 μs) and digitalized. A very low background, which can be reached by the use of a complete coaxial setup of the scanners, is of great importance due to the small currents of maximum 500 μA.

The two beam profiles shown in figure 6 were recorded with the first wire scanner in the transfer line. Because the emitting surface is a full circle, a homogeneous and constant current density is expected. The measurements are in good agreement with the expected profile (red curves) and legitimate the assumptions for the simulation. The slight charge redistribution is caused by inhomogeneous fringe fields of a permanent magnet of an ion getter pump.

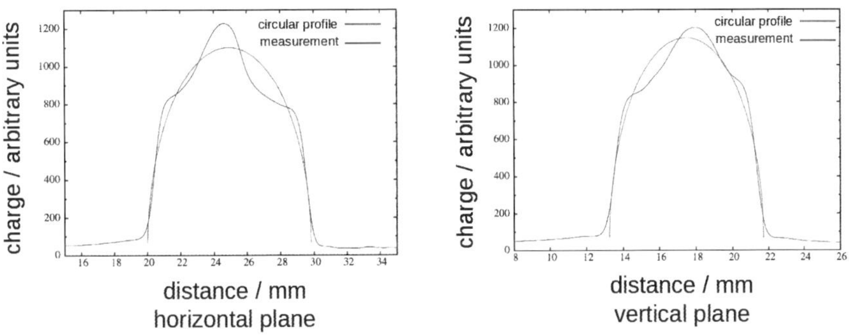

Fig. 6. Beam profiles in both planes recorded with the first wire scanner in the transfer line.

4. Conclusion

Since 2000, a source of polarized electrons has been in operation providing an 80 % polarized beam of 100 mA emitted in space-charge limitation. Measurements of the photo-emission current and the numerical simulation of the space-charge dominated beam-transport show that an intensity upgrade to 200 mA is feasible.

References

1. W. Hillert, *Eur. Phys. J. A* **28 S01**, 139 (2006).
2. A. Sokolov and I. Ternov, *Sov. Phys. Dokl.* **8**, 1203 (1964).
3. W. Hillert, The 50 kv inverted source of polarized electrons at ELSA, in *The 14th International Spin Physics Symposium, SPIN2000*, eds. W. Hillert and *et al.*, AIP Conf. Proc., Vol. 570 (AIP, New York, 2001).

4. M. E. *et al.*, The 50 kev Source of Polarized Electrons at ELSA: Past and Future, in *The 17th International Spin Physics Symposium, SPIN2007*, eds. K. Imai and *et al.*, AIP Conf. Proc., Vol. **915** (AIP, New York, 2006).
5. J. Buon, *Beam Phase Space and Emittance*, tech. rep., CERN Yellow Report (1994), CERN-94-01.
6. A. Septier, *Focusing of Charged Particles* (Academic Press, New York, 1969).

LIFETIME MEASUREMENTS OF DBR AND NONDBR PHOTOCATHODES AT HIGH LASER INTENSITIES

E. Riehn*, V. Tioukine and K. Aulenbacher

Institut für Kernphysik, Johannes Gutenberg–Universität,
55099 Mainz, Germany
**E-mail: riehn@kph.uni-mainz.de*
www.kph.uni-mainz.de/B2

The effects of intense laser irradiation on the lifetime of superlattice photocathodes with and without Distributed Bragg Reflector (DBR) have been studied. Both types were exposed to different laser intensities in the range of 30 mW to 800 mW at a wavelength of 808 nm without producing any photocurrent. The observed lifetime is dependent on the laser power and also on the history of every cathode. It was demonstrated that the lifetime of DBR photocathodes at a laser intensity of 300 mW is, by a factor of ≈ 7, higher in comparison to the nonDBR ones.

Keywords: Distributed Bragg reflector; lifetime; superlattice photocathodes.

1. Introduction

Experiments at future electron beam facilities like MERHIC[1] may require highly polarized currents in the range of 30 mA to 100 mA, which is orders of magnitude larger than presently available. Since the photocathode in such a source should be operational for at least one lifetime (i.e. quantum yield (QY) drops to $\frac{1}{e}$ of its primary value), the required average laser powers may reach levels of >50 W if we assume state of the art QY of about 1 %. The cesium activation layer of NEA (negative electron affinity) cathodes only tolerates a very moderate increase of temperature above room temperature, hence requiring sufficient cooling, which in turn poses additional technical challenges for the construction of such an electron source.

In this article we provide evidence that the problem may be alleviated by avoiding unnecessary absorption in the structural components of the photocathode, which do not actively contribute to photoemission. For today's highly polarized structures such additional losses take place in the substrate, since absorption in the active layer is only of the order of 10 % to 20 %. One solution to this problem is to extract unabsorbed light from the cathode by back-reflection, which can be achieved by growing the active layer on top of a Distributed Bragg Reflector (DBR).[2,3]

In the following, we compare the laser power tolerance of InGaAlAs/GaAlAs superlattice active regions with (DBR-type) and without (nonDBR-type) such a reflector. These cathodes were produced by the Joffe Institute in St. Petersburg, Russia.

2. Structure of SuperLattice (SL) photocathodes

2.1. *nonDBR-type*

As shown in figure 1 (left side) SL photocathodes typically consist of five layers with different functions. At the very bottom there is a GaAs-substrate as the supporting material. It is followed by a Buffer Layer (BL) which prevents dislocations and other defects from spreading from the substrate into the Active Layer (AL). In the present case, the AL itself is a strained superlattice structure of 25 alternating $In_{0.2}Al_{0.19}Ga_{0.61}As$ and $Al_{0.4}Ga_{0.6}As$ layers. The AL is covered with a 6 nm wide, highly Be-doped layer which provides a strong band bending within its width. This fact, in conjunction with the work function lowering by the Cs:O[a] layer, allows to achieve NEA conditions, which is the prerequisite for efficient extraction of conduction band edge electrons from the structure. The GaAs layer also prevents oxygen from the surrounding residual gas from reaching the AL and reacting with its aluminum compounds. When irradiated with a laser only a small amount of the light is absorbed in the AL, whereas most of the photons get absorbed in the substrate.

We believe that only a small fraction (6 %)[4] of the absorbed energy leaves the crystal by radiative recombination of the electrons (photoluminescence), whereas the reminder is transferred to crystal lattice as heat. At elevated temperatures the Cs:O structure will undergo irreversible structural changes which may be visualized as "evaporation". This leads to an accelerated decrease in QY until no reasonable operation is possible.

2.2. *DBR-type*

The reduction of photo absorption within the substrate can be realized by embedding a so called DBR-structure within the cathode while the remain-

[a] Cs:O represents any combination Cs_xO_y of cesium and oxygen.

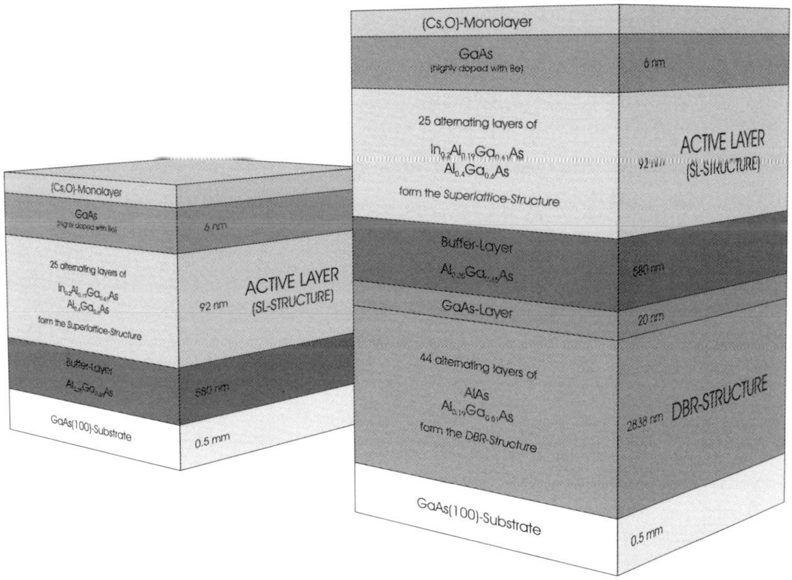

Fig. 1. Left: Structure of the nonDBR sample SL 7-395 (layer thickness not to scale). The undermost layer consists of GaAs and has a thickness of 0.5 mm. The buffer layer, a compound of $Al_{0.35}Ga_{0.65}As$, is 580 nm thick and adapts the active layer to the substrate. The active layer consists of 25 alternating layers of $In_{0.2}Al_{0.19}Ga_{0.61}As$ and $Al_{0.4}Ga_{0.6}As$ and has a thickness of 92 nm. The following 6 nm thin Be-doped GaAs layer seals the AL from the residual gas. The monolayer of cesium and oxide is crucial for electron emission and needs to be renewed periodically. Right: Structure of the DBR sample SL 7-396. The structure is similar to the nonDBR sample but two additional layers have been added: The 2838 nm thick DBR structure is a stack of 44 alternating layer of AlAs and $Al_{0.19}Ga_{0.81}As$ and acts as a mirror with reflectivity close to 1. A layer of GaAs with thickness of 20 nm links the DBR with the nonDBR structure.

ing layers stay unchanged. This DBR structure then acts as a mirror with almost 100 % reflectivity. Figure 1 (right side) shows the composition of a SL cathode akin to the one described in the previous section but with the additional DBR, realized by a stack of layers with thickness lambda-fourth.

The original idea behind the DBR concept was to develop a Fabry-Pérot-Interferometer (FPI) by reflecting light at the BL-DBR interface and again at the AL-vacuum interface, trapping the light between these boundaries in a resonator-like structure and therefore increasing the photo absorption within the AL if the exciting laser beam meets the resonator resonance condition. The intensity reflectivity coefficient of the DBR can be taken as 1, the second one can be calculated from the involved refractive indices and is close to 0.31.

3. Reflectivity measurements

Since the reflectivity of a DBR is close to 1 only in a bounded wavelength region, measurements have been done to determine the lower and upper limit of the DBR working range. These measurements were performed *in vacuo* but with an uncesiated surface. Figure 2 shows data for both the DBR and the nonDBR cathode.

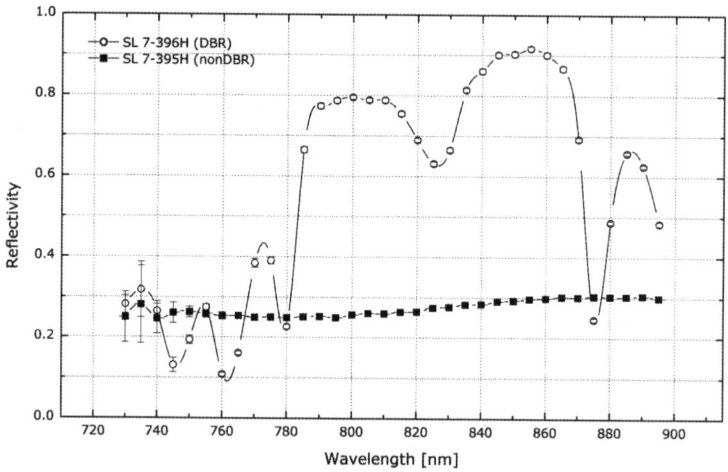

Fig. 2. Reflectivity measurement for SL 7-395 (nonDBR) and SL 7-396 (DBR). For wavelengths over 760 nm, the size of the error bars is smaller than the size of the symbols. Splines are just to guide the eye. The values of the nonDBR sample (filled squares) vary only slightly over the whole measured range. The DBR data (unfilled circles) show two important structures: The rise at 780 nm and the drop at 870 nm mark the working range of the DBR mirror, the oscillating substructure is caused by the FPI. When the incident wavelength is an integer value of the resonator length, the light is trapped in the structure and a larger amount is absorbed in the AL.

As expected, the nonDBR curve displays a mostly constant behaviour while one can clearly see a plateau between 790 nm and 860 nm in the DBR case. Assuming complete reflectivity of the mirror, our data allow us to estimate the absorption in the active region to be between <20 % (off resonance) and <40 % at resonance. In a subsequent spectrally resolved QY measurement we found that the maxima of QY were corresponding to the minima in figure 2 which demonstrates the realization of a resonator. The resulting advantage in QY in the maximum of polarization (P) with respect to a nonDBR structure is <30 %. Whereas this gain is only marginal, the real advantage of a DBR structure lies in the laser power handling capability.

4. Irradiation experiments

High power irradiations were performed with a commercial diode laser at 808 nm which is very close to the optimum working point (i.e. maximum value of P with highest QY) of the photocathode.[5] DBR and nonDBR cathodes were activated[b], transferred to the gun chamber and then irradiated at different laser powers. During irradiation the accelerating high voltage was switched off so no photo current was produced. In this mode of operation most otherwise lifetime limiting effects like ion back-bombardment, or back-streaming from gases released by the electron beam are absent. Hence, lifetime is limited only by the stability of the cathode under the given vacuum conditions and by specific other interactions which are mediated by the laser irradiation and include an increased cathode temperature. Every 30 minutes the main laser was switched off and high voltage of 100 kV was applied to the cathode. Then, a second (much weaker) laser operating at only 25 μW was turned on and photo current was extracted for about 30 seconds to determine the QY. This procedure was periodically repeated for at least one lifetime. In addition to the expected exponential decay in QY nearly every dataset showed one out of three superposed structures dependent on its history and therefore had to be purged. These structures, shown in figures 3 a) - c), are:

(a) There is no decrease in QY during the first hours/days after activation which leads to an apparent extension in lifetime.
(b) A temporary rise of QY occurs after increasing the irradiation power which again leads to an apparent lifetime extension.
(c) After decreasing the irradiation power a continuing (even accelerated) decrease of QY simulates a shorter lifetime.

We define "lifetime" as the time constant of the exponential decay which sets in after the transient phenomena in cases (a) and (b) have died out. We have avoided the situation (c) by performing a reactivation of the photocathode each time before changing from high to low intensities, in order

[b]The process of heating the cathode to 550 °C for about 30 minutes and the following application of Cesium and Oxygen is called (re)activation.

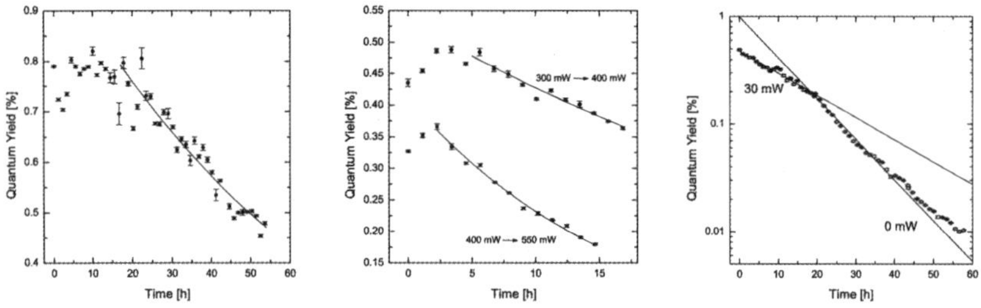

Fig. 3. Three effects adulterate the lifetime: (a) The QY remains constant for the first hours/days after activation. This effect leads to an apparent extension in lifetime and therefore first data have been discarded for extracting the lifetime. (b) After increasing the irradiation power, a temporary rise of the QY occurs. The upper curve shows the QY after switching the laser power from 300 mW to 400 mW; the lower curve shows the following measurement in which the laser power has been increased from 400 mW to 550 mW. Like above, the first data points have been refused for the fit. (c) The decrease of QY continues (is even accelerated) if the irradiating laserpower is decreased. In the shown experiment, the sample was irradiated with 30 mW for 20 hours; then the laser was switched off. The observed lifetimes do not reflect the true lifetime that would have been observed if the intensity would have been constant from the beginning.

to remove the memory-effect. Taking all this into account, the laserpower-dependent lifetime could be extracted. The results of the fits are presented in figure 4. We find that there is negligible difference between hydrogen-cleaned[7] and non-hydrogen-cleaned photocathodes. We have determined the laser induced temperature growth for our cathode holders by observing the shift of the photoluminescence spectrum with increasing irradiation power.[6] From this we get a calibration of the temperature scale (upper x-axis in fig. 4) as $\Delta T \approx 350\frac{\mathrm{K}}{\mathrm{W}}$ which is only valid for the nonDBR structure, since the DBR structure absorbs much less power. The reason for the strong increase of T is the high thermal resistance of the mechanical connection between the photocathode and its molybdenum holder. Higher thermal conductivity can be provided by choosing appropriate thermal connections and active cooling, for instance Weigel *et al.* demonstrated $\Delta T < 20\frac{\mathrm{K}}{\mathrm{W}}$ in a cryogenic environment.[8] Taking this result as "state of the art", we find that it will be technically difficult to avoid substantial increases of cathode temperature in projects requiring several tens of Watts of average power. Operation may then reach a regime where cathode lifetime is influenced — or maybe even dominated — by thermal effects, a situation which we have investigated by our irradiation experiments.

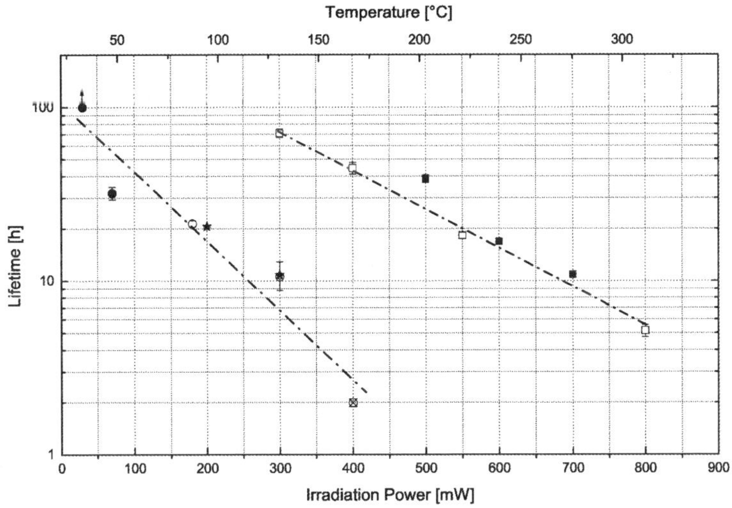

Fig. 4. Lifetimes for different cathodes and different activations at different irradiation powers. Squares show the lifetime of sample SL 7-396 (DBR), circles represent data from a SL 7-395H (nonDBR) sample whose surface was cleaned with atomic hydrogen before activation and stars are the results from a SL 7-395 (nonDBR) sample. The temperature scale on the upper axis was taken from reference 6 and then extrapolated. Please note that it is only valid for the nonDBR sample. The arrow at the leftmost data point indicates that the end of the plateau mentioned in section 4 has not been reached before the measurement was aborted. Lines are just to guide the eye.

5. Conclusion

It has been demonstrated that the presence of a DBR layer within the cathods structure has a major benefit on the lifetime of the cathode in the irradiation dominated regime. Besides, other attributes like QY[c] and P remain mostly the same.[5] With a reflectivity of close to 1, the DBR prevents the light from entering the substrate and reduces the effects ascribed to cathode heating. The advantages of this effect are two-fold:

(a) The DBR cathode allows for a factor of ≈ 3 more laserpower (i.e. beam current) at given lifetime. This factor roughly coincides with the reduced absorption in the DBR structure.

[c]QY curves show some additional (resonator caused) structure but the overall value stays equal.

(b) For a given laser power — i.e. equal laser power and photocurrent, but different temperatures — the effect is even more dramatic: If we compare the lifetimes at 300 mW, the DBR lifetime is larger by a factor of ≈ 7.

Using a DBR cathode, all of the previously mentioned improvements can be realized without any constructional change of the electron gun and the existing infrastructure can remain unaltered. Likewise, the treatment for activation is completely equivalent to the nonDBR case. Especially on facilities that cannot apply active cooling to their photo cathodes, structures containing a distributed Bragg reflector are a promising option.

Acknowledgments

We especially want to thank Y. Mamaev, L. Gerchikov and Y. Yashin (Ioffe PTI, Russia) for the stimulating experimental and theoretical input. This work was supported by the SFB 443 of the Deutsche Forschungsgemeinschaft.

References

1. I. Ben-Zvi, Polarized Electron Photocathode R&D at BNL these proceedings.
2. J. C. Gröbli, D. Oberli, F. Meier, A. Dommann, Y. Mamaev, A. Subashiev and Y. Yashin, *Phys. Rev. Lett.* **74**, 2106 (1995).
3. T. Saka, T. Kato, T. Nakanishi, M. Tsubata, K. Kishino, H. Horinaka, Y. Kamiya, S. Okumi, C. Takahashi, Y. Tanimoto, M. Tawada, K. Togawa, H. Aoyagi and S. Nakamura, *Jpn J. Appl. Phys.* **32**, L1837 (1993).
4. L. Gerchikov, *private communication.*
5. Y. Yashin, *private communication.*
6. K. Aulenbacher, V. Tioukine, M. Wiessner and K. Winkler, Status of the polarized source at mami, in *Proc. 15th Int. Spin Physics Symposium and Workshop on Polarized Electron Sources and Polarimeters (SPIN 2002)*, eds. Y. I. Makdisi, A. U. Luccio and W. W. MacKay, AIP Conf. Proc., Vol. 675 (AIP, New York, 2003).
7. V. Tioukine, K. Aulenbacher and E. Riehn, Hydrogen cleaning of superlattice photocathodes, in *Proc. 18th Int. Spin Physics Symposium*, eds. D. G. Crabb, Y. Prok, M. Poelker, S. Liuti, D. B. Day and X. Zheng, AIP Conf. Proc., Vol. 1149 (AIP, New York, 2009).
8. U. Weigel, D. Orlov, S. Kosolobov, D. Schwalm, A. Terekhov and A. Wolf, *Nucl. Instr. Meth. A* **536**, 323 (2005).

POLARIZED ELECTRON SOURCE BASED ON FZD SRF GUN

R. Xiang*, A. Arnold, P. Michel, P. Murcek and J. Teichert

Research Centre of Dresden-Rossendorf
Bautzner Landstr. 400, 01328 Dresden, Germany
**E-mail: r.xiang@fzd.de*

The state of the art DC guns with GaAs, type photocathodes have been successfully operated as polarized electron sources for accelerators, the beam emittance is regretfully very high because of the bunch compressing after the gun. Though not all of the high energy physics experiments using polarized beams are sensitive to the source emittance, but a new source with lower emittance can simplify the injector system and lower radiation load during the beam transport. A normal conducting RF gun can produce beams with low emittance, but the limit on vacuum is still an open question for the currently designed RF guns. In this paper a new type of polarized source is reported: an FZD polarized SRF gun, i.e. an FZD Superconducting RF gun (SRF gun) with a GaAs-type photocathode. The SRF gun is able to produce a 1 mA CW beam with 9.5 MeV energy and 1 mm mrad emittance. It has higher a accelerating field and thus lower emittance than DC guns, at the same time vacuum condition much better than RF guns. Based on the successful running experience in last two years, an SRF gun applied with GaAs-type cathode is considered as a promising alternative for current polarized guns. Some interesting questions will be discussed here, including the back bombardment on cathode, cathode dielectric loss in strong RF field and the cathode time-response.

Keywords: Polarized electron source; SRF gun; GaAs photocathode.

1. Motivation

The International Linear Collider (ILC) and the electron-ion collider eRHIC will give physicists a new cosmic doorway to explore energy ranges beyond the reach of today's accelerators, both of which require polarized electron sources.[1,2] The state of the art polarized electron sources are based on DC guns. For example, in Mainz (MAMI),[3] Thomas Jefferson National Laboratory (TJNAF),[4] Stanford Linear Accelerator Centers (SLAC),[5] MIT Bates,[6] ELSA,[7] and Darmstadt,[8] DC polarized guns have routinely provided beam, been commissioned for medium and high energy physics experiments.

These DC guns use GaAs-type photocathodes as electron sources, and apply DC voltage from 100 kV to 250 kV for acceleration. The vacuum in the DC guns can be kept in a range of 10^{-11} mbar, which makes GaAs photocathodes' charge lifetime up to 200 C.[9] But because the accelerating field in the gun is limited under 8 MV/m in order to avoid the field emission on the electrodes, the principal limitation of the DC-biased source is the low energy of the space-charge dominated electrons and thus large transverse emittance. Although not all of the high energy physics experiments using polarized beams are sensitive to the source emittance, a new source with lower emittance can simplify the injector system and lower radiation load during the beam transport.

A polarized RF gun, i.e. a normal conducting radio frequency gun with the GaAs photocathode, has the potential to produce electron beams with better transverse and longitudinal emittance than traditional DC guns, because of the higher surface field compared with DC guns; so RF gun is a simple, efficient injector system without any compression, and eliminates the need for RF or magnetic bunchers.[10] Up to now, tests in the RF gun have been done by SLAC[11] and BINP,[12] but the GaAs potocathode was damaged after several RF cycles because of the bad vacuum in the Cu cavity and the electron back bombardment on the photocathode. So the vacuum problem is the biggest barrier faced by the injector experts. To improve the vacuum environment, SLAC made a new design of cavity with a bigger port,[13] but there is no successful process reported up to now.

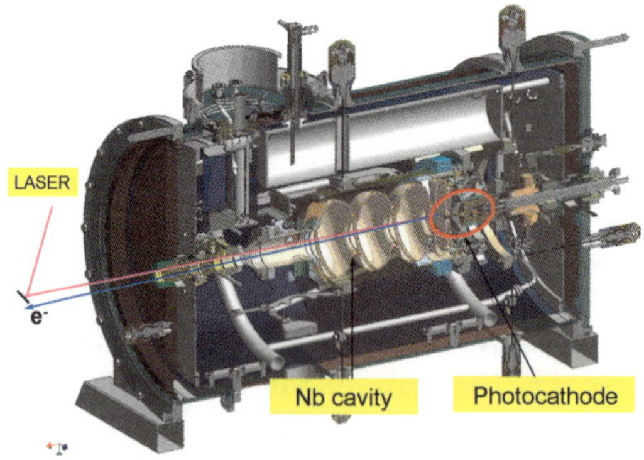

Fig. 1. The design of the FZD SRF gun cryo-module.

The idea of the polarized superconducting RF gun appears under this background. The SRF cavity itself, sinking in the liquid helium tank, is a perfect cryo-pump, which provides the best vacuum environment for the photocathode in the cavity. At the same time, in the SRF gun a peak electric field up to 50 MV/m can be achieved, which is much higher than that in DC guns (8 MV/m). Besides FZD, Brookhaven National Lab (BNL) has its SRF polarized-gun plan too. BNL tests its half-cell SRF gun with GaAs photocathode attached on the back wall.[14,15] But the dielectric RF power loss on normal conducting GaAs crystal is the potential barrier for the realization of this type of gun. In FZD-type SRF gun, the cathode is electrical and thermal isolated to the cavity, and the cathode itself is cooled with liquid Nitrogen, so the RF loss power on cathode has only ignorable effect to the cavity. Thus the FZD SRF gun combined with GaAs photocathode is a promising polarized electron gun to provide polarized electron beam with high bunch charge and low emittance.

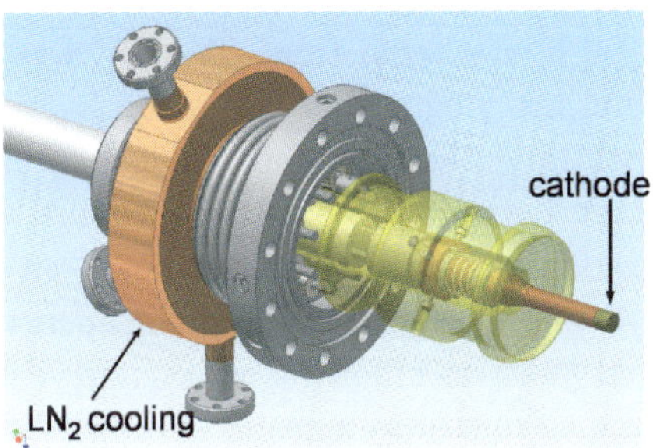

Fig. 2. Cathode supporter to isolate the cathode from the superconducting cavity.

2. FZD SRF gun introduction

The SRF gun developed within the collaboration of HZB, DESY, MBI and FZD has been in operation in FZD since 2007.[16,17] This new type of injector is a promising candidate of high current and high brightness electron source for the new light sources and big accelerator facilities, such as FEL, ERL and so on. It is able to produce 1 mA CW beam with 9.5MeV energy and 1 mm mrad emittance. At first it is based on SRF cavity technology which allows continued wave mode operation, so the repetition rate and thus the average current can be much higher than in the normal conducting RF gun; secondly, good niobium cavity treatment technique makes it possible to operate the cavity at a very high acceleration field in the cavity, and on the cathode surface, to reduce the influence of the space-charge effect in the low energy electron bunch; thirdly, the electron bunches are produced from the photocathode with the fast response time driven by the pico-second laser, so the bunch length can be as short as several ps, and the structure of the bunches could be controlled by shaping the laser pulses.

To combine the normal conducting Cs_2Te photocathode and the superconducting niobium cavity, the cathode supporter in the centre of the half cell cavity (fig. 2) is designed.[18] The cathode is isolated from the SC cavity by a vacuum gap and cooled further with liquid nitrogen. The gun operation proves that this design is efficient and successful. There is no obvious degradation found in the cavity quality, since the RF measurement result shows that the cavity with the cathode inside has the quality factor Q_0 the same as the virgin cavity and there is no visible degradation of the cavity performance after ~1000 hours of operation.[19]

Up to now, four Cs_2Te photocathodes have been applied in the SRF gun for more than 100 hours beam time. The QE was measured at more than 1 % and the lifetime longer than two months.[20] The bunch charge reached 300 pC/bunch, and the laser pulse length is 60 ps after the gun. The maximum average current achieved was 18 μA at 125 kHz repetition rate. At the gradient of 6 MV/m, the beam energy was measured with dipole magnet as maximum 3 MeV and the energy spread as 25 keV. With the gun solenoid and two following screens downstream, the normalized transverse emittance has been measured, which is 3 mm mrad for 77 pC/bunch. For the higher bunch charge, the slit masks method is in the developing.

3. Polarized gun based on the FZD SRF gun

In principle the NEA GaAs-type photocathode can be applied directly in the FZD SRF gun instead of Cs_2Te. The sample holder showed in figure 3 will be used to install the GaAs crystal onto the cathode body. The indium wire in between is used for more flexible mechanical structure and better thermal contact. The bulk GaAs(Cs, O) crystal is planned to be tested because it is much cheaper and easier than the strained semiconductor and the superlattice crystal. But there are some physics and technique problems that arise from the application of the GaAs crystal in the superconducting RF cavity.

3.1. *Ion bombardment*

In the SRF gun cavity, the accelerated photoelectrons will hit and ionize the rest gas molecules, which results in more charged particles in the RF field. Among them, some ions and electrons appearing in proper RF phase will fly back to the cathode surface and hit the cathode film with large momentum, which is harmful for the photocathode material. In the DC gun with GaAs photocathode the ion back bombardment is a serious problem.[21,22] To research the ion bombardment of GaAs photocathode in the FZD SRF gun, a simulation has been done for the 3.5 cell gun. The discussion is only based on the electron-ion impacting. The rest gas in the gun is assumed as uniform. H are the only ions in this simulation, because H is the most dangerous element in the gun, which can be ionized from H_2, water and alcohol molecules. Only the ions near the cathode surface are found to fly back to the cathode.

If the electron beam is 77 pC/bunch, 13 MHz repetition rate, the H^+ current on the cathode is 0.28 fA, which means 1750 ions will hit the cathode surface in one second. The mean momentum of these ions is about 1 MeV/c, or kinetic energy 0.533 keV.

For the polarized DC gun, the ion current on the cathode can be estimated in the same way. The partial gas pressure of H_2 in the DC polarized-gun is assumed as 1.6×10^{-11} Torr,[21] so the H_2 gas density is 5.1×10^{11} m^{-3}. The ion current on the cathode is 3.57 fA, which is over 10 times more than the ion current in the SRF gun and the impact energy is 0$\sim$100 keV.

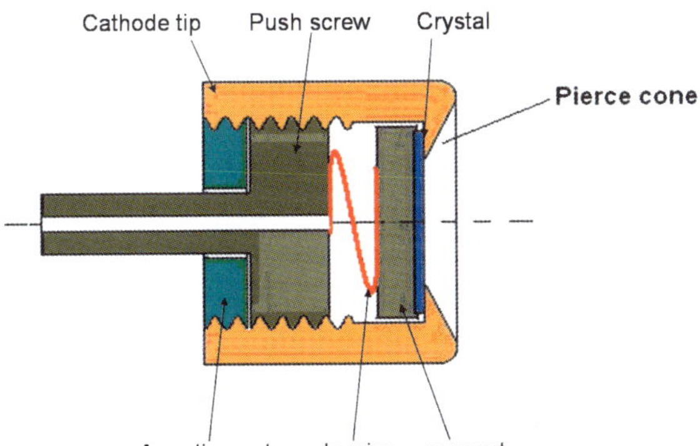

Fig. 3. The designed GaAs crystal holder. Indium wire is used for more flexible mechanical structure and better thermal contact.

3.2. *Dielectric RF power loss on GaAs film*

When the FZD gun is applied with a GaAs photocathode, the RF power will dissipate on the dielectric cathode material. For the dielectric material, complex dielectric permittivity is

$$\varepsilon = \varepsilon' + i\varepsilon'' = |\varepsilon| e^{i\delta}, \tag{1}$$

where ε' is the ac capacitivity, ε'' the dielectric loss factor, and δ the dielectric loss angle. AC capacitivity $\varepsilon' = \varepsilon_r \varepsilon_0$, $\varepsilon_0 = 8.85\times10^{-12} A \cdots / V{\cdot}m$, is the permittivity of free space. For the bulk GaAs, $\varepsilon_{\mathrm{r,GaAs}} = 12.74@75.6$ K, $tan\delta_{\mathrm{GaAs}} = 0.0004$, then the RF power loss

$$P = E^2 \omega A d \varepsilon_r \cdot \varepsilon_0 {\cdot} tan\delta, \tag{2}$$

where E is the electric field on the cathode surface, $\omega = 2\pi f$ the RF angular frequency, A the cathode area, and d the thickness of the dielectric material.

Table 1. The calculation results of dielectric loss on GaAs film and the radiation to cavity.

Thickness (μm)	Diameter (mm)	E cathode (MV/m)	Power loss (W)	Radiation to cavity (mW)
50	5	20	14.6	10
100	5	20	29.2	30
150	5	20	43.8	65
350	5	20	102	450

Table 1 gives the calculation results of the dissipation on the GaAs semiconductor for the variable cathode size and the electric field in the cavity. If the cathode contacts the superconducting cavity directly without extra cooling, like the design of the BNL SRF gun,[15] the cavity will quench and is not a superconductor any longer. However, in the FZD SRF gun, the cathode is isolated from the SC cavity and cooled with LN_2. This power loss can only result in the temperature rising of the cathode and hence very little radiation to the SC cavity. From the relationship of the cathode temperature and the radiation power to cavity,[23] this radiation power can be calculated as in the last column in table 1. Because the liquid helium system can tolerate tens of watt power dissipation, this radiation will not affect the cavity operation.

3.3. *Response time of the GaAs photocathode*

As known, GaAs bulk material has a slow response time in the level of 30 ps[24] and a long tail[25] which means the bunch will occupy 14° phase space in the 1.3 GHz acceleration RF field. Because the electrons in the head and in the tail of the bunch see much different field, the beam parameters will be much different from the ones of beam with short bunches. In order to optimize the beam parameters, the electron bunch length must be reduced and the bunch temporal profile must be shaped. The first method is to use thinner crystal, i.e. 50 μm, which reduces the depth of the photoelectric effect and of course the time taken for the photoelectrons to disperse. A second method is to add a buffer layer with a higher band gap (for example $Al_xGa_{1-x}As$) under the active layer, then the photoelectric effect will be limited only in the thin active layer and thus reduce the dispersal time. The disadvantage of both methods is that the quantum efficiency of the cathode will be reduced too.

A simulation has been done to watch the change of the emittance versus the variation of the bunch's temporal structure.[26] The transverse emittance ε_x becomes worse when the tail is longer, but the curve of the longitudinal emittance ε_y versus the tail length has the binomial shape. The best emittance ε_y is 37π mm mrad with bunch charge of 0.77 nC, bunch length of 20 ps, and beam radius of 1.3 mm. The corresponding ε_x is 1.6π mm mrad.

4. Summary and outlook

The SRF gun developed in FZD has been successfully operated with a Cs_2Te photocathode. If combined with GaAs photocathode, the FZD SRF gun will be a promising polarized electron gun to provide polarized electron beam with high bunch charge and low emittance. The relevant physics problems have been discussed in this paper. In the near future, the bulk p-type GaAs (Zn highly doped) semiconductor will be activated and tested in the SRF gun. For this goal, the vacuum in the present preparation chamber for Cs_2Te will be improved to eXtreme High Vacuum (XHV) 10^{-11} mbar, a light source in the range of 600 nm$\sim$800 nm with polarizer will be built, and a Mott polarimeter with a Wien filter will be developed in FZD.

References

1. `http://www.linearcollider.org/cms/` .
2. `http://quark.phy.bnl.gov/~raju/eRHIC.html` .
3. K. Aulenbacher *et al.*, *Nucl. Instr. Meth. A* **391**, 498 (1997).
4. M. Poelker *et al.*, Status of 100 kV DC High Voltage Polarized Electrons Guns at CEBAF, in *Proc. ERL07, Daresbury Laboratory, UK, 2007.*
5. R. Alley *et al.*, *Nucl. Instr. Meth. A* **365**, 1 (1995).
6. M.Farkhondeh *et al.*, MIT-Bates polarized source, in *Proc. Workshop on*

Polarized Electron Sources and Polarimeters PESP2002, AIP Conf. Proc. **588** (AIP, New York, 2003).
7. Hillert, *Eur. Phys. J. A* **28**, 139 (2006).
8. C. Hessler *et al.*, Status of the polarized-electron gun at the s-dalinac, in *Proc. EPAC 2006, Edinburgh, Scotland.*
9. C. Sinclair *et al.*, *Phys. Rev. ST Accel. Beams* **10**, 023501 (2007).
10. D. Edwards, Polarized RF guns for Linear Colliders, in *Proc. Workshop ICFA, Fermi National Accelerator Lab., USA, 2001.*
11. K. Aulenbacher *et al.*, Initial test of an RF gun with a GaAs cathode installed, in *Proc. 12th Int. Sym. On High-Energy Spin Physics, NIKHEF, Amsterdam, 1996.*
12. A. Aleksandrov *et al.*, High power test of GaAs photocathode in RF gun, in *Proc. EPAC98, Stockholm, 1998.*
13. J. Clendenin *et al.*, RF guns for generation of polarized electron beams, in *Proc. 36th ICFA Advanced Beam Dynamics Workshop (NANOBEAM 2005), Kyoto, Japan.*
14. D. Holmes *et al.*, Superconducting RF photocathode gun for low emittance polarized electron beams, in *Proc. PAC07, Albuquerque, New Mexico, 2007.*
15. I. J. Kewisch *et al.*, The polarized SRF gun experiment, in *Proc. PESP2008, CEBAF Center, JLab, Newport News, 2008.*
16. J. Teichert *et al.*, Initial Commissioning Experience with the Superconducting RF Photoinjector at ELBE, in *Proc. FEL08, 2008.*
17. J. Teichert *et al.*, First Operation Results of the Superconducting Photoinjector at ELBE, in *Proc. EPAC08, 2008.*
18. F. Staufenbiel *et al.*, *Physica C* **441**, 216 (2006).
19. R. Xiang *et al.*, Running experience of FZD SRF photoinjector, in *Proc. FEL09, Liverpool, UK, 2009.*
20. R. Xiang *et al.*, Properties of normal conducting cathodes in the FZD Sperconducting gun, in *Proc. SRF2009, Berlin, Germany, 2009.*
21. J. Grams *et al.*, Ion back-bombardment of GaAs photocathodes inside DC high voltage electron guns, in *Proc. PAC05, Knoxville, Tennessee 2005.*
22. T. Siggins *et al.*, *Nucl. Instr. Meth. A* **475**, 549 (2001).
23. F. Staufenbiel *et al.*, Test of the photocathode cooling system of the 3 1/2 cell SRF gun, in *Proc. SRF 2005.*
24. P. V. Logatchev *et al.*, Measurement of time response of laser-triggered GaAs photocathode, in *Proc. EPAC1994, 1994.*
25. J. Schuler *et al.*, in *Proc. of the 14th International Spin Physics Symposium, 2000.*
26. R. Xiang *et al.*, Low emittance polarized electron source based on superconducting RF gun, in *Proc. SRF 2007.*

MAJOR ADVANCES IN SEOP OF ^{3}HE TARGETS

P. Dolph*, K. Mooney, V. Nelyubin, J. Singh, A. Tobias and G. Cates

Department of Physics, University of Virginia,
Charlottesville, VA 22903, USA
**E-mail: pad8c@virginia.edu*
www.virginia.edu

Major advances in Spin-Exchange Optical Pumping (SEOP) of polarized ^{3}He targets include the introduction of line-narrowed lasers, hybrid-alkali alloys, and convection driven recirculation.

Keywords: ^{3}He; SEOP; hybrid alkali.

1. Introduction

Spin-Exchange Optical Pumping (SEOP) uses circularly-polarized laser light to polarize an alkali metal; the alkali metal in turn transfers its polarization to a noble-gas nucleus such as ^{3}He or ^{129}Xe.[1] Until recently SEOP usually employed rubidium (Rb) and broadband lasers (2.0 nm Full Width Half Maximum (FWHM)), which can achieve 40 % ^{3}He polarization in large target cells (1-3 liters). Recent advances, including the introduction of hybrid alkali mixtures and spectrally narrow lasers (0.2 nm FWHM), have produced polarizations in excess of 70 %.

Because both the electrons and protons in ^{3}He are paired, polarized ^{3}He provides what is effectively a polarized neutron target when it is used in the context of electron scattering.[2] Such targets can be used to study a wide variety of topics including neutron form factors and the internal spin structure of the neutron. In our case we use a two chambered cell in which gas can freely diffuse between pumping chamber (where SEOP occurs) and target chamber (where the electron beam interaction occurs). This talk will outline recent improvements in SEOP and future directions in target cell design.

2. Spin-Exchange Optical Pumping

Traditional (pure Rb) optical pumping uses circularly polarized light at 795 nm to induce D1 transitions (see fig. 1). Through collisional mixing, the two excited states populate evenly; collisions with nitrogen molecules non-radiatively relax the excited electrons.[3] Although both ground states are being repopulated, only one of the two states is being depopulated; consequently, the Rb vapor becomes polarized.

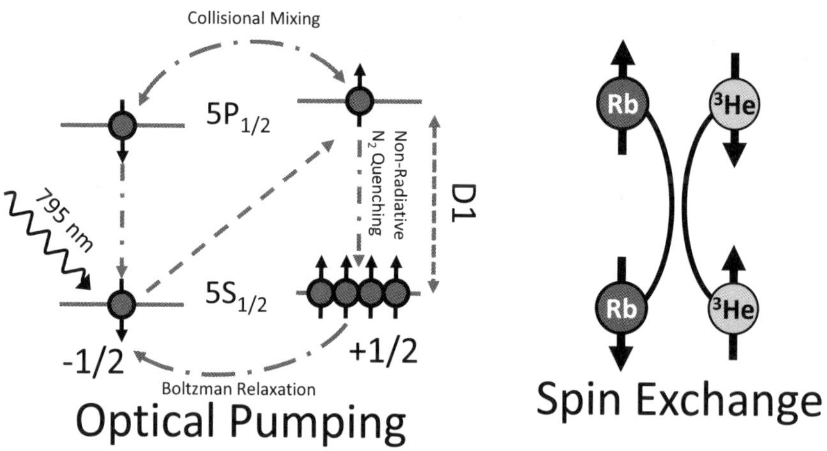

Fig. 1. Spin-Exchange Optical Pumping (SEOP) of rubidium.

The alkali polarization (assuming a build up from zero polarization) is given by

$$P_A(t) = P_{light}\frac{R}{R+\Gamma_{Rb}}\left(1-e^{-(R+\Gamma_{Rb})t}\right), \tag{1}$$

where P_{light} is the polarization of the laser light, R is the optical pumping rate

$$R = \int \phi(\nu)\cdot\sigma(\nu)d\nu \approx 35\text{ kHz per }25\text{ W}/23\text{ cm}^2/2.0\text{ nm (broadband)}, \tag{2}$$

and Γ is the Rb relaxation rate

$$\Gamma_{Rb} = k_{\mathrm{Rb}}[\mathrm{Rb}] + k_{^3\mathrm{He}}[^3\mathrm{He}] + k_{\mathrm{N}_2}[\mathrm{N}_2] \approx 0.8\text{ kHz }(200^\circ\text{C \& }[^3He] = 7\text{ amg}). \tag{3}$$

In the above equations, $\phi(\nu)$ is the photon flux, $\sigma(\nu)$ is the photon absorption cross section, k_X is the Rb-X relaxation rate constant, and $[X]$ is the number density of species X. From these equations, it appears as though the alkali polarization approaches 1; however, because the photon flux decreases as the laser light passes through the cell (by transfering its angular momentum to the alkali vapor), the optical pumping rate (and therefore the alkali polarization) decreases as the light passes through the cell. It should also be noted that the alkali polarization reaches its equilibrium value on a time scale of tens of microseconds.

The incident photon flux, $\phi(\nu)$ of the laser light has a gaussian spectral profile as well as a gaussian spatial profile.

$$\Phi(\nu, \vec{r}) = \Phi_0(\vec{r})G(\nu), \tag{4}$$

$$\Phi_0(\vec{r}) = \frac{P}{h\nu}\frac{2}{w^2\pi}e^{-2r^2/w^2}, \tag{5}$$

$$G(\nu) = \frac{1}{\sqrt{2\pi}\sigma_l}e^{-(\nu-\nu_l)^2/2\sigma_l^2}, \tag{6}$$

where ν_l is the laser central frequency; σ_l is the gaussian width of the laser, related to the FWHM by $\sigma_l = \Delta\nu/2\sqrt{\ln 4}$; P is the power of the laser; and w is the beam waist. The absorption cross section, $\sigma(\nu)$ has a lorentzian spectral profile whose width is pressure-broadened by collisions with ^{3}He: 4

$$\sigma(\nu) = \frac{\pi r_e c}{3}\frac{1}{\pi}\left(\frac{\frac{\Gamma_A}{2}}{(\nu-\nu_0)^2 + \left(\frac{\Gamma_A}{2}\right)^2}\right) \tag{7}$$

The absorption line width, $\Gamma_A \approx .04$ nm/amg $\cdot$ [^{3}He] is ≈ 0.3 nm under our operating conditions.

The ^{3}He polarization is limited by the alkali polarization. Techniques for increasing the alkali polarization will be discussed.

3. Improvements in optical pumping

The optical pumping rate (eq. 2) can be increased by better matching the photon-absorption cross section of the alkali metal to the spectral width of the laser light; this can be accomplished by using line-narrowed lasers. In the past, we've used broad-band Coherent FAP diode lasers with a FWHM of approximately 2.0 nm. Recently, we've acquired several Volume Bragg Grating (VBG) locked Spectra-Physics line-narrowed Comet lasers, each with a FWHM of approximately 0.2 nm. The two systems each provide approximately 25 W of laser light; however, because the Comet laser spectrum is better matched to the alkali absorption cross section, the initial optical pumping rate of the Comet is about five times larger than the FAP's. When line-narrowed lasers are used in our lab, ^{3}He polarizations increase from 40 % to 60 %.

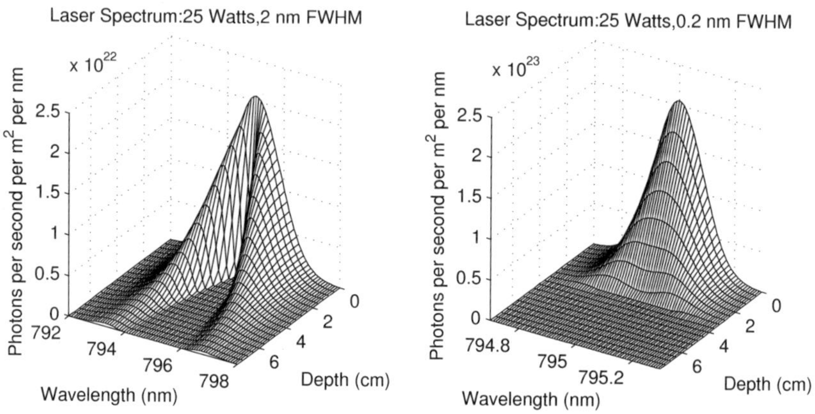

Fig. 2. Comparison between line-narrowed (right) and broadband (left) laser spectra. The simulation was carried out on a spherical cell (7.5 cm diameter) with 25 W of laser light. Note the different Y- and Z-scales.

The simulated plots (fig. 2) show the difference in $\phi(\nu)$ between line-narrowed (right) and broadband lasers (left). Although the two spectra have the same total number of photons, the light from the line-narrowed laser is more resonant (note the different Z-scale); consequently, all of the Comet light is absorbed, while the off-resonance tail ends of the FAP light exit the cell unabsorbed. Since the line-narrowed light is being used more efficiently, a higher alkali polarization can be achieved.

4. Improvements in spin exchange

Rb-^{3}He spin exchange is a relatively inefficient process [5] – under our typical operating conditions, it takes roughly 50 polarized Rb atoms to polarize one ^{3}He atom. Spin exchange with a lighter alkali metal is more efficent – it would take roughly ten polarized potassium (K) atoms or six polarized sodium (Na) atoms to polarize one ^{3}He atom; however, optical pumping of lighter alkali metals is more difficult. Aside from the need to invest in new laser systems, lighter alkali metals have lower vapor pressure curves and therefore must be heated to a higher temperature (which is experimentally difficult). Additionally, whereas the D1-D2 fine structure splitting for Rb

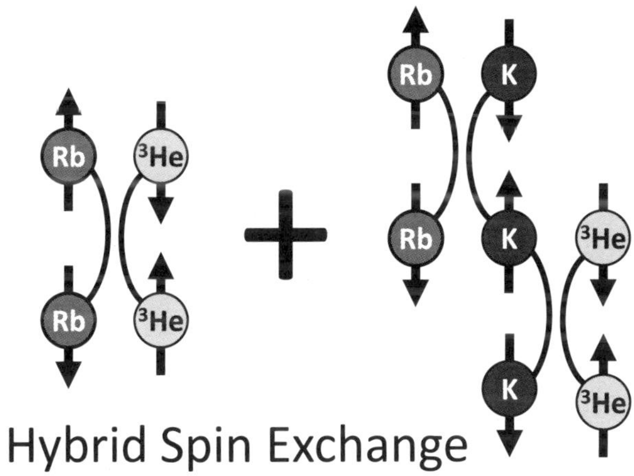

Fig. 3. Hybrid spin exchange.

is 15 nm, for K it's 3.5 nm, and for Na it's 0.4 nm. Such small splittings can lead to inefficient optical pumping. The hybrid technique attempts to solve this problem.[6]

The hybrid technique employs a mixture of Rb and K; the lasers optically pump the former and it shares its polarization with the latter with minimal losses. Potassium's smaller electron cloud enables it to more efficiently transfer its polarization to ^{3}He. Both atoms spin exchange with ^{3}He (see fig. 3); however, for a fixed amount of laser light, a lower Rb density can be used which allows the laser light to penetrate deeper into the cell. In order to compensate for the lower Rb density, the K density is increased and therefore, K carries out the majority of spin exchange. The laser polarized Rb indirectly transfers its spin more efficiently, and higher ^{3}He polarizations (≈ 55 %) are attained with shorter spinup times.

To make use of the hybrid technique, an appropriate balance between light-absorbing Rb density and spin-exchange K density needs to be made. To that end, we define the number density ratio, $D = [\mathrm{K}]/[\mathrm{Rb}]$. We've found that a D between three and seven (see fig. 4) gives the best results (highest ^{3}He polarization).

Figure 5 illustrates the improvements line-narrowed lasers and hybrid-alkali mixtures bring to SEOP. Qualitatively, hybrid-alkali mixtures stretch out the polarization roll-off curve (a lower Rb density allows for deeper laser penetration). Line-narrowed lasers change the shape to a sharper roll-off

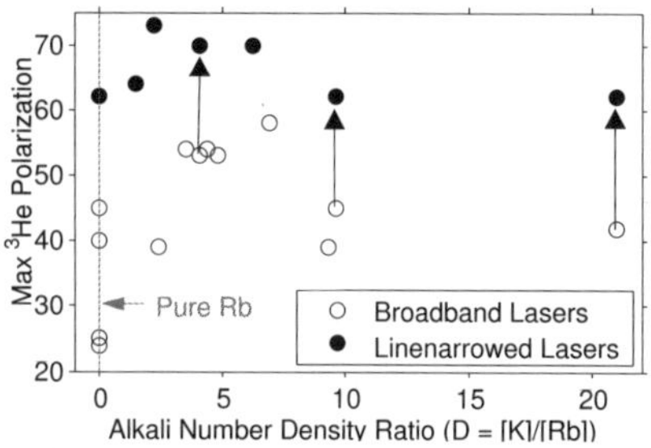

Fig. 4. Target cell performance.

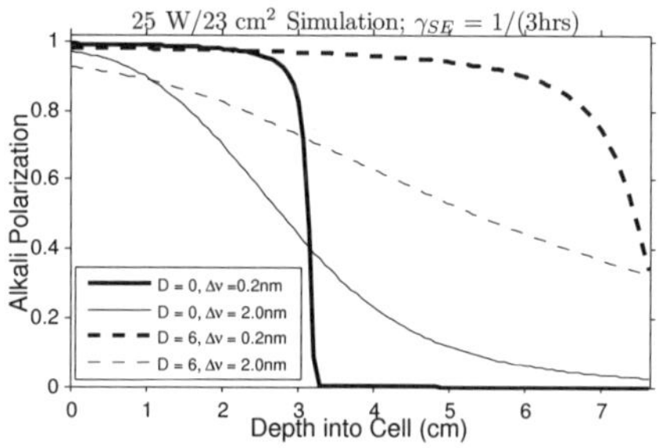

Fig. 5. Theoretical alkali polarization.

(the more resonant laser light is absorbed quicker). When used in conjunction, a larger cell can be more highly polarized. It should be noted, however, that all four methods perform the same when used on relatively large (when all the light has been absorbed) or small cells (before much of the light can be absorbed).

5. Future directions in ^{3}He targets

The figure of merit in a ^{3}He-electron-scattering-asymmetry experiment is proportional to the square of the ^{3}He polarization. Consequently, doubling the ^{3}He polarization decreases the required beam time by a factor of four. Often, beam time can also be reduced by increasing beam current; however, since the electron beam strongly relaxes the ^{3}He, the maximum beam polarization is limited by the time it takes for freshly polarized gas to diffuse from the pumping chamber to the target chamber. We plan to increase the beam current by an order of magnitude without sacrificing ^{3}He polarization by drastically increasing the mixing time. We've successfully tested a new target cell in which we've driven a convection current that recirculates the gas two orders of magnitude faster than diffusion (see fig. 6).

Like current incarnations of target cells, the next generation target cell will have two chambers; however, instead of a single transfer tube connecting them, the new target cell will have two transfer tubes. One of the transfer tubes will be heated to a controlled temperature, while the other will be cooled. This temperature difference will lead to a density difference, which in turn will drive the convection current,

$$\Delta\rho g V = F_{retarding}(v) \tag{8}$$

$$\Delta\rho = \frac{m_{\mathrm{He}} P}{k_B}\left(\frac{1}{T_{drive}} - \frac{1}{T_{cold}}\right) \tag{9}$$

In the above equations, V is the volume of transfer tube gas that is being heated to T_{drive}; $F_{retarding}(v)$ is the retarding frictional forces exerted on the gas. The convection current can be increased by increasing the driving temperature (see fig. 6).

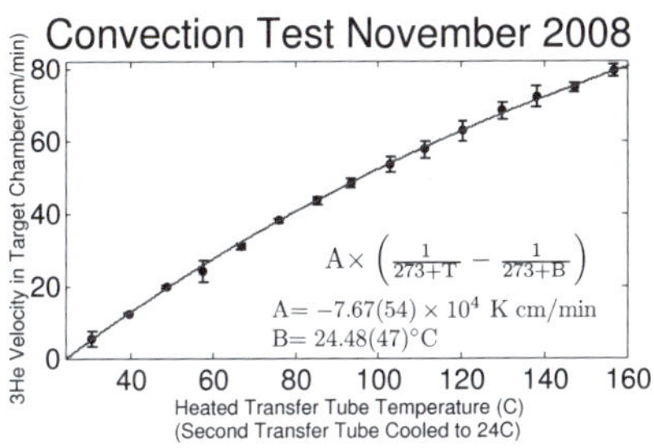

Fig. 6. Test cell performance.

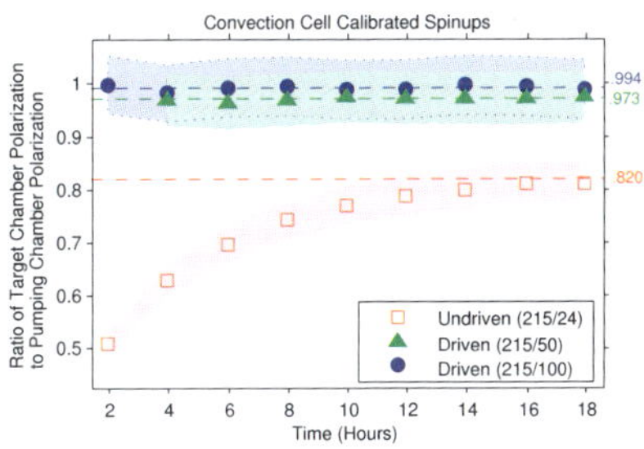

Fig. 7. Convection vs diffusion.

A consequence of rapidly recirculating the gas is that both the pumping and target chambers achieve nearly the same polarization. In traditional diffusion cells, the target chamber polarization was often 95 % that of the pumping chamber. Since the nuclear physics experiment occurs in the target chamber, it is important that this polarization be as high as possible. Figure 7 illustrates the improvement that convection brings to target chamber polarization. In addition to the target chamber polarization saturating closer to the pumping chamber polarization, it does so faster (on the time scale of minutes). In this case, the non-driven (diffusion based) polarization build-up saturates much lower than a typical target cell because the cell used had very long transfer tubes.

The proposed higher electron beam current will significantly stress our traditional all-glass target cell design. We are working on a new design that incorporates a gold-coated aluminum target chamber alongside our robust glass pumping chamber. ^{3}He relaxation rates of various metals (Al = 6 hrs, Au = 20 hrs) have been measured.[7] We've tested a gold-coated glass sphere, but have been unable to measure a lifetime much longer than an hour.

References

1. T. G. Walker and W. Happer, *Rev. Mod. Phys.* **69**, 629 (1997).
2. B. Blankleider and R. M. Woloshyn, *Phys. Rev. C* **29**, 538 (1984).
3. S. Applet, A. B.-A. Baranga, C. Erickson, M. Romalis, A. Young and W. Happer, *Phys. Rev. A* **58**, 1412 (1998).
4. M. Romalis, E. Miron and G. Cates, *Phys. Rev. A* **56**, 4569 (1997).
5. E. Babcock, I. Nelson, S. Kadlecek, B. Driehuys, L. W. Anderson, F. W. Hersman and T. G. Walker, *Phys. Rev. Lett.* **91**, 123003 (2003).
6. W. Happer, G. Cates, M. Romalis and C. Erikson., U.s. patent no. 6318092 (2001).
7. A. Deninger, W. Heil, E. W. Otten, M. Wolf, R. K. Kremer and A. Simon, *Eur. Phys. J. D* **38**, 439 (2006).

A STUDY OF POLARIZED METASTABLE ^{3}HE BEAM PRODUCTION

Yu. A. Plis*, E. D. Donets, V. V. Fimushkin, I. V. Gapienko,
Yu. V. Prokofichev and V. P. Vadeev

Joint Institute for Nuclear Research,
Dubna, Russia
**E-mail: plis@nusun.jinr.ru*
www.jinr.ru

The problems of metastable ^{3}He beam production are considered. There is an essential difference between metastable ^{3}He and hydrogen or deuterium atoms concerning the radio-frequency transitions of atomic states. The Schroedinger equations are given in the uncoupled basis $|\psi_e, \psi_h>$ and also in the basis of stationary states. The results of computer simulations agree with the published data. This paper has shown a possibility to get positive and negative values of helion polarization by using two types of the weak field transitions in the metastable ^{3}He atoms.

Keywords: Polarization; ion beams; ^{3}He; atomic beams.

1. Introduction

Our goal is to make a source of polarized $^3He^{++}$ ions (helions) on the basis of the polarized deuteron source for the JINR Accelerator Complex. The RF dissociator is fed with helium-3 gas to produce ^{3}He atoms in the metastable 2^3S_1 state. Stern-Gerlach separation with a sextupole magnet and RF transitions in a weak magnetic field are used for nuclear polarization of the metastable atoms. Ionization to $^3He^{++}$ and accumulation of the polarized helions may be carried out by the Electron Beam Ion Source (EBIS).[1]

Earlier, the Laval University group (Canada)[2] polarized ^{3}He atoms in the metastable state 2^3S_1 (a lifetime of 7860 s) and then ionized them to $^3He^+$. A cold cathode discharge source of metastable atoms produced a flux of 6×10^{15} atoms/s sterad with an average velocity of 2.5×10^5 cm/s. It is known that the ionization potential of metastable atoms is quite low, 4.6 eV, compared to 24.6 eV for atoms in the ground state. The ionizer of the Laval University group discriminated the metastables with respect to the ground-state atoms, thus producing $^3He^+$ from the nuclear polarized metastable atoms. The subsequent ionization to $^3He^{++}$ was effected by stripping in the base of a Van de Graaf accelerator at 7.5 MV.

We have one more inspiring example from Saclay. The SATURNE group[3] reported the results of the experiments performed with the known HYPERION polarized ion source. Helium gas was passing through a dissociator of the polarized proton source where in the RF discharge its small part was excited into the 2^3S_1 metastable state. The ionizer with a reflex electron beam yielded, as it was supposed, $^3He^+$ ions with a pulsed beam current of 50 μA and a pulse duration of 1 ms. The difference between the sextupole magnet "on" and "off" modes was 10 μA. We used this value to estimate the helion intensity in the helion ion source being offered in this paper.

The spin-dependent part of the hamiltonian for helium-3 atoms in the metastable state 2^3S_1 with the electron spin moment equal to 1, is

$$\hat{H} = -\mu_J \vec{J}\vec{B}(t) - \mu_h \vec{\sigma_h}\vec{B}(t) - \frac{1}{3}\Delta W \vec{\sigma_h}\vec{J}, \tag{1}$$

where $\vec{\sigma_h}$ are the Pauli spin matrices of a helion, $\vec{J}$ are the electron spin matrices ($J = 1$), $B(t)$ is magnetic field power, ΔW is hyperfine splitting; $\Delta W = 4.4645 \times 10^{-24}$ J $= \hbar \times 4.2335 \times 10^{10}$ rad/s, $\mu_J = 2\mu_e = -1.85695275 \times 10^{-23}$ J/T $= -\hbar \times 1.76085977 \times 10^{11}$ rad/s T, $\mu_h = -1.07455 \times 10^{-26}$ J/T $= -\hbar \times 1.0189 \times 10^8$ rad/s T.

Now we consider a case of static magnetic field B directed along the z-axis. The wave functions of the hyperfine states $\Psi(F, M_F)$ at this magnetic field and for a high field limit are as follows:

$$\begin{aligned}
\Psi_1(1/2,+1/2) &= -\sin\beta\,\psi_h^+\psi_J^0 + \cos\beta\,\psi_h^-\psi_J^+ \;\Rightarrow \psi_h^-\psi_J^+,\\
\Psi_2(3/2,+3/2) &= \psi_h^+\psi_J^+,\\
\Psi_3(1/2,-1/2) &= -\sin\alpha\,\psi_h^+\psi_J^- + \cos\alpha\,\psi_h^-\psi_J^0 \;\Rightarrow \psi_h^-\psi_J^0,\\
\Psi_4(3/2,+1/2) &= \cos\beta\,\psi_h^+\psi_J^0 + \sin\beta\,\psi_h^-\psi_J^+ \;\Rightarrow \psi_h^+\psi_J^0,\\
\Psi_5(3/2,-1/2) &= \cos\alpha\,\psi_h^+\psi_J^- + \sin\alpha\,\psi_h^-\psi_J^0 \;\Rightarrow \psi_h^+\psi_J^-,\\
\Psi_6(3/2,-3/2) &= \psi_h^-\psi_J^-,
\end{aligned} \tag{2}$$

$\sin\beta = \sqrt{A_+}$, $\cos\beta = \sqrt{1 - A_+}$; $\sin\alpha = \sqrt{A_-}$, $\cos\alpha = \sqrt{1 - A_-}$;

$$A_+ = \frac{1}{2}\Bigl(1 - \frac{x + 1/3}{\sqrt{1 + \frac{2}{3}x + x^2}}\Bigr),\quad A_- = \frac{1}{2}\Bigl(1 - \frac{x - 1/3}{\sqrt{1 - \frac{2}{3}x + x^2}}\Bigr);$$

$$x = \frac{B}{B_c}, \quad B_c = \frac{\Delta W}{-\mu_J/J + \mu_I/I} = \frac{\Delta W}{-\mu_J + 2\mu_h} = 0.2407\,\mathrm{T}.$$

The Breit-Rabi diagram of the six Zeeman hyperfine components of this metastable state is shown in figure 1, where the numbers correspond to those of the wave functions.

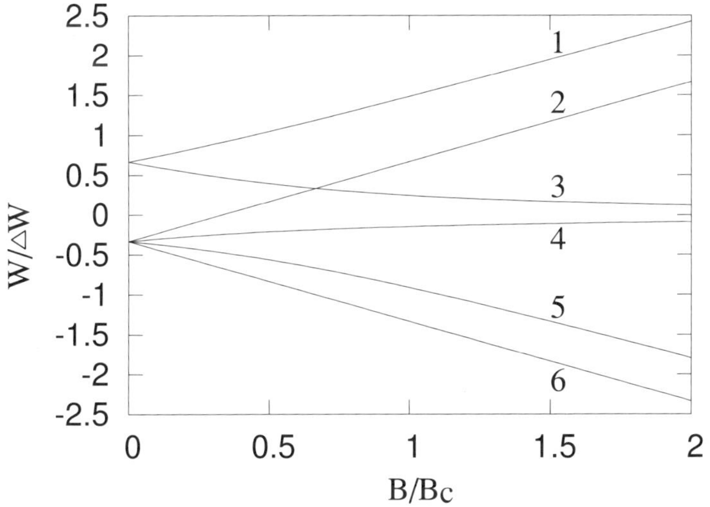

Fig. 1. The scheme of ^{3}He 2^3S_1 hyperfine structure and Zeeman splitting, B_c = 0.2407 T.

2. The Schroedinger equations in the uncoupled state basis

For hydrogen, the task of the adiabatic transitions of atomic spin states was considered by Antishev and Belov.[4] Oh[5] published detailed results for weak field transitions in deuterium. Here we solve this problem for Weak Field Transitions (WFT) in ^{3}He in the metastable state 2^3S_1.

In the uncoupled $|m_h, m_J>$ state basis:

$$\begin{aligned}\Psi(t) = {} & C_1(t)\psi_h^+\psi_J^+ + C_2(t)\psi_h^+\psi_J^0 + C_3(t)\psi_h^+\psi_J^- + C_4(t)\psi_h^-\psi_J^+ \\ & + C_5(t)\psi_h^-\psi_J^0 + C_6(t)\psi_h^-\psi_J^-,\end{aligned} \quad (3)$$

$dC_1/dt = -i/\hbar\{C_1[-(\mu_h + \mu_J)B_z + \Delta W/3] - C_2\mu_J B_x/\sqrt{2} - C_4\mu_h B_x\}$,
$dC_2/dt = -i/\hbar\{-C_1\mu_J B_x/\sqrt{2} - C_2\mu_h B_z - C_3\mu_J B_x/\sqrt{2} + C_4\sqrt{2}\Delta W/3 - C_5\mu_h B_x\}$,

$dC_3/dt = -i/\hbar\{-C_2\mu_J B_x/\sqrt{2} + C_3[(-\mu_h + \mu_J)B_z - \Delta W/3] + C_5\sqrt{2}\Delta W/3 - C_6\mu_h B_x\}$,
$dC_4/dt = -i/\hbar\{-C_1\mu_h B_x + C_2\sqrt{2}\Delta W/3 + C_4[(\mu_h - \mu_J)B_z - \Delta W/3] - C_5\mu_J B_x/\sqrt{2}\}$,
$dC_5/dt = -i/\hbar\{-C_2\mu_h B_x + C_3\sqrt{2}\Delta W/3 - C_4\mu_J B_x/\sqrt{2} + C_5\mu_h B_z - C_6\mu_J B_x/\sqrt{2}\}$,
$dC_6/dt = -i/\hbar\{-C_3\mu_h B_x - C_5\mu_J B_x/\sqrt{2} + C_6[(\mu_h + \mu_J)B_z + \Delta W/3]\}$.

3. The Schroedinger equations in the basis of the stationary states

Another way is to solve the Schroedinger equation in the basis of the stationary states, as it was made by Beijers[6] for hydrogen. But he used the "static" hamiltonian slowly changing with time because atoms move through the changing magnetic field. Hasuyama and Wakuta[7] used the "static" hamiltonian for strong field transitions in deuterium. This approach is not always correct.

We use the stationary states as the basis existing at the value of the magnetic field, $B_z = B_0$, at the input into RF field. For WFT 2–6 we consider only a four-level system of the 2, 4, 5, 6 substates of the $F = 3/2$ state, since the levels of the $F = 1/2$ state are sufficiently distant to have no significant effect on our problem. For the amplitudes of these states, we have used notations c_2–c_6.

The Schroedinger equation is written as follows:

$$i\hbar\frac{d\Psi}{dt} = [\hat{H} + \hat{H}'(t)]\Psi, \tag{4}$$

where $\hat{H}$ is a time-independent hamiltonian whose eigenfunctions satisfy the equations $\hat{H}\psi_n = W_n\psi_n$. The exact wave function is written in the following form:

$$\Psi = \sum c_n(t)\psi_n e^{-iW_n t/\hbar}. \tag{5}$$

Coefficients $c_n(t)$ must satisfy differential equations:

$$i\hbar\frac{dc_k(t)}{dt} = \sum H'_{kn}(t)c_n(t)e^{i\omega_{kn}t}, \tag{6}$$

where $\omega_{kn} = (W_k - W_n)/\hbar$, and $H'_{kn}(t) =< \psi_k|\hat{H}'(t)|\psi_n >$.

For WFT one may use

$$\hat{H}'(t) = -\mu_h\sigma_{hx}B_x(t)\sin\omega t - \mu_J S_{Jx}B_x(t)\sin\omega t - \mu_h\sigma_{hz}b_z(t) - \mu_J S_{Jz}b_z(t), \tag{7}$$

where $b_z(t) = (dB_z/dx)vt$.

The matrix elements are:

$$H'_{22} = (-\mu_h - \mu_J)b_z(t),$$

$$H'_{24} = [-\mu_h \sin\beta - (\mu_J/\sqrt{2})\cos\beta]B_x(t)\sin\omega t,$$

$$H'_{44} = [\mu_h(\sin^2\beta - \cos^2\beta) - \mu_J \sin^2\beta]b_z(t),$$

$$H'_{45} = [-\mu_h \sin\alpha\cos\beta - \mu_J/\sqrt{2}(\sin\alpha\sin\beta + \cos\alpha\cos\beta)]B_x(t)\sin\omega t \quad (8)$$

$$H'_{55} = [\mu_h(\sin^2\alpha - \cos^2\alpha) + \mu_J \cos^2\alpha]b_z(t),$$

$$H'_{56} = [-\mu_h \cos\alpha - (\mu_J/\sqrt{2})\sin\alpha]B_x(t)\sin\omega t,$$

$$H'_{66} = (\mu_h + \mu_J)b_z(t).$$

The values of $\sin\alpha$, $\cos\alpha$, $\sin\beta$, $\cos\beta$, ω_{ik} are taken at the initial value of $B_z(x_{init})$. To find the final amplitude, we have to multiply the resulting wave function by Ψ_n^* at the final value of the field, $B_z(x_{final})$.

Also, we need the matrix elements for the transitions 1–3. We use notations c_1 and c_3.

$$H'_{11} = [\mu_h(\cos^2\beta - \sin^2\beta) - \mu_J \cos^2\beta]b_z(t),$$

$$H'_{13} = [\mu_h \sin\beta\cos\alpha - \mu_J/\sqrt{2}(\sin\alpha\sin\beta + \cos\alpha\cos\beta)]B_x(t)\sin\omega t, \quad (9)$$

$$H'_{33} = [\mu_h(\cos^2\alpha - \sin^2\alpha) + \mu_J \sin^2\alpha]b_z(t).$$

The energies of the states Ψ_1–Ψ_6 are

$$W_1 = \frac{\Delta W}{6} - \mu_J\frac{B}{2} + \frac{\Delta W}{2}\sqrt{1 + \frac{2}{3}x + x^2},$$

$$W_2 = -\frac{\Delta W}{3} - \mu_J B - \mu_h B,$$

$$W_3 = \frac{\Delta W}{6} + \mu_J\frac{B}{2} + \frac{\Delta W}{2}\sqrt{1 - \frac{2}{3}x + x^2},$$

$$W_4 = \frac{\Delta W}{6} - \mu_J\frac{B}{2} - \frac{\Delta W}{2}\sqrt{1 + \frac{2}{3}x + x^2},$$

$$W_5 = \frac{\Delta W}{6} + \mu_J\frac{B}{2} - \frac{\Delta W}{2}\sqrt{1 - \frac{2}{3}x + x^2},$$

$$W_6 = -\frac{\Delta W}{3} + \mu_J B + \mu_h B.$$

We note that at the weak magnetic field ($x \ll 1$) the level distance W_1–$W_3 \approx -\frac{4}{3}\mu_J B$, and W_2–$W_4 = W_4$–$W_5 = W_5$–$W_6 \approx -\frac{2}{3}\mu_J B$.

This differs from the case of deuterium where all distances between the levels at $F = 3/2$ and $F = 1/2$ at the weak magnetic field are equal to $\approx -\frac{2}{3}\mu_e B$.

If we have a system of two sextupoles with the space between them then, realizing the transition $1 \to 3$ in the space between the sextupoles, after the second sextupole we get the pure state 2 with $F = 3/2$, $m_F = 3/2$ and after ionization in the strong magnetic field $P \approx +1$.

If we add the transition $2 \to 6$ after the second sextupole, we produce pure state 6 with $F = 3/2$, $m_F = -3/2$ and $P \approx -1$ after ionization.

So, for the polarized metastable ^{3}He atomic beam, we need two WFT units that should be placed between and after the sextupole magnets.

It is interesting to note that the pure states of ^{3}He(2S) may be transferred into the pure states of $^3He^+$ after stripping one electron in 2S-state. The results of Slobodrian *et al.*[2] have confirmed this point. In their scheme, the sextupole produces an axial field dropping with distance, so, using the RF field perpendicular to the beam trajectory, it is possible to reverse the orientation of components 1 and 2, transforming them into 3 and 6.

In the magnetic field the wave function of the hyperfine substate 3 of the ^{3}He(2S) atom is:

$$\psi(F = 1/2, m_F = -1/2) = -\sin\alpha\,\psi_h^+\psi_J^- + \cos\alpha\,\psi_h^-\psi_J^0 \Rightarrow \psi_h^-\psi_J^0.$$

With $B = 0.2$ T, ($x = 0.8309$) $\cos\alpha = 0.8564$ and $\sin\alpha = 0.5164$. The substate 6, $\psi(F = 3/2, m_F = -3/2) = \psi_h^-\psi_J^-$, does not change.

It can be easily shown that if the second ionization is effected in zero magnetic field, the expected value of P for the pure state of $^3He^+$ would be $P = -0.68$, and for the mixed state $^3He^+$ $P = -0.44$. The measured value is $P = -(0.6 \div 0.8)$.

A tapered electromagnet produces the static magnetic field $B_z(x)$ perpendicular to the beam path with field gradient dB_z/dx along $x = vt$.

$$B_z(x) = B_0 + \frac{dB_z}{dx}x, \; B_x(x) = B_1(x)\sin\omega x. \tag{10}$$

We supposed that some parameters would have the same values as in the paper by Oh:[5]

$B_0 = 1.17 \times 10^{-3}$ T, $dB_z/dx = -1.4 \times 10^{-2}$ T/m for the negative static field gradient (or $B_0 = 4.7 \times 10^{-4}$ T, $dB_z/dx = 1.4 \times 10^{-2}$ T/m for the positive gradient),

$l = 5 \times 10^{-2}$ m, $\omega = 9.63 \times 10^7$ rad/s for the $2 \to 6$ transition and $\omega = 1.93 \times 10^8$ rad/s for the $1 \to 3$ transition. The atomic beam velocity $v = 1.2 \times 10^3$ m/sec. The RF amplitude $B_1(x)$ is a quadratic function of x with zero values at $x = 0$ and $x = l$; $B_1^{max} = B_1(l/2) = (1-2) \times 10^{-4}$ T.

The results of the computer calculations for the atom velocity of 1200 m/s give practically 100 % probability of the transition.

4. Ionizer

The problems of ionization and depolarization in the ionizer were discussed earlier.[8] An evident way for producing polarized helions is to follow the way of the Laval University group: to ionize the polarized helium-3 2^3S_1 atoms in an ionizer into $^3He^+$ ions and inject them into the electron beam ion source (EBIS) for subsequent ionization to get helions.

But there is a possibility of ionizing the polarized metastable atoms directly to $^3He^{++}$ and accumulating them in the ion trap of the EBIS with subsequent 8-μs-pulse extraction and injection into the JINR Accelerator Complex.

Earlier, the pulsed extraction of ions from the trap was carried out for 7 μs with a current of 1 mA, which corresponded to 4×10^{10} charges.[1]

The experiments[9] with the ionizer of the polarized deuteron source POLARIS have shown a feasibility of accumulating up to 4×10^{11} charges in the ion trap.

5. Depolarization

The time between the metastability exchange collisions is $\tau = 1/\sigma v N$, where σ is a cross section for the metastability exchange, v is a velocity of the metastables, and N is a density of ground state atoms. With $v = 1.2 \times 10^5$ cm/s and $\sigma = 4 \times 10^{-16}$ cm^2,[10] the condition for $\tau \gg T_{acc}$, where T_{acc} is the time of accumulation, is:

$$\sigma v N \ll T_{acc}^{-1} \text{ or } N \ll 2 \times 10^{10} T_{acc}^{-1},$$

or

$$p \ll 6 \times 10^{-7}/T_{acc} \text{ Torr}.$$

For $T_{acc} = 15$ ms, the required value of pressure is $p \ll 4 \times 10^{-5}$ Torr. Also, a dangerous process is the symmetric resonant charge transfer

$$^3He^{++} + ^3He \to ^3He + ^3He^{++}.$$

The cross section of this process is estimated to be $\simeq 7 \times 10^{-16}$ cm^2,[11] even larger than the cross section for metastability exchange.

Let the metastable flux be 3×10^{15} atoms/s sterad. If we assume that the flux of atoms in the ground state is by $\simeq 10^3$ times higher than the metastable flux, the pressure of the ground state atoms in the ionizer at a distance of 120 cm from the nozzle, is $\simeq 10^{-7}$ Torr.

The experience of the SATURNE group[12] has shown that depolarization processes seem to be unimportant. They ionized $^6Li^+$ polarized ions to bare nuclei $^6Li^{3+}$ in the EBIS at the field of 5 T without depolarization during ionization, accumulation and 3 ms extraction.

There is an interesting possibility for ionization in the EBIS to use a nuclear-polarized ground-state ^{3}He beam produced with a permanent sextupole magnet at the energy of 8 meV and intensity 1.4×10^{14} atoms/s in the focused 2 mm diameter beam.[13]

6. Conclusion

A possibility for developing a polarized helion source for the JINR Accelerator Complex, has been discussed. It is feasible to provide a polarized beam with rather high polarization and helion intensity up to 2×10^{11} ions/pulse of 8 μs. The depolarizing effects in the polarized ion source are expected to be low. For helion acceleration at the NUCLOTRON-M it is necessary to provide conditions for low depolarization. Installation of the polarized helion source at the JINR Accelerator Complex is sure to extend the program of spin physics experiments.

References

1. E. D. Donets *et al.*, *Rev. Sci. Instr.* **71**, 887 (2000).
2. R. J. Slobodrian *et al.*, *Nucl. Instr. and Meth. A* **244** 127 (1986).
3. P. Y. Beauvais *et al.*, in *Proc. Int. Symposium Dubna-Deuteron-93*, 278 (JINR, Dubna, 1994).
4. E. P. Antishev and A. S. Belov, in *Proc. 12th Int. Workshop on Polarized Ion Sources, Targets and Polarimetry (PSTP2007)*, AIP Conf. Proc. **V.980**, 263 (AIP, New York, 2008).
5. S. Oh, *Nucl. Instr. Meth.* **82**, 189 (1970).
6. J. P. M. Beijers, *Nucl. Instr. Meth. A* **536**, 282 (2005).
7. H. Hasuyama and Y. Wakuta, *Nucl. Instr. Meth. A* **260**, 1 (1987).
8. Yu. A. Plis *et al.*, in *Proc. 19th Int. Baldin Seminar on High Energy Physics Problems (ISHEP)*, 3 (JINR, Dubna, 2008).
9. V. V. Fimushkin *et al.*, *Czech. J. Phys. A* **51**, 319 (2001).

10. F. D. Colegrove *et al.*, *Phys. Rev.* **132**, 2561 (1963).
11. H. Schrey and B. Huber, *Z. Phys. A* **273**, 401 (1975).
12. A. Courtois *et al.*, *Rev. Sci. Instr.* **63**, 2815 (1992).
13. A. P. Jardine *et al.*, *Rev. Sci. Instr.* **72**, 3834 (2001).

POLARIZED ^{3}HE TARGETS FOR REAL AND VIRTUAL PHOTONS

J. Krimmer*, J. Ahrens, P. Aguar Bartolomé, M. Distler, W. Heil, S. Karpuk and Z. Salhi

Institut für Physik & Institut für Kernphysik,
Johannes Gutenberg-Universität Mainz,
55128 Mainz, Germany
**E-mail: krimmer@uni-mainz.de*

Polarized ^{3}He from metastability exchange optical pumping is used for various double polarized experiments at the MAinz MIcrotron (MAMI). More than 70 % initial target polarization has been obtained at the experimental area. Besides the description and the performance of the target for the electron beam, details of the newly developed target for the photon beam inside the Crystal Ball (CB) detector will be given.

Keywords: Polarized ^{3}He; polarized target; MEOP.

1. Introduction

Polarized ^{3}He gas has a wide spectrum of applications ranging from basic research to medical applications. Due to the strong spin dependence of the neutron absorption cross section, polarized ^{3}He is used as a neutron spin filter.[1] Free spin precession of ^{3}He in a magnetically shielded room serves as a test of Lorentz invariance.[2] Medical applications of polarized ^{3}He involve MRI of the lung where nowadays measurements of the apparent diffusion coefficient allow probing of the pulmonary microstructure *in vivo*.[3] The application of polarized ^{3}He focused on in this paper is the use as an effective polarized neutron target. Due to the large S-state probability[4] the two protons saturate their spin and the spin of the neutron is aligned with the spin of the nucleus. This property can be used, for example, to extract the electric formfactor of the neutron $G_{e,n}$ via a double polarized electron scattering experiment.[5] The setup described in this paper has been used for measurements of $G_{e,n}$ at $Q^2 = 1.5(\text{GeV/c})^2$.[6] Furthermore, via double polarized photoabsorption measurements one has experimental access to the Gerasimov Drell Hearn (GDH) sum rule.[7,8] For the neutron, only data from polarized deuterium targets exist so far.[9,10] Here, due to the spin structure, measurements on ^{3}He will give a complementary and more direct access to the GDH sum rule for the neutron. Gaseous ^{3}He is polarized via the method of Metastability Exchange Optical Pumping (MEOP).[11,12] The metastable 2^3S_1 state is reached via a weak gas discharge at pressures of 0.8-1.0 mb and can then be optically pumped by circularly polarized laser light at 1083 nm ($2^3S_1 \rightarrow 2^3P_0$). The nuclear polarization of the 2^3S_1 state is transferred to unpolarized ground state atoms via metastable exchange collisions. After polarization buildup the gas is compressed to the desired pressures of about 5 bar by means of a nonmagnetic piston where less than 2 % of the polarization is lost.[13] Up to 76 % nuclear polarization can be obtained at a flux of 2 bar·l/h.[14,15] The target cells are filled with polarized ^{3}He gas at the polarizing facility at the Institute of Physics and are then brought in an auxiliary magnetic field to the experimental area at MAMI. This remote type of operation requires long relaxation times of the polarized gas inside the target cells and a minimal loss of polarization during the transport.

The paper is organized as follows. In section 2 the target setup at the electron beam is given together with its performance during the $G_{e,n}$ run in July 2008. section 3 deals with the assembly for the photon beam which has been used for a first measurement in July 2009.

2. Polarized ^{3}He target for the electron beam

2.1. *Setup*

The principle target setup for electron scattering experiments is given in the left part of figure 1. The target cell with the polarized ^{3}He is located in the middle of a box[16,17] providing a homogeneous magnetic holding field of B_0=0.4 mT. Provided that the relative field gradient $|dB/dr|/B_0$ can be kept below $5{\cdot}10^{-4}$ cm^{-1} in the target cell region, the corresponding partial relaxation time T_1^{grad} is larger than 1000 hours at a pressure of 5 bar.[18]

Holes in the box for the primary electron beam and for the scattered electrons, as well as missing coil windings in the corners of the box, spoil the perfect homogeneity of the magnetic field. These effects can, however, be compensated for by means of correction coils. The field in the target cell region along the z-axis is shown in the right part of figure 1, where the raw field is given by the dashed line, and the solid line denotes the field with the correction coils in use. For display purposes, a constant offset of +0.22

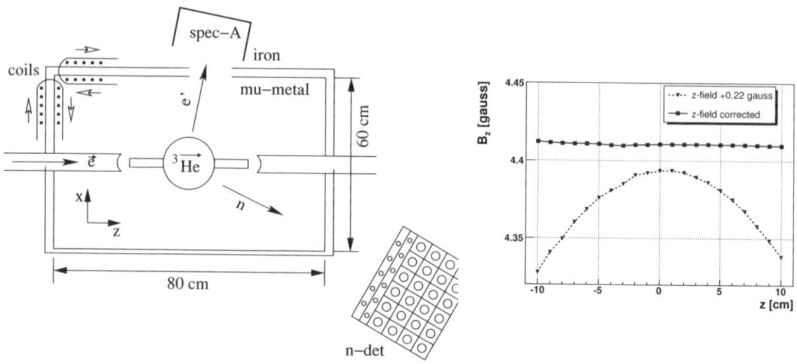

Fig. 1. Left: Target setup for electron scattering experiments. Right: Magnetic field along the z-direction in the target cell region.

G has been added to the uncorrected field data. The relative gradient can be kept well below the demanded $5 \cdot 10^{-4}$ cm^{-1}.

2.2. *Target cells*

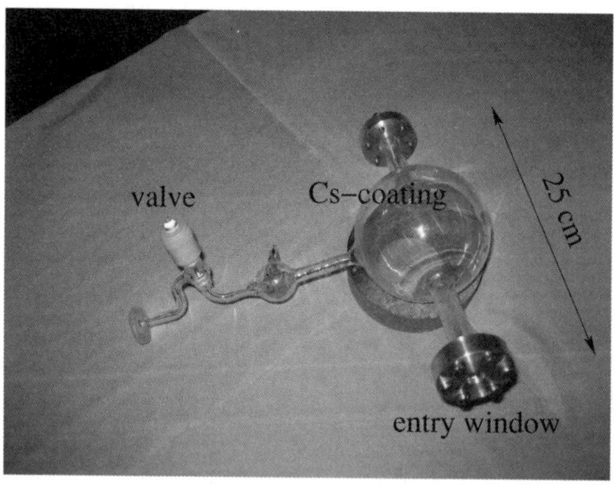

Fig. 2. Target cell used for electron scattering experiments.

The target cells used for electron scattering experiments (see fig. 2) comprise a spherical part with an outer diameter of 10 cm and two cylindrical side arms giving rise to a total length of 25 cm. The cells are made from quartz glass. After coating with cesium, they exhibit wall relaxation times T_1^{wall} between 100 and 200 hours. The entry and exit windows for the electron beam consist of 50 μm beryllium, covered with 0.4 μm aluminum.

2.3. *Polarization measurement*

The polarization can be measured online during the experiment by two means. A relative polarization information is obtained from the Free Induction Decay (FID) signal measured in pickup coils after tipping the B-field axis nonadiabatically by 2°. The measured FID signal is displayed in the

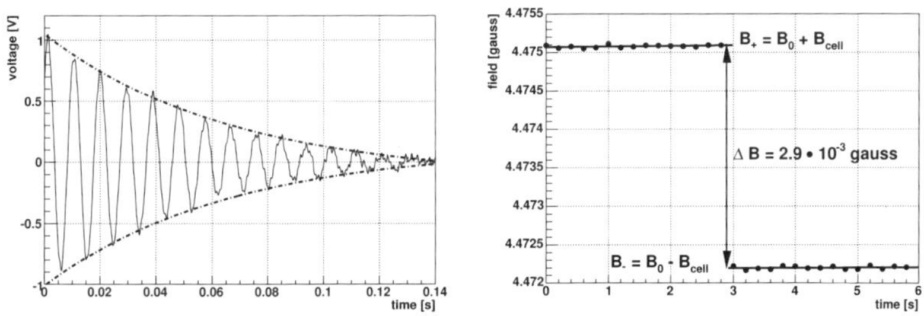

Fig. 3. Left: FID signal. Right: Total magnetic field before and after a 180° spin-flip via AFP.

left part of figure 3. This polarization measurement does not interfere with the data taking for the experiment and the relative polarization loss due to this measurement is 0.02 %.

The second method is based on the precise measurement of the magnetic field produced by a dense sample of polarized gas.[19] The field B_{cell} produced by the target cell filled with polarized gas, is in the order of 1 mG which is more than a factor 1000 smaller than the holding field $B_0 = 4.48$ G. The total magnetic field is measured with the cell magnetization parallel ($B_+ = B_0 + B_{cell}$) and antiparallel ($B_- = B_0 - B_{cell}$) to the holding field B_0 (see right part of fig. 3). The difference $\Delta B = B_+ - B_- = 2B_{cell}$ is independent of B_0 if the sequence of measurement is fast enough so that drifts in B_0 don't play any role. The reversal of the magnetization with respect to the guiding field via Adiabatic Fast Passage (AFP) guarantees a complete and non-destructive spin-reversal even in a slightly inhomogeneous B_0. The polarization P_{He} can then be obtained via

$$P_{He} = \frac{r^3}{2\mu_{^3He}} \cdot \frac{1}{N} \cdot \Delta B \qquad (1)$$

where $\mu_{^3He} = 1.0746 \cdot 10^{-26}\text{Am}^2$ denotes the magnetic moment of the ^{3}He nucleus. The number of atoms N in a target cell with volume V at a temperature T and a pressure p is given by $\frac{p \cdot V}{k_B \cdot T}$ with k_B the Boltzmann constant. Due to the cubic dependence of the distance r between the field sensor and the center of the cell, a proper calibration is needed which is described in detail in reference20. The absolute polarization can be determined with a relative uncertainty of 3 %. The relative polarization loss due to this polarization determination can be kept below 0.2 %. In contrast to the FID method described previously, the method described here cannot be used in parallel to experimental data taking as a 180° spin-flip is involved. It is performed during pauses, when the spin has been rotated to a different direction anyway.

2.4. *Performance in the electron beam*

In the left part of figure 4, the polarization decay of a target cell in the electron beam under running conditions is displayed. Here, additional relaxation mechanisms become relevant, like the dipole-dipole interaction[21] and the production of $^3\text{He}^+$-ions in the electron beam.[22] The parametrization (black line in fig. 4) takes into account all relaxation terms plus the polarization loss due to the polarization measurements. The FID values and the parametrization are scaled by a constant factor in order to match the AFP values. The T_1 times obtained with all methods agree among themselves. In the right part of figure 4 the polarization values during the complete

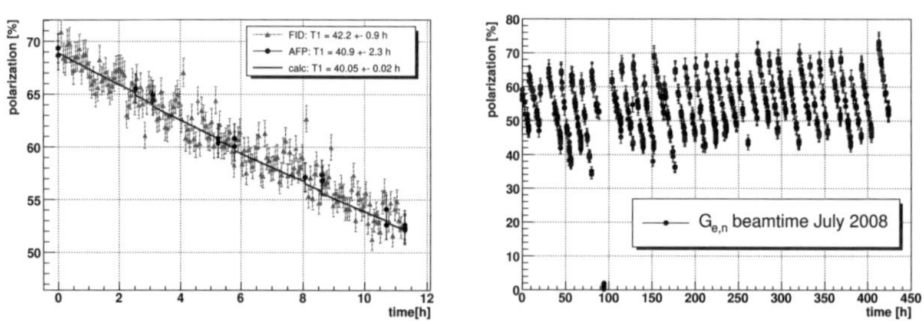

Fig. 4. Left: Polarization decay in the electron beam. The FID values and the parametrization (black line) are scaled by a constant factor in order to match the AFP values. Right: Polarization values during the $G_{e,n}$ beamtime in July 2008.

$G_{e,n}$ beamtime at MAMI in July 2008 are shown. The initial polarization steadily increased during the run due to an improved cell exchange procedure, reaching up to 72 %. With a target cell exchange twice per day, the mean polarization resulted in 55-60 %.

More details about the polarized ^{3}He target for the electron beam can be found in reference 20.

3. Polarized ^{3}He target for the photon beam

3.1. *Setup*

A sketch of the target setup used for experiments in the tagged photon beam of MAMI is given in figure 5. During the experiment the target cell with the polarized ^{3}He is situated in the middle of the CB detector. The holding field along the beam axis is provided by a solenoid[23] with 1500 windings, an outer diameter of 8 cm, and a length of 80 cm. The relative field gradient $(dB/dz)/B_0$ along the axis is less than $5 \cdot 10^{-4}$ cm^{-1} in the target cell region.

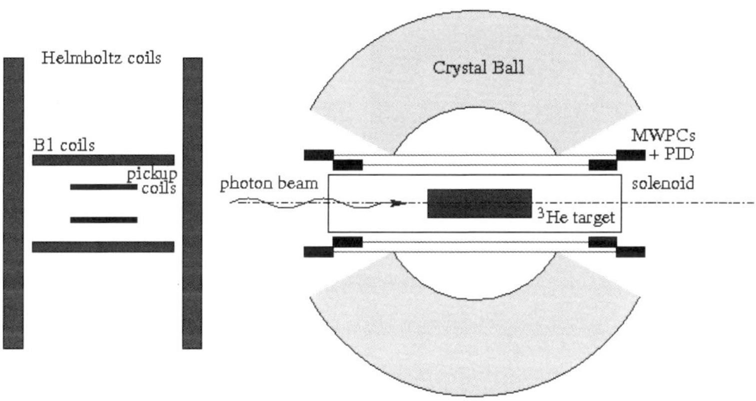

Fig. 5. Target setup for the tagged photon beam of MAMI.

3.2. *Polarimetry*

Due to space limitations inside the CB detector, polarimetry is performed outside the detector on the upstream site. Here, a pair of Helmholtz coils with 100 windings and a diameter of 1.6 m is set up, together with the B_1 and pickup coils, for a polarization measurement according to the FID method (see sec. 2.3). An automatic transport system has been installed, which allows a remote controlled movement of the target cell between the position inside the CB detector and the Helmholtz coil. During the movement the solenoid is off in order to avoid the sharp gradients at the end of the solenoid. The time needed to complete the FID measurement procedure, including the movement of the cell, is three minutes and the relative polarization loss amounts to 1 %.

3.3. *Target cells and performance in the beam*

The target cells used for the photon beam obey a cylindrical geometry with a length of 20 cm and an outer diameter of 6 cm (see left part of fig. 6). Various materials for the entry windows for the photon beam have been tested, the results are given in table 1.

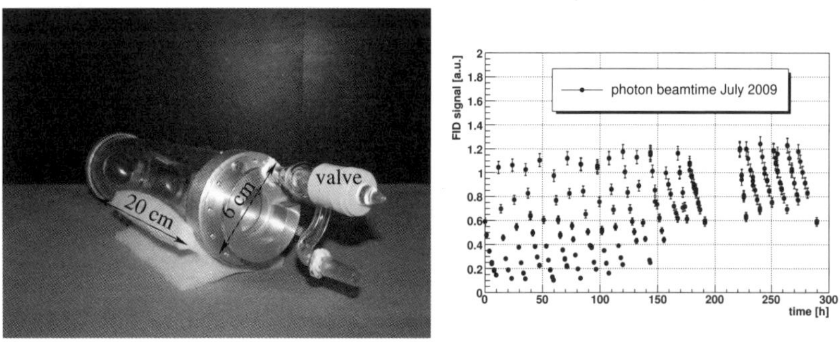

Fig. 6. Left: target cell used for experiments in the photon beam. Right: performance in the beam during the first measurement at MAMI in July 2009.

Table 1. Properties of various window materials.

material	d [μm]	$\rho \cdot d$ [g/cm^2]	T_1^{wall}	comments
Mylar(Al)	50	0.007	18	not 100 % tight (H_2O)
Be	150	0.027	14	thickness, short T_1 times
Havar	10	0.008	15	short T_1 times
Ti	50	0.022	40	preferred solution

To find the optimum material, a compromise has to be made between long T_1 times and a suppression of background events coming from the windows. This background scales as $\rho \cdot d$, where ρ denotes the density and d the thickness of the window material. A good compromise seemed to be Mylar, metallized with aluminium, but these foils are not 100 % tight, in particular water vapour diffuses inside the cell, which destroys the cesium coating within hours. Regarding background, Havar would be ideal, but it is not ideal for keeping the ^{3}He gas polarized. The preferred solution so far is titanium which resulted in the longest T_1 times during the test.

In July 2009 the first measurement with a polarized ^{3}He gas target in a circularly polarized photon beam has been done at MAMI. In the right part of figure 6 the FID signal in arbitrary units is given for the complete beamtime.

More details about the target for the photon beam will be given in an upcoming paper.

References

1. W. Heil *et al.*, *Nucl. Instr. Meth. A* **485**, 551 (2002).
2. C. Gemmel *et al.*, submitted to *Eur. Phys. J. D.*
3. W. G. Schreiber *et al.*, *Respir. Physiol. Neurobiol.* **148**, 23 (2005).
4. B. Blankleider and R. M. Woloshyn, *Phys. Rev. C* **29**, 538 (1984).
5. J. Bermuth *et al. Phys. Lett. B* **564**, 199 (2003).
6. S. Schlimme, PhD thesis, University of Mainz (Mainz, Germany, in preparation).
7. S. Gerasimov, *Sov. J. Nucl. Phys.* **2**, 430 (1966).
8. S. D. Drell and A. C. Hearn, *Phys. Rev. Lett.* **16**, 908 (1966).
9. H. Dutz *et al.*, *Phys. Rev. Lett.* **94**, 162001 (2005).
10. J. Ahrens *et al.*, *Phys. Rev. Lett.* **97**, 202303 (2006).
11. F. Colegrove *et al.*, *Phys. Rev.* **132**, 2561 (1963).
12. P. Nacher and M. Leduc, *J. Physique* **46**, 2057 (1985).
13. J. Schmiedeskamp, PhD thesis, University of Mainz (Mainz, Germany, 2005).
14. E. Otten, *Europhysics News* **35**, 16 (2004).
15. M. Batz *et al.*, *J. Res. Natl. Inst. Stand. Technol.* **110**, 293 (2005).
16. D. Rohe, PhD thesis, University of Mainz (Mainz, Germany, 1998).
17. Y. Gussev *et al.*, submitted to *J. Magn. Res.*
18. L. Schearer and G. Walters, *Phys. Rev. A* **139**, 1398 (1965).
19. E. Wilms *et al.*, *Nucl. Instr. Meth. A* **401**, 491 (1996).
20. J. Krimmer *et al.*, *Nucl. Instr. Meth. A* **611**, 18 (2009).
21. N. Newbury *et al.*, *Phys. Rev. A* **48**, 4411 (1993).
22. T. Chupp *et al.*, *Phys. Rev. C* **45**, 915 (1992).
23. P. Aguar-Bartolomé, Diploma Thesis, University of Mainz (Mainz, Germany, 2006).

SPIN-FILTERING STUDIES AT COSY AND AD

F. Rathmann for the PAX-collaboration

Institut für Kernphysik
Forschungszentrum Jülich
52425 Jülich, Germany
**E-mail: f.rathmann@fz-juelich.de*

Polarized antiprotons provide access to a wealth of single- and double-spin observables, thereby opening a window to physics uniquely accessible with the High Energy Storage Ring (HESR) at FAIR. The physics program proposed by the PAX collaboration includes a first measurement of the transversity distribution of the valence quarks in the proton, a test of the predicted opposite sign of the Sivers-function, related to the quark distribution inside a transversely polarized nucleon in Drell-Yan as compared to semi-inclusive DIS, and a first measurement of the moduli and the relative phase of the time-like electric and magnetic form factors $G_{\mathrm{E,M}}$ of the proton. In polarized and unpolarized $\mathrm{p\bar{p}}$ elastic scattering, open questions like the contribution from the odd charge-symmetry Landshoff-mechanism at large $|t|$, and spin-effects in the extraction of the forward scattering amplitude at low $|t|$ can be addressed.

Provided that antiproton beams with a polarization in excess of 20 % can be obtained with the APR, the HESR at FAIR could be converted into a double-polarized asymmetric $\bar{p}p$ collider by installation of an additional COSY-like ring. In this setup, antiprotons of 3.5 GeV/c collide with protons of 15 GeV/c at c.m. energies of $\sqrt{s} \approx \sqrt{200}$ GeV with a luminosity in excess of 10^{31} $\mathrm{cm^{-2}s^{-1}}$.

A recent experiment at COSY revealed that ep spin-flip cross sections are too small to cause a detectable depolarization of a stored proton beam. This measurement rules out a proposal to use polarized positrons to polarize an antiproton beam by $\mathrm{e^{+}\bar{p}}$ spin-flip interactions. The most promising approach to provide a beam of polarized antiprotons, adopted by the PAX collaboration, is based on spin-filtering using an internal polarized hydrogen gas target – a method that has been shown to work with stored protons. We expect to produce a polarized antiproton beam with ten orders of magnitude higher intensity than secondary polarized antiproton beams previously available.

Keywords: Polarized antiprotons; ep interaction; polarized beams and targets.

1. Physics case

Polarized antiproton-proton interactions at the High Energy Storage Ring (HESR) at the future Facility for Antiproton and Ion Research (FAIR) at Darmstadt, Germany, will provide unique access to a number of new fundamental physics observables, which can be studied neither at other facilities, nor at HESR without transverse polarization of protons and antiprotons.

1.1. *The transversity distribution*

This is the last leading-twist missing piece of the QCD description of the partonic structure of the nucleon. It describes the quark transverse polarization inside a transversely polarized proton.[1] Unlike the more conventional unpolarized quark distribution $q(x, Q^2)$ and the helicity distribution $\Delta q(x, Q^2)$, the transversity $h_1^q(x, Q^2)$ can neither be accessed in inclusive deep-inelastic scattering of leptons off nucleons, nor can it be reconstructed from the knowledge of $q(x, Q^2)$ and $\Delta q(x, Q^2)$. It may contribute to some single-spin observables, but always coupled to other unknown functions. The transversity distribution is directly accessible uniquely via the double transverse spin asymmetry A_{TT} in the Drell-Yan production of lepton pairs. The theoretical expectations for A_{TT} in the Drell-Yan process with transversely polarized antiprotons interacting with transversely polarized protons at HESR are in the 0.3–0.4 range.[2,3] With the expected antiproton beam polarization of $P \approx 0.3$, achieved by spin-filtering in a dedicated low-energy Antiproton Polarizer Ring (APR), and the luminosity available with the HESR, the PAX experiment[a] is uniquely suited for the definitive observation of $h_1^q(x, Q^2)$ of the proton for the valence quarks. The determination of $h_1^q(x, Q^2)$ will open new pathways to the QCD interpretation of Single-Spin Asymmetry (SSA) measurements. In conjunction with the data on SSA from the HERMES collaboration,[4] the PAX measurements of the SSA in Drell-Yan production on polarized protons can for the first time provide a test of the theoretical prediction[5] of the reversal of the sign of the Sivers function[6] from semi-inclusive DIS to Drell-Yan production.

1.2. *Magnetic and electric form factors*

The origin of the unexpected Q^2-dependence of the ratio of the magnetic and electric form factors of the proton as observed at the Jefferson laboratory[7] can be clarified by a measurement of their relative phase in the time-

[a]PAX collaboration, `http://www.fz-juelich.de/ikp/pax`.

like region, which discriminates strongly between the models for the form factor. This phase can be measured via SSA in the annihilation $\bar{p}p^{\uparrow} \to e^+e^-$ on a transversely polarized target.[8,9] The first ever measurement of this phase at PAX will also contribute to the understanding of the onset of the pQCD asymptotics in the time-like region and will serve as a stringent test of dispersion theory approaches to the relationship between the space-like and time-like form factors.[10–12] The double-spin asymmetry will allow independently the $G_E - G_M$ separation and serve as a check of the Rosenbluth separation in the time-like region which has not been carried out so far.

1.3. *Hard scattering*

Arguably, in $p\bar{p}$ elastic scattering the hard scattering mechanism can be checked beyond $|t| = \frac{1}{2}(s - 4m_p^2)$ accessible in t-u-symmetric pp scattering, because in the $p\bar{p}$ case the u-channel exchange contribution can only originate from the strongly suppressed exotic dibaryon exchange. Consequently, in the $p\bar{p}$ case the hard mechanisms[13–15] can be tested at t almost twice as large as in pp scattering. Even unpolarized large angle $p\bar{p}$ scattering data can shed light on the origin of the intriguing oscillations around the s^{-10} behavior of the 90 ° scattering cross section in the pp channel and put stringent constraints on the much disputed odd-charge conjugation Landshoff mechanism.[16–19] If the Landshoff mechanism is suppressed, the double transverse asymmetry in $p\bar{p}$ scattering is expected to be as large as the one observed in the pp case.

2. A polarized asymmetric antiproton-proton collider

The possibility of testing the nucleon structure via double spin asymmetries in polarized proton-antiproton reactions of FAIR was suggested by the PAX collaboration in 2005.[20] Since then, there has been much progress, both in understanding the physics potential of such an experiment[2,3,9] and in studying the feasibility of efficiently producing polarized antiprotons.[21]

The accelerator setup proposed by PAX is shown in figure 1. Its main features are:

1. An Antiproton Polarizer Ring (APR)[22,23] built inside the HESR area with the crucial goal of polarizing antiprotons, which are subsequently transferred to the other rings, and accelerated.
2. A second Cooler Synchrotron Ring (CSR, COSY–like) in which protons or antiprotons can be stored with a momentum up to 3.5 GeV/c. This

ring shall have a straight section, where the PAX detector[20] will be installed, running parallel to the experimental straight section of HESR.

3. By deflection of the HESR beam into the straight section of the CSR and back, both the collider and the fixed-target mode become feasible. In the collider mode, protons at 15 GeV/c collide with polarized antiprotons at 3.5 GeV/c.

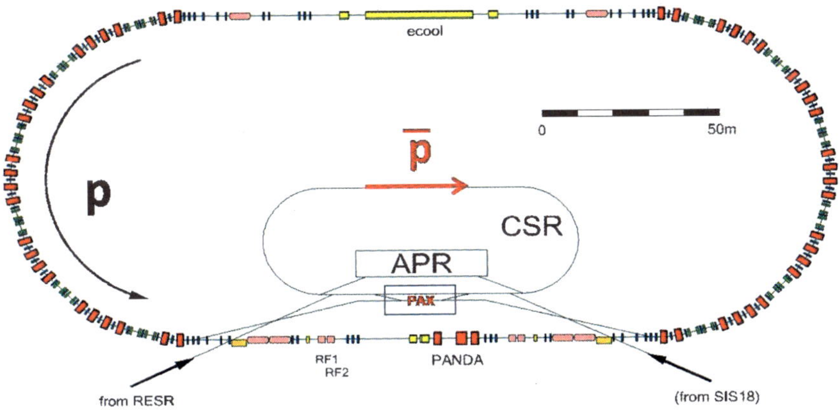

Fig. 1. Double-polarized asymmetric collider proposed by the PAX collaboration on the basis of the HESR. Necessary for fixed target operation during phase I are: CSR, APR, beam transfer lines, and polarized proton injector. For the asymmetric collider operation in phase II, one needs, in addition, two transfer lines.

3. Spin-filtering experiments at COSY and AD

For more than two decades, physicists have tried to produce beams of polarized antiprotons,[24] generally without success. Conventional methods like Atomic Beam Sources (ABS), appropriate for the production of polarized protons and heavy ions cannot be applied, since antiprotons annihilate with matter. So far, the only polarized antiproton beam has been produced from the decay in flight of $\bar{\Lambda}$ hyperons at Fermilab. At polarizations $P > 0.35$, the achieved intensities never exceeded $1.5 \cdot 10^5$ s^{-1}.[25] Scattering of antiprotons off a liquid hydrogen target could yield polarizations of $P \approx 0.2$, with beam intensities of up to $2 \cdot 10^3$ s^{-1}.[26] Unfortunately, both abovementioned approaches do not allow for an efficient accumulation in a storage ring, which would be needed to enhance the luminosity. Spin-splitting, using the Stern-Gerlach separation of magnetic substates in a stored antiproton beam was already proposed in 1985.[27] Although the theoretical understanding has much improved since then,[28] spin-splitting using a stored beam has yet to be observed experimentally. In contrast to that, a proof of the spin-filtering principle has been produced by the FILTEX experiment at the TSR-ring in Heidelberg.[22]

The experimental basis for predicting the polarization buildup in a stored antiproton beam is practically non-existent. The AD-ring at CERN is a unique facility at which stored antiprotons in the appropriate energy range are available and whose characteristics meet the requirements for the first ever antiproton polarization buildup studies. Therefore, it is of the highest priority in the quest for polarized antiprotons to make use of this opportunity, and to perform spin-filtering experiments using stored antiprotons at the AD-ring at CERN. In preparation for the experiment at the AD, a number of dedicated spin-filtering experiments will be carried out with protons at the COoler SYnchrotron COSY at Jülich, Germany, in order to commission the equipment needed, and to gain additional insight into accelerator physics aspects of the project. To this end, two proposals have been submitted in 2009, one to the COSY PAC,[29] and one to the SPS committee of CERN.[30]

3.1. *Experimental setup for the AD-ring*

At present, the AD at CERN is actually the only place worldwide, where the proposed measurements can be performed. The effort involved is substantial. Although we will perform most of the design and commissioning work outside of CERN, it is obvious, that many aspects in the design require a close collaboration with the CERN machine group. The main components that need to be installed in the AD are shown in figure 2. All components will be tested and commissioned at COSY in Jülich.

The measurements require implementing a storage cell for Polarized Internal Target (PIT) in the straight section between injection and electron cooling of the AD (see fig. 3). PITs nowdays represent a well established technique with high performance and reliability shown in many different experiments with hadronic and leptonic probes.[31] Targets of this kind have been operated successfully at TSR in Heidelberg,[32] later on they were used at HERA/DESY,[33] at Indiana University Cyclotron Facility (IUCF), and at MIT-Bates. A new PIT is presently operated at ANKE-COSY.[34,35] Typical target densities range from a few 10^{13} to 2×10^{14} cm^{-2}.[33] The target density depends strongly on the transverse dimension of the storage cell. In order to provide a high target density at AD, the β-function at the

Fig. 2. Full installation foreseen at the AD for the straight section between injection and electron cooling (see fig. 3). The beam moves from right to left. The outer AD quadrupole magnets define the up- and downstream boundaries of the low-β insertion. The magnets next in line are COSY arc (short) and straight section (long) quadrupoles. The two inner qaudrupoles next to the target chamber have been recuperated from the CELSIUS ring. The atomic beam source is mounted above the target chamber that houses the detector system and the storage cell. Three sets of Helmholtz coils providing magnetic holding fields along x, y, and z are mounted on the edges of the target chamber. The Breit-Rabi target polarimeter and the target-gas analyzer are mounted outwards of the ring. Fast shutters are used on the target chamber on all four main ports. The complete section can be sealed off from the rest of the AD by valves.

storage cell should be about $\beta_x = \beta_y = 0.3$ m. In order to achieve that, a special insertion includes additional quadrupoles around the storage cell (see fig. 2). The low-β section is designed in such a way that the storage cell limits the machine acceptance only marginally. A careful machine study has been carried out in order to maintain the machine performance at injection energy and at low energies for the other AD experiments. The section which houses the PIT has to be equipped with a powerful differential pumping system, that is capable of maintaining good vacuum conditions in the other sections of the AD.

We utilize the former HERMES ABS to feed the storage cell.[36,37] The target will be operated in a weak magnetic guide field of a about 10 G. The

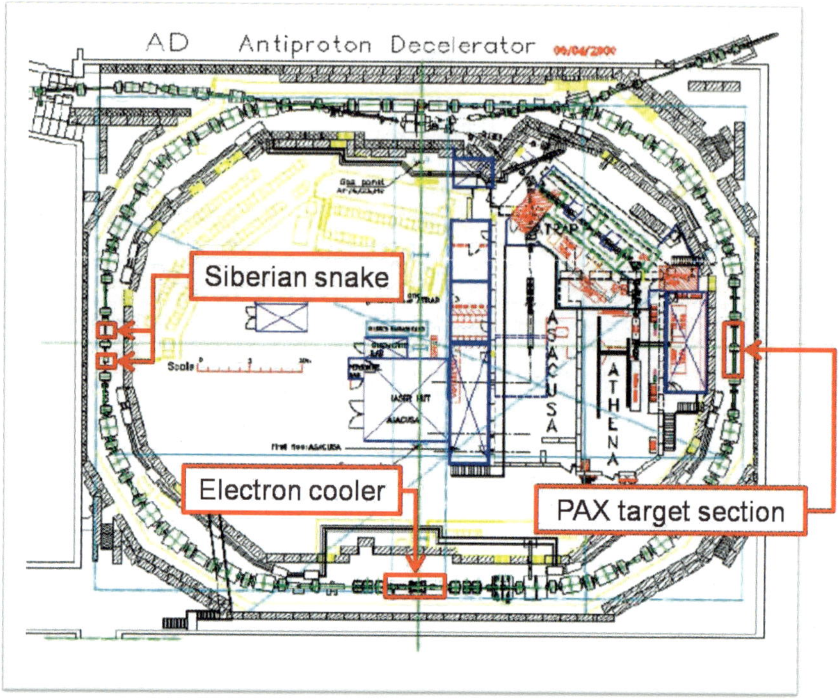

Fig. 3. Floorplan of the AD ring at CERN. The PAX target is located in the straight section on the right (for details, see fig. 2). In order to explore spin-filtering at energies higher than about 50 MeV, the AD electron cooler has to be upgraded. A Siberian snake needs to be installed in the straight section opposite the target to investigate the longitudinal spin-dependence of the $\bar{p}p$ interaction.

orientation of the target polarization is maintained by a set of Helmholtz coils in transverse and longitudinal direction.

4. Conclusion

To summarize, we note that the storage of polarized antiprotons at HESR will open unique possibilities for testing QCD in hitherto unexplored domains. This will provide another cornerstone to the antiproton program at FAIR.

The depolarization study carried out by the ANKE and PAX collaborations constitutes the first step of investigations at COSY, shedding light on the ep spin-flip cross sections when the target electrons are unpolarized.

The experimental finding rules out the practical use of polarized leptons to polarize a beam of antiprotons with present-day technologies.[21] This leaves us with the only proven method to polarize a stored antiproton beam *in situ*, namely spin-filtering by the strong interaction. The experimental basis for predicting the polarization buildup in a stored antiproton beam by spin-filtering is practically non-existent. Therefore, a series of dedicated spin-filtering experiments using stored antiprotons needs to be carried out at the AD ring at CERN. At that facility stored antiprotons in the appropriate energy range are available with characteristics that meet the requirements for the first-ever antiproton polarization buildup studies. The equipment required for the spin-filtering experiments at the AD, i.e., the polarized internal target and the new low-β section, efficient polarimeters to determine target and beam polarizations, and a Siberian snake to maintain the longitudinal beam polarization, are presently commissioned and tested at COSY. Only through the investigations at the AD, one can obtain direct access to the spin dependence of the total $\bar{\text{p}}\text{p}$ cross sections. Apart from the obvious interest for the general theory of $\bar{\text{p}}\text{p}$ interactions, the knowledge of these cross sections is necessary for the interpretation of unexpected features of the $\bar{\text{p}}\text{p}$, and other antibaryon-baryon pairs, contained in final states in J/Ψ and B-decays. Of course, once these experiments have provided an experimental data base, the design of a dedicated APR can be targeted.

References

1. A review on the transverse spin structure of the proton can be found in: V. Barone, A. Drago and P. Ratcliffe, *Phys. Rep.* **359**, 1 (2002).
2. M. Anselmino, V. Barone, A. Drago and N. Nikolaev, *Phys. Lett. B* **594**, 97 (2004).
3. A. Efremov, K. Goecke and P. Schweitzer, *Eur. Phys. J* **35**, 207 (2004).
4. K. Rith, in *Proc. 18th Int. Spin Physics Symp., 6–11 Oct. 2008, Charlottesville, Viginia, USA*, eds. D. G. Grabb *et al.*, AIP Conf. Proc. **1149**, 21 (AIP, New York, 2009).
5. J.C. Collins, *Phys. Lett. B* **536**, 43 (2002).
6. D. Sivers, *Phys. Rev. D* **41**, 83 (1990); *Phys. Rev. D* **43**, 261 (1991).
7. M. K. Jones *et al.*, *Phys. Rev. Lett.* **84**, 1398 (2000); O. Gayou *et al.*, *Phys. Rev. Lett.* **88**, 092301 (2002).
8. A. Z. Dubnickova *et al.*, *Nuovo Cimento* **109**, 241 (1966).
9. S.J. Brodsky *et al.*, *Phys. Rev. D* **69**, 054022 (2004).
10. For a discussion on the validity of continuing space-like form factors to the time–like region, see, B. V. Geshkenbein, B. L. Ioffe, and M. A. Shifman, *Sov. J. Nucl. Phys.* **20**, 66 (1975) [*Yad. Fiz.* **20**, 128 (1974)].
11. H.-W. Hammer *et al.*, *Phys. Lett. B* **385**, 343 (1996); H.-W. Hammer; U.-G. Meißner, *Eur. Phys. J.* A **20**, 469 (2004).

12. E. Tomasi-Gustafsson and M.P. Rekalo, *Phys. Lett. B* **504**, 291 (2001); *Nuovo Cimento* **109**, 241 (1996).
13. V. Matveev *et al.*, *Lett. Nuovo Cimento* **7**, 719 (1972).
14. S. Brodsky and G. Farrar, *Phys. Rev. Lett.* **31**, 1153 (1973); *Phys. Rev. D* **11**, 1309 (1973).
15. M. Diehl *et al.*, *Phys. Lett. B* **460**, 204 (1999).
16. P. Landshoff, *Phys. Rev. D* **10**, 1024 (1974); P. Landshoff and D. Pritchard, *Z. Phys.* **C6**, 69 (1980).
17. J.P. Ralston and B. Pire, *Phys. Rev. Lett.* **61**, 1823 (1988); ibid. **49**, 1605 (1982); *Phys. Lett. B* **117**, 233 (1982).
18. G. P. Ramsey and D. W. Sivers, *Phys. Rev. D* **52**, 116 (1995); *Phys. Rev. D* **47**, 93 (1993); *Phys. Rev. D* **45**, 79 (1992).
19. D. Dutta and H. Gao, *Phys. Rev. C* **71**, 032201 (2005).
20. Technical Proposal for *Antiproton-Proton Scattering Experiments with Polarization*, PAX Collaboration, spokespersons: P. Lenisa (Ferrara University, Italy) and F. Rathmann (Forschungszentrum Jülich, Germany), `http://www.fz-juelich.de/ikp/pax/`.
21. D. Oellers *et al.*, *Phys. Lett. B* **674**, 269 (2009).
22. F. Rathmann *et al.*, *Phys. Rev. Lett.* **71**, 1379 (1993).
23. F. Rathmann *et al.*, *Phys. Rev. Lett.* **94**, 014801 (2005).
24. *Proc. Workshop on Polarized Antiprotons, Bodega Bay, CA, 1985*, eds. A. D. Krisch, A. M. T. Lin, and O. Chamberlain, AIP Conf. Proc. **145** (AIP, New York, 1986).
25. D.P. Grosnick *et al.*, *Nucl. Instr. Meth. A* **290**, 269 (1990).
26. H. Spinka *et al.*, *Proc. 8th Int. Symp. on Polarization Phenomena in Nuclear Physics, Bloomington, Indiana, 1994*, eds. E. J. Stephenson and S. E. Vigdor, AIP Conf. Proc. **339**, 713 (AIP, New York, 1995).
27. T.O. Niinikoski and R. Rossmanith, *Nucl. Instr. Meth. A* **255**, 460 (1987).
28. P. Cameron *et al.*, *Proc. 15th Int. Spin Physics Symp., Upton, New York, 2002*, eds. Y. I. Makdisi, A. U. Luccio, and W. W. MacKay, AIP Conf. Proc. **675**, 781 (AIP, New York, 2003).
29. AD Proposal SPSC-P-337, *Measurement of the Spin-Dependence of the pp Interaction at the AD-Ring*; PAX Collaboration, spokespersons: P. Lenisa (Ferrara University, Italy) and F. Rathmann (Forschungszentrum Jülich, Germany), `http://www.fz-juelich.de/ikp/pax` .
30. COSY Proposal #199, *Spin-Filtering Studies at COSY*; PAX Collaboration, spokespersons: M. Nekipelov and Chr. Weidemann (both Forschungszentrum Julich, Germany), `http://www.fz-juelich.de/ikp/pax` .
31. E. Steffens and W. Haeberli, *Rep. Prog. Phys.* **66**, 1887 (2003).
32. K. Zapfe *et al.*, *Rev. Sci. Instr.* **66**, 28 (1995).
33. A. Airapetian *et al.*, *Nucl. Instr. Meth. A* **540**, 68 (2005).
34. M. Mikirtychyants, these proceedings.
35. R. Engels, these proceedings.
36. A. Nass, these proceedings.
37. C. Barschel, these proceedings.

EXPERIMENTAL SETUP FOR SPIN-FILTERING STUDIES AT COSY AND AD

A. Nass* for the PAX-collaboration

Physikalisches Institut II, Universität Erlangen-Nürnberg,
91058 Erlangen, Germany
**E-mail: a.nass@fz-juelich.de*

The high physics potential of experiments with stored high-energy polarized antiprotons led to the proposal of PAX (Polarized Antiproton eXperiment) for the High Energy Storage Ring (HESR) of the new FAIR facility at GSI (Darmstadt/Germany). It is proposed to polarize a stored antiproton beam by means of spin-filtering with a polarized hydrogen (deuterium) gas target. The feasibility of spin-filtering with protons has been demonstrated in the FILTEX experiment. In an additional $\vec{e}\bar{p}$ depolarization experiment at COSY no influence of electron scattering on the proton polarization was found. Several experimental studies with protons (at COSY/Jülich) as well as antiprotons (at AD/CERN) will be carried out to measure spin-dependent $\bar{p}\vec{p}$ and $\bar{p}\vec{d}$ cross sections. A Polarized Internal gas Target (PIT) surrounded by silicon detectors and immersed into a low-β section has to be set up.

Keywords: Polarized targets; antiproton-induced reactions.

1. Principle of spin-filtering

Several methods to polarize antiprotons for a future polarized antiproton experiment[1] were reviewed at workhops held in Bodega Bay, 1985,[2] Daresbury, UK, 2007[3] and Bad Honnef, Germany, 2008.[4] The only successfully tested method to produce a polarized beam is spin filtering. It is based on the effect of selective removal of (anti)protons of a stored beam by a polarized target. The total cross section

$$\sigma_{tot} = \sigma_0 + \sigma_\perp \vec{P} \cdot \vec{Q} + \sigma_\parallel (\vec{P} \cdot \vec{k})(\vec{Q} \cdot \vec{k}), \tag{1}$$

consists of a transverse $\sigma_\perp$ and longitudinal part $\sigma_\parallel$, where $\vec{P}$ is the proton beam polarization, $\vec{Q}$ the target polarization and $\vec{k}$ the proton beam direction. For initially equally populated states $\uparrow$ ($m = +\frac{1}{2}$) and $\downarrow$ ($m = -\frac{1}{2}$)

the total cross sections for the transverse and longitudinal cases are

$$\sigma^{\perp}_{tot\pm} = \sigma_0 \pm \sigma_{\perp}\vec{Q} \quad \text{and} \quad \sigma^{\parallel}_{tot\pm} = \sigma_0 \pm (\sigma_{\perp} + \sigma_{\parallel}) \cdot \vec{Q}\,, \tag{2}$$

respectively. Therefore an initially unpolarized (anti)proton beam will become polarized.[5]

Experiments at the Test Storage Ring (TSR) at Heidelberg in 1993 (fig. 1) showed that the spin-filtering technique works.[6] A polarization buildup was observed (fig. 1, right panel) with an effective polarization buildup cross section of $\sigma_{\perp} = 73 \pm 6$ mb. The current interpretation[7] shows that by considering only elastic pp-scattering, a total cross section of $\sigma_{\perp} = 86 \pm 2$ mb is obtained. Depolarization experiments were already carried out at COSY[7] to investigate the influence of electron scattering on the polarization of a stored proton beam. Spin-filtering experiments[8] will be carried out at COSY with protons, followed by spin-filtering experiments with antiprotons at the Antiproton Decelerator ring (AD/CERN).[9]

Required for the spin-filtering experiments is a highly polarized internal gas target with areal densities of up to $5 \cdot 10^{13}$ atoms/cm^2 using a storage cell. A low-β section is necessary to pass the stored (anti)proton beam through the storage cell and to reduce the Coulomb losses, in order to achieve long storage times of several hours. It is expected that nuclear polarized deuterium could be equally effective for spin-filtering as hydrogen. Therefore, the target should run with hydrogen *and* deuterium with nuclear polarization along x,y, and z using variable target holding fields. Because no analyzing power measurements for $\bar{p}\vec{d}$ scattering exist in this energy range, the deuterium target gas has to be quickly replaced by hydrogen in order to measure the antiproton beam polarization. For longitudinal spin-filtering a Siberian snake has to be implemented in order to preserve the longitudinal polarization of the beam at the location of the internal target.

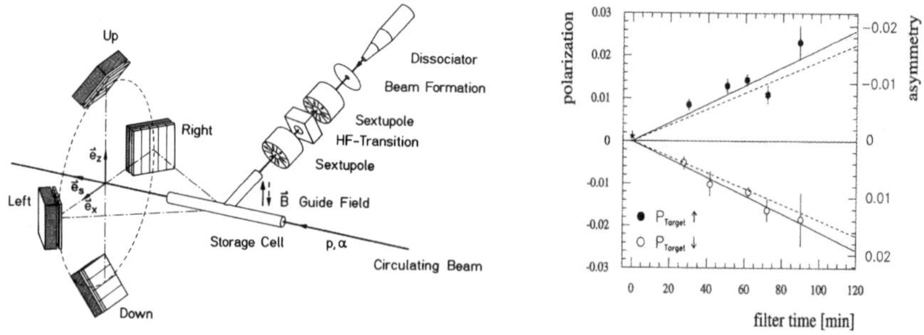

Fig. 1. The setup of the test experiment at TSR and the results.[6]

2. Experimental setup

The setup for the spin-filtering experiments (fig. 2) consists of an Atomic Beam Source (ABS) to produce polarized target gas, a target chamber with storage cell and a detector system to detect forward and recoil (anti)protons. A so-called Breit-Rabi Polarimeter (BRP) is used to measure the polarization of the target gas. A low-β section consisting of four(six) magnets is necessary for the measurements at COSY(AD). This is designed for the Interaction Point IP 1 at COSY, and for one of the straight sections at AD. For longitudinal spin-filtering at COSY, the solenoids of WASA and the electron cooler, located in the opposite straight section, can be used as a siberian snake at injection energy of 45 MeV. At the AD, and for higher energies at COSY, an additional snake is required.

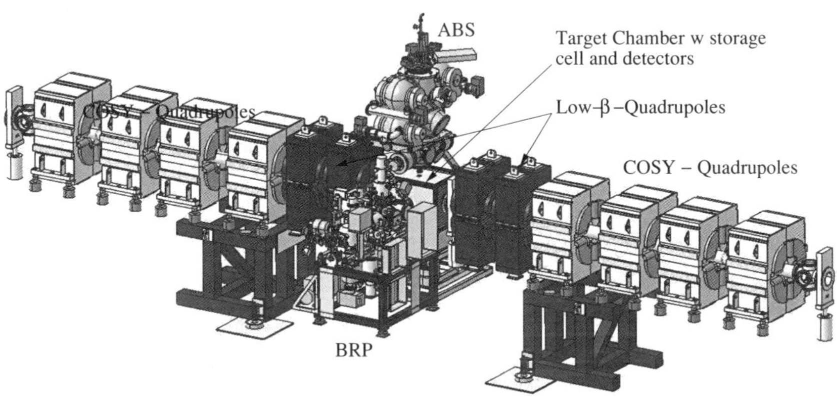

Fig. 2. Overview over the setup for the spin-filtering experiments at COSY.

2.1. *The ABS*

The former HERMES-ABS was set up in Jülich with a modified vacuum system, mounted on a new support. The cryogenic pumps were replaced by turbo molecular pumps and an oil-free forevacuum system. The source was completely recabled to allow for a fast assembly and disassembly at COSY and AD. The control system was renewed to allow for a full remote control via computer. The vacuum system with the microwave dissociator is operating well.

After construction of a new analysis chamber with Quadrupole Mass Spectrometer (QMS) and a calibrated compression tube, the first intensity

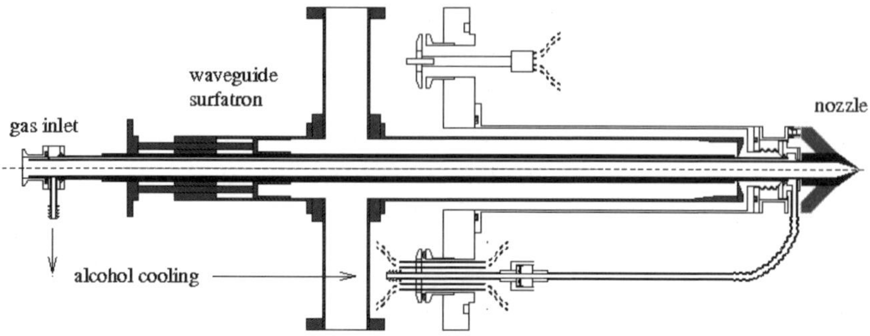

Fig. 3. Sketch of the new liquid (alcohol) cooled microwave dissociator.

measurements were carried out. The measurements showed atomic beam intensities of up to $6 \cdot 10^{16}$ atoms/s for hydrogen in two hyperfine states. As a new development, a liquid cooled microwave dissociator (fig. 3) was installed. First tests showed a stable behaviour with microwave powers ranging up to 1200 W. The dependence of the ABS output intensity on the MW-power was linear, without showing any sign of saturation. The dependence on the temperature of the nozzle is shown in the left panel of figure 4. It shows a well understood behaviour with a maximum beam intensity at around 100 K. In contrast to this, the ABS intensity showed a strong dependence on the temperature of the cooling liquid (fig. 4, right panel). It seems that the surface recombination is high and lower temperatures will reduce this effect.

The new and unique feature of the present setup is that the ABS will be able to produce nuclear polarized hydrogen or deuterium beams in short sequence (5 minutes) without any mechanical modifications.

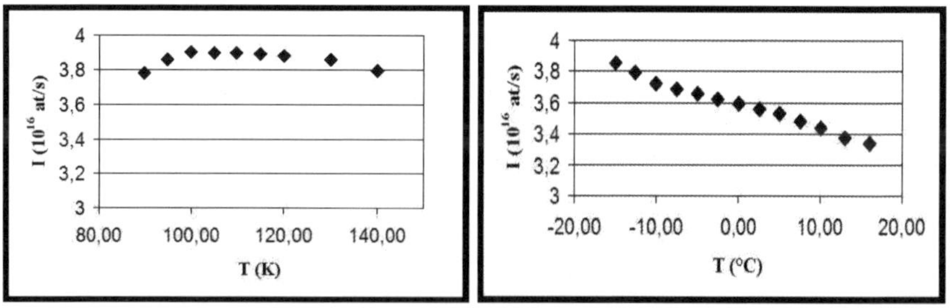

Fig. 4. The dependence of ABS intensity vs. temperature of the nozzle (left panel) and vs temperature of the cooling liquid (right panel). The hydrogen flux was 82 sccm (1.37 mbarl/s), the oxygen flux 0.2 sccm ($3 \cdot 10^{-3}$ mbarl/s), and the microwave power 1200 W.

2.2. *The target chamber with storage cell*

Since spin-filtering requires areal densities of up to $5 \cdot 10^{13}$ atoms/cm^2, the use of a storage cell is mandatory. The present cell design consists of 5 μm Teflon walls supported by an aluminum frame. Thin walls allow low energy recoil particles to pass through and be detected by the Silicon Tracking Telescopes (STT). Teflon also suppresses depolarization and recombination of the target gas inside the cell. The cell has to be openable to provide enough space for the beam during injection at AD. The cell will be closed after the beam is decelerated and cooled. Subsequently, the target gas is injected. Longitudinal and transverse weak holding field coils, added on the outside of the target chamber, provide the quantization axis either x, y, or z for the polarized atoms.

The vacuum system of the target section comprises one large cryogenic pump with a pumping speed of about 20000 l/s, and two turbo molecular pumps, backed with smaller turbo molecular pumps, and a dry forevacuum pump (fig. 5). This will ensure that most of the target gas exiting the storage cell is pumped away in the target chamber. Flow limiters will be installed between the target chamber and the magnet chambers in order to reduce the gas load.

Fig. 5. The planned vacuum system of the target chamber. See also figure 2.

2.3. *The BRP*

The BRP is necessary to determine the nuclear polarization of deuterium at the AD because the analyzing powers for $\bar{p}\vec{d}$ scattering are unknown. In addition, it will provide information about the status of the polarized target and the efficiencies of the high frequency transitions. The former HERMES-BRP was rebuilt on a new support stucture with modifications due to the new configuration within the experiment. Tracking calculations led to a modified sextupole magnet configuration in the BRP to adjust for the higher temperature of 300 K of the effusive hydrogen/deuterium beam out of the uncooled storage cell; at HERMES the cell was at 100 K. In addition, a new strong field transition "dual cavity" was designed in order to induce transitions between hyperfine levels of hydrogen or deuterium. The quadrupole mass spectrometers and the high frequency transitions are working properly. The start-up of the data aquisition system is on the way.[10]

To match the tight spacial conditions at AD, the lower part of the BRP has to be rebuilt (fig. 6) using a new 90° cryogenic pump and a modified lower BRP chamber. The latter will also contain a cooling inset for the chamber walls around the Ti-ball to increase the pumping capability in this area.

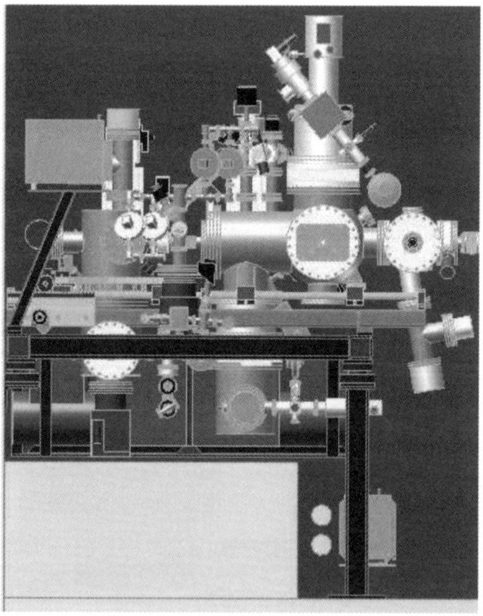

Fig. 6. The Breit-Rabi polarimeter in the AD environment.

2.4. *The detection system*

The beam polarization at COSY and AD will be measured using pp (p$\bar{p}$)-elastic scattering. To this aim, a detector system consisting of 12 STTs will be implemented around the target cell (fig. 7). The STTs will detect both the low energy recoil particles ($< 8\,\text{MeV}$) as well as the forward scattered particles with a large angular coverage and high resolution. In addition, the proton polarization of the target can be measured using an initially unpolarized (anti)proton beam. This allows for the calibration of the BRP.

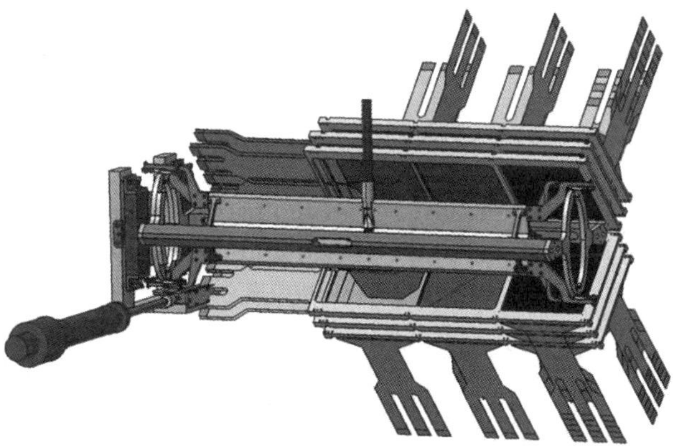

Fig. 7. The designated detector setup.

2.5. *The low-beta section*

The low-β-section will consist of four(six) normal conducting quadrupole magnets (fig. 2). They will be implemented into the COSY (AD) lattice prior to the installation of the target. Calculations show that the dependence of the beam envelope along the target section will match the requirements.[11]

3. Planned measurements

A first series of spin-filtering measurements is planned to be carried out at COSY/Jülich with an initially unpolarized proton beam and a nuclear polarized target in a weak holding field. It will determine the polarization buildup[8] and commission the experiment for the AD. With the installation of the magnets of the low-β section, the first step towards the realization of the experiment was done in summer 2009. Now the COSY team has to understand and operate the system. The installation of the polarized target is foreseen to take place in summer 2010, and subsequently, a first spin-filtering experiment will be carried out.

A second series of measurements is planned at AD/CERN. These measurements will provide data to estimate the achievable polarization in spin-filtering with an antiproton beam.[9]

References

1. *Antiproton-Proton Scattering Experiments with Polarization*, Technical Proposal for the HESR at FAIR, Jülich, e-Print Archive: hep-ex/0505054 (2005).
2. A. D. Krisch *et al.* (eds.), *Polarized antiprotons*, AIP Conf. Proc. **145** (1986).
3. D. P. Barber *et al.* (eds.), *Polarized Antiproton Beams - How?*, AIP Conf. Proc. **1008** (2008).
4. *Polarized Antiprotons*, WE-Heräus Seminar, Bad Honnef, `http://www.fe.infn.it/heraeus/index.html` (2008).
5. F. Rathmann *et al.*,*Spin-filtering studies at COSY and AD*, these proceedings.
6. F. Rathmann *et al.*, *Phys. Rev. Lett.* **71**, 1379 (1993).
7. D. Oellers *et al.*, *Phys. Lett. B* **674**, 269 (2009).
8. *Spin-Filtering Studies at COSY*, Proposal for COSY, available at `http://www.fz-juelich.de/ikp/pax/` (2009).
9. *Measurement of the Spin-Dependence of the* $\bar{p}p$ *Interaction at the AD-Ring*, Proposal for AD/CERN, available at `http://www.fz-juelich.de/ikp/pax/` (2005).
10. C. Barschel *et al.*,*Target section for spin-filtering studies at COSY and AD*, these proceedings (2010).
11. A. Nass *et al.*, *spin-filtering Studies at COSY and AD*, AIP Conf. Proc. **1149**, 781 (AIP, New York, 2009).

POLARIZING A STORED PROTON BEAM BY SPIN-FLIP? — A REANALYSIS

D. Oellers* on behalf of the PAX-Collaboration

IKP-2, Forschungszentrum Jülich,
52428, Jülich, Germany
**E-mail: di.oellers@fz-juelich.de*
www.fz-julich.de/ikp/

We discuss polarizing a proton beam in a storage ring, either by selective removal or by spin-flip of the stored ions. Prompted by recent, conflicting calculations, we have carried out a measurement of the spin-flip cross section in low-energy electron-proton scattering. The experiment uses the cooling electron beam at COSY as an electron target. A re-analysis of the data leeds to reduced statistical errors, resulting in reduction by a factor of four of the upper limit for the spin-flip cross section. The measured cross sections are too small for making spin-flip a viable tool in polarizing a stored beam.

Keywords: Polarized beams; storage rings; electron-proton scattering; antiprotons.

1. Introduction

Usually, polarized ions in a storage ring are provided by injecting an already polarized beam from a suitable ion source. Alternatively, it is conceivable to polarize an initially unpolarized beam while it is stored in the ring. In the case of a spin-1/2 beam (with two spin states) this would be achieved by either selectively discarding particles in one spin state ("filtering"), or by selectively reversing the spin of particles in one spin state ("flipping").

After summarizing ideas and experimental results concerning the *in situ* polarization of a stored proton beam, we report in this paper a direct experimental evaluation of spin-flip in electron-proton scattering, and its contribution to polarizing the proton beam. The experiment, which is making use, for the first time, of the electron cooler as an electron target, has been carried out to resolve the discrepancy between two recently published calculations,[1,2] and to settle the question of whether, in the future, spin-flip will play a role in polarizing stored beams.

We are interested in *in situ* polarization of a stored proton beam because we hope to be able to apply the same technique to antiprotons. The need for polarized antiproton beams is well recognized prerequisite for addressing several important topics in particle physics, including a first measurement of the transversity distribution of the valence quarks in the proton, a test of the predicted opposite sign of the Sivers-function (related to the quark distribution inside a transversely polarized nucleon) and a first measurement of the moduli and the relative phase of the time-like electric and magnetic form factors of the proton.[3]

2. *In situ* polarization of a stored beam

2.1. *Evolution of the beam polarization*

Let us consider a storage ring that contains $N = N_\uparrow + N_\downarrow$ spin-1/2 particles in the two allowed substates, $\uparrow$ and $\downarrow$. The arrows indicate spins pointing along or opposite the quantization axis. The beam polarization is given by $P_B = \frac{N_\uparrow - N_\downarrow}{N}$. The beam interacts with an internal spin-1/2 target with polarization P_T and area number density d_T. The orbit frequency is f_R.

A particle traversing the target may be removed from the stored beam by a reaction or by scattering by an angle larger than the ring acceptance. The removal cross section, integrated over the appropriate solid angle, is defined as $\sigma_R \equiv 1/2(\sigma_R(\downarrow\uparrow) + \sigma_R(\uparrow\uparrow))$. In principle, it is possible to change the polarization of the stored beam by spin-flip of particles that interact with the target but remain in the ring. The cross section for the spin-flip of a beam particle is defined as $\sigma_S \equiv 1/2(\sigma_S(\downarrow\uparrow) + \sigma_S(\uparrow\uparrow))$. The arrows indicate whether the spins of projectile and target are opposite or parallel. The spin-dependent part of these two cross sections are given by $\Delta\sigma_R \equiv 1/2(\sigma_R(\downarrow\uparrow) - \sigma_R(\uparrow\uparrow))$ and $\Delta\sigma_S \equiv 1/2(\sigma_S(\downarrow\uparrow) - \sigma_S(\uparrow\uparrow))$. Scattering within the ring acceptance, but without a spin-flip, does not affect the beam polarization at all and can be ignored.

The time evolution equations for the beam polarization P_B and the number of stored particles N have been discussed repeatedly, see for example 4. Here only two special cases are discussed. The first case deals with polarizing an initially unpolarized beam ($P_B = 0$). As long as P_B is still small, the rate of change of polarization is constant and given by

$$\frac{dP_B}{dt} = f_R d_T P_T \left[2\Delta\sigma_S + \Delta\sigma_R\right]. \tag{1}$$

We define the "polarizing cross section", σ_{pol}, as the sum of the two terms in the bracket.

The second special case describes the effect of an unpolarized target ($P_T = 0$) on an already polarized beam,

$$\frac{dP_B}{dt} = -2 f_R d_T \sigma_S P_B, \tag{2}$$

which shows that the "de-polarizing cross section" is equivalent to twice the spin-flip cross section σ_S. Since it is always true that $\sigma_S \geq \Delta\sigma_S$, it follows from equations (1) and (2) that if a polarized target is capable of polarizing an unpolarized beam by spin-flip, an unpolarized target will de-polarize an already polarized beam. The experiment described in this paper makes use of this principle.

2.2. *Spin-filtering*

The first (and so far only) evidence that a stored hadron beam can be polarized *in situ* was presented in 1993 by the FILTEX group.[5] The experiment was carried out in the TSR at Heidelberg with a 23 MeV proton beam, orbiting with $f_R = 1.177\,\text{MHz}$ in the presence of a polarized atomic hydrogen target. The target atoms were in a single spin state, i.e. protons and electrons were both polarized. The polarization buildup of an initially unpolarized beam was measured; the result was $\frac{dP_B}{dt} = (1.29 \pm 0.06) \cdot 10^{-2}$ per hour.[5]

In the FILTEX experiment, the target thickness was $d_T = (5.3 \pm 0.3) \cdot 10^{13}\ \text{cm}^{-2}$ and the target polarization was $P_T = 0.795 \pm 0.024$. Inserting these numbers into equation (1), one finds for the polarizing cross section[5,6]

$$\sigma_{\text{pol}} = (73 \pm 6)\,\text{mb}. \tag{3}$$

Theoretical calculations[4,7,8] based on pp interaction result in

$$\sigma_{\text{pol,theo}} = (86 \pm 2)\,\text{mb}. \tag{4}$$

The fact that experiment and theory (eqs. (3) and (4)) disagree by two standard deviations was the original motivation for investigating the role of spin-flip.

2.3. *Spin-flip*

During the analysis of the FILTEX result, it became clear that small-angle scattering, for which the ion remains in the beam, is a significant part of the total cross section.[9] It was argued that this scattering, without loss, may be accompanied by spin-flip. This would include scattering not only from the polarized protons of the atomic beam target, but also from the electrons,[10] which are also polarized. Because of their much larger mass, protons scattering from electrons always stay within the acceptance. Evaluating the spin transfer cross section (as defined for example in ref. 11) at small angles between 10 and 100 MeV, sizeable effects were predicted.[9] A decade later, Milstein and co-workers[12] showed that the relevant quantity to evaluate is the spin-flip cross section, which is much smaller that the spin transfer cross section and is in fact negligible for the proton energy used in the FILTEX experiment.

More recently, Arenhövel[1] predicted that the spin-flip cross section in electron-proton scattering at low energy (a few eV in the center-of-mass system) is very large because of the mutual attraction of the two oppositely charged particles. Walcher and co-workers adopted this idea for a proposal to polarize stored antiprotons with a co-moving beam of polarized positrons.[13] The proper low interaction energy would be achieved by making the two beam velocities almost the same. Even though the achievable positron beam intensities are quite low, the predicted spin-flip cross sections are so large that the scheme would still be feasible. For instance, at a center-of-mass energy of 0.93 eV (corresponding to a proton energy in the lepton rest frame of $T_h = 1.7\,\text{keV}$) Arenhövel predicts a spin-flip cross section of $\sigma_S = 4 \cdot 10^{13}\,\text{b}$. However, a calculation of the same quantity by Milstein and co-workers[2] resulted in $\sigma_S = 0.75\,\text{mb}$. The goal of the experiment described in the following is to resolve this discrepancy of 16 orders of magnitude.

3. Experiment

The goal of this experiment is to determine the depolarization of a polarized proton beam by its interaction with the electrons of the cooler beam. The measurement is carried out with a proton beam in the COSY ring,[14] using the detector setup in the target chamber of the ANKE spectrometer.[15] The proton energy is $T_p = (49.3 \pm 0.1)\,\text{MeV}$, corresponding to a velocity of $v_p = 0.312 \cdot c$, and the usual relativistic parameter $\gamma_p = 1.053$.

3.1. *Cooler beam as an electron target*

In this experiment, the COSY electron cooler[16] serves two functions. On the one hand, as usual, it provides the phase-space cooling of the stored proton beam, while on the other hand it plays the role of an electron target for the actual measurement of the low-energy spin-flip cross section in *ep* scattering.

In the cooling mode, the electron velocity is adjusted to the velocity v_p of the stored protons. When the cooler is used as a target, a relative motion between the proton and the electron beam is achieved by "detuning" the accelerating voltage by ΔU, changing the electron velocity by Δv_e, and inducing an average relative "detune" velocity u_0.

Besides this induced velocity, there are additional contributions to the relative motion between protons and electrons. The dominant effect arises from the transverse thermal motion of the electrons. Other contributions include the betatron motion of the protons, the velocity spread of both beams, and the ripple on the electron high-voltage supply.

3.2. *Cycle scenario*

The scenario of our experimental cycle is shown in figure 1. At the beginning of the cycle, the ring is filled with vertically polarized protons (typically, the beam polarization is $P_B \approx 0.5$). During the first half of the cycle, the coasting beam is interacting with the electrons in the cooler. During the second half, while cooling the beam, the internal deuteron target is turned on to measure the beam polarization.

The first half of the cycle contains 49 sub-cycles of 10 s length. During such a sub-cycle the electron velocity is first tuned to the beam velocity to cool the beam for 5 s, then the electron beam velocity is detuned for another 5 s. This is the time when the actual experiment takes place with a total "interaction" time in the detuned mode of $t_{\rm int} = 245$ s per cycle. The scenario just described shall be called "E-cycle". To reduce systematic

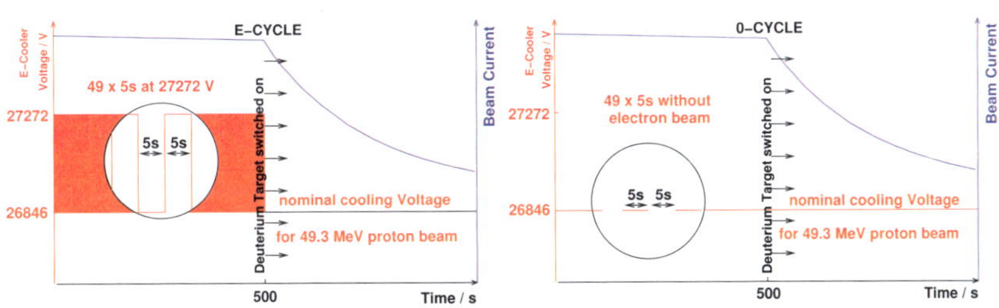

Fig. 1. Time sequence of the experimental "E-" and "0-cycles". Both electron accelerating voltage ($\Delta U = 320.42$ V) and number of protons orbiting in the ring are plotted.

uncertainties, E-cycle polarization measurements are compared to those observed in a reference cycle, or "0-cycle". Reference cycles are identical in every respect, except that during the interaction time (in the second half of the sub-cycles, for a total time $t_{\rm int}$ in each cycle) the cooler beam is *turned off*. During the experiment, E-cycles and 0-cycles are alternated, first with beam polarization up (↑), then with an unpolarized beam and finally with polarization down (↓). The deduced polarization ratio $R \equiv P_E/P_0$ (see sec. 3.3.3) reflects the effect of an electron target on the beam polarization.

3.3. *Polarimetry*

3.3.1. *Hardware*

The beam polarization is measured using *pd* elastic scattering. Precise analyzing power data are available at $T_p = 49.3\,\text{MeV}$[17] and cross sections have been measured at a nearby energy ($T_p = 46.3\,\text{MeV}$).[18] The beam energy for this experiment was chosen partly because of this.

The target consists of a deuterium cluster jet with about $5 \cdot 10^{14}$ deuterons per cm^2.[19]

The detector system consists of two silicon tracking telescopes[20] placed symmetrically to the left and right of the beam, as shown in figure 2. Each telescope features three position-sensitive detectors, oriented parallel to the beam direction. The first two layers are 300 μm thick with an active area of 51 mm by 66 mm. They are located 28 mm and 48 mm from the beam axis. The third, 5 mm thick detector, 68 mm from the beam axis is not used in this experiment. Within the mechanical constraints of the detector support, the telescope positions with respect to the interaction region are chosen to optimize the figure of merit for the pd analyzing reaction. The position resolution of the detectors is about 200 μm, both, vertically (y axis) and along the beam direction (z axis).

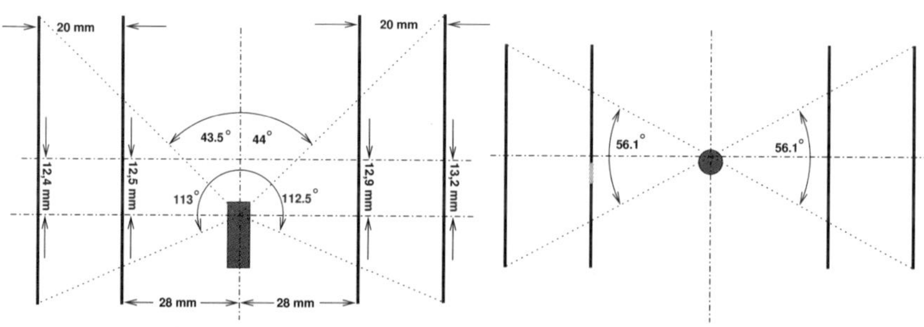

Fig. 2. Detector setup in the target chamber of the ANKE spectrometer, seen from the top (left) and in beam direction (right). The detector telescopes are mounted to the left and right of the interaction region. The beam target overlap is as well indicated as a region in the left detector, which due to radiation damage gives no data.

3.3.2. *Event selection*

A first analysis described in reference 21, is based on clearly identified pd elastic scattering events. Additionally, two different samples have been extracted from the data. A minimum bias sample "MB" was selected by choosing the complete deuteron region in the energy loss spectra and applying an additional cut on the scattering angle (fig. 3). In case no deuteron was found by "MB", the event was analyzed for events with exactly one track and in case this track stored in the "OT" sample. By this the "MB" and the "OT" samples are completely disjunct and therefore statistical independent.

3.3.3. *Determination of asymmetry*

Making use of the cross ratio method, we calculate the asymmetry for bins in the deuteron scattering angle (a detailed description can be found in ref. 21).

$$\epsilon_n = \frac{1}{\langle\cos\phi\rangle}\frac{\delta_n - 1}{\delta_n + 1}, \text{ where } \delta_\mathrm{n} = \sqrt{\frac{\mathrm{Y_{L\uparrow}(n) \cdot Y_{R\downarrow}(n)}}{\mathrm{Y_{L\downarrow}(n) \cdot Y_{R\uparrow}(n)}}}. \tag{5}$$

Taking the weighted average for all bins, one arrives at the beam polarization. This procedure is carried out separately for E-cycles and 0-cycles, resulting in the respective polarizations P_E and P_0, with or without electron beam during the "interaction" part of the cycle. The ratio $R \equiv P_E/P_0$ then constitutes the final result of the polarization measurement. The systematic errors of this measurement can be neglected.

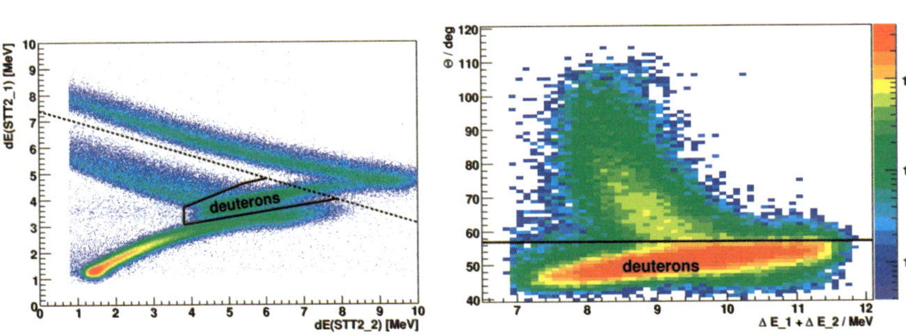

Fig. 3. Left: This energy loss spectrum for STT2 (right STT) indicates the minimum bias cuts, which are rsed to reconstruct deuterons. Right: The additional cut $\theta < 57°$ strongly reduces the background from breakup protons.

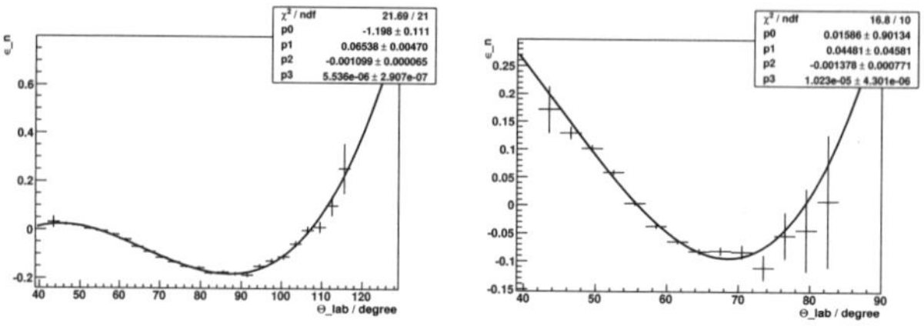

Fig. 4. Left(Right): Figure for the "OT"("MB") sample. The asymmetries ϵ_n with its statistical errors *vs* the scattering angle in laboratory θ. The horizontal bars indicate the bin width of 3°. The solid curve represents a polynomial fit to the measured asymmetry.

Similar to the polarization, the asymmetries ϵ_E and ϵ_0 of the "MB" and "OT" samples have been individually used to evaluate the ratio R of the polarizations, as it is independent of the analyzing powers.

$$R = \frac{P_E}{P_0} = \frac{\epsilon_E \cdot A_y}{\epsilon_0 \cdot A_y} = \frac{\epsilon_E}{\epsilon_0}. \tag{6}$$

The solid curve in figure 4 is a polynomial fit to that part of the data from the E-cycle with 426 V detuning potential, scaled to fit the individual datasets. The fitted functions follows the shown data points with a detuning potential $\Delta U = 246\,\mathrm{V}$. This is true for the asymmetries ϵ_0 and ϵ_E for all data points and therefore a proof of the stability of the event selection. The scaling factors α_0 and α_E are directly proportional to the measured asymmetries and their ratio gives R.

3.4. *Results*

For each of the six detune potentials $\Delta k (k = 1 \ldots 6)$, the result of the measurement consists of the ratios $R_k \equiv \left(\frac{P_E}{P_0}\right)_k$ as described in the previous section. Due to the random thermal movement of the electrons, two cross sections with spin along (transverse) σ_λ (σ_T) the relative motion contribute and one obtains[21]

$$y_k \equiv \frac{-\ln R_k}{2ct_{\mathrm{int}} n_{e,k} u^{\star 2} (L_C / L_R)} = \sigma^{\star}_{S,\tau} I_{\tau,k} + \sigma^{\star}_{S,\lambda} I_{\lambda,k}. \tag{7}$$

The denominator contains the speed of light, the interaction time $t_{\mathrm{int}} = 245\,\mathrm{s}$, the electron density n_e, a reference velocity, arbitrarily set to $u^\star = 0.002$, the active length $L_C = (1.75 \pm 0.25)\,\mathrm{m}$ of the cooler, and the ring circumference $L_R = 183.47\,\mathrm{m}$. The cooler length is uncertain because of details of inflection and extraction of the electron beam, and the electron density is affected by uncertainties of the electron beam current $I_e = 170\,\mathrm{mA}$ and its area $A_e = 5\,\mathrm{cm}^2$. We estimate that the overall systematic uncertainty of the denominator is $\pm 20\,\%$.

The asymmetry ratios R_k (fig. 5) are consistent with unity, i.e. the polarization differences between E-cycles and 0-cycles are of the order of their statistical errors.

The depolarizing cross sections, $\sigma^\star_{S,\tau}$ and $\sigma^\star_{S,\lambda}$ (at the reference velocity $u^\star$) appear as unknowns in equation (7). Since our experiment fails to find a depolarization effect, we instead derive an upper limit for the two cross sections that is compatible with our data. Following the usual treatment, we define the likelihood function

$$L(\overrightarrow{y}|\sigma^\star_{S,\tau},\sigma^\star_{S,\lambda}) \equiv \prod_k \exp\left(-\frac{(y_k - \sigma^\star_{S,\tau} I_{\tau,k} - \sigma^\star_{S,\lambda} I_{\lambda,k})^2}{2\delta y_k^2}\right). \qquad (8)$$

The experimental result, y_k, is defined in equation (7); Following the bayesian approach, we calculate the posterior probability density function

$$p(\overrightarrow{y}|\sigma_{S,\tau},\sigma_{S,\lambda}) = \frac{L(\overrightarrow{y}|\sigma^\star_{S,\tau},\sigma^\star_{S,\lambda})h(\sigma^\star_{S,\tau},\sigma^\star_{S,\lambda})}{\int L(\overrightarrow{y}|\hat{\sigma}^\star_{S,\tau},\hat{\sigma}^\star_{S,\lambda})h(\hat{\sigma}^\star_{S,\tau},\hat{\sigma}^\star_{S,\lambda})d\hat{\sigma}^\star_{S,\tau}d\hat{\sigma}^\star_{S,\lambda}}. \qquad (9)$$

The function h reflects our prior knowledge (cross sections are positive numbers) and is set to a constant for all non-negative values of $\sigma^\star_{S,\tau}$ and $\sigma^\star_{S,\lambda}$, and to zero otherwise.

The probability p is evaluated numerically. The upper cross section limits, shown in figure 5, are contours of constant p.

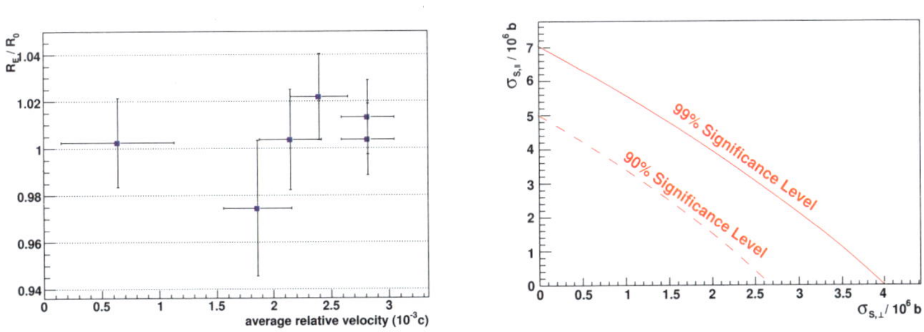

Fig. 5. Left: Ratio R of Polarizations with and without electron beam. Right: Upper limit allowed by the data of this experiment for transverse and longitudinal spin-flip cross sections at a relative velocity of $u^\star = 0.002$.

As mentioned earlier, the spin-flip cross sections are proportional to the inverse square of the relative velocity $u^\star$. The values shown in figure 5 are for $u^\star = 0.002$, corresponding to a center-of-mass energy of about 1 eV, or to a proton kinetic energy in the electron rest system of $T_h = 1.2\,\text{keV}$.

The present result is in agreement with the calculation of Milstein *et al.*,[2] but clearly rules out the validity of the prediction of $\sigma_{S,\lambda} = 4 \cdot 10^{13}\,\text{b}$ claimed in references 1 and 13. Since the completion of this experiment, the calculation presented in these two references has been withdrawn[22,23]).

Acknowledgments

We are grateful to the operators of the COSY facility for their help in setting up the accelerator and the unusual performance parameters of the cooler. We also appreciate the occasional contributions of a number of members of the ANKE and PAX collaborations, who were not directly involved in the experiment. D. Oellers would like to thank the organizers of the conference for the invitation.

References

1. H. Arenhövel, *Eur. Phys. J. A* **34**, 303 (2007).
2. A. I. Milstein, S. G. Salnikov and V. M. Strakhovenko, *Nucl. Instrum. Meth. B* **266**, 3453 (2008).
3. Technical Proposal for Antiproton-Proton Scattering Experiments with Polarization, PAX Collaboration, `http://arxiv.org/abs/hep-ex/0505054` (2005). An update can be found at the PAX website `http://www.fz-juelich.de/ikp/pax` .
4. N. Nikolaev and F. Pavlov, `http://arXiv.org/abs/hep-ph/0701175` .
5. F. Rathmann *et al.*, *Phys. Rev. Lett.* **71**, 1379 (1993).
6. F. Rathmann, Ph. D. thesis, Phillips-Universität Marburg, Jan. (Marburg, Germany, 1994).
7. V. Strakhovenko, in *Proc. Int. Workshop on Polarized Antiproton Beams - how?*, AIP Conf. Proc. **1008**, 44 (AIP, New York, 2008).
8. N. Nikolaev and F. Pavlov, in *Proc. Int. Workshop on Polarized Antiproton Beams - how?*, AIP Conf. Proc. **1008**, 34 (AIP, New York, 2008).
9. H. O. Meyer, *Phys. Rev. E* **50**, 1485 (1994).
10. C. Horowitz and H. O. Meyer, *Phys. Rev. Lett.* **72**, 3981 (1994).
11. J. Bystricky, F. Lehar and P. Winternitz, *J. Physique* **39**, 1 (1978).
12. A. I. Milstein and V. M. Strakhovenko, *Phys. Rev. E* **72**, 066503 (2005).
13. Th. Walcher *et al.*, *Eur. Phys. J. A* **34** , 447 (2007).
14. R. Maier *et al.*, *Nucl. Instr. Meth. A* **390**, 1 (1997).
15. S. Barsov *et al.*, *Nucl. Instr. Meth. A* **462**, 364 (2001).
16. H. J. Stein *et al.*, *Atomic Energy* **94**, 24 (2003) and in *Proc. XVIII Conference on Accelerators of Charged Particles, RUPAC-2002, Obninsk, Russia*, eds. I. N. Meshkov *et al.*, 220 (NRCRFf, Obninsk, 2004).
17. N. S.King *et al.*, *Phys. Lett. B* **69**, 151 (1977).

18. S. N. Bunker *et al.*, *Nucl. Phys. A* **113**, 461 (1968).
19. A. Khoukaz *et al.*, *Eur. Phys. J. D* **5**, 275 (1999).
20. R. Schleichert *et al.*, *IEEE Trans. Nucl. Sci.* **50**, 301 (2003).
21. D. Oellers *et al.*, *Phys. Lett. B* **674**, 269-275 (2009).
22. H. Arenhövel, *Eur. Phys. J. A* **39**, 133 (2009).
23. Th. Walcher *et al.*, *Eur. Phys. J. A* **39**, 137 (2009).

TRACKING STUDIES OF SPIN COHERENCE IN COSY IN VIEW OF EDM POLARIZATION MEASUREMENTS

A. U. Luccio[a,*], F. Lin[a], C. J. G. Onderwater[b] and E. J. Stephenson[c]

[a]*Brookhaven National Laboratory, Upton, NY 11973, USA*
[b]*KVI & Uni. of Groeningen, The Netherlands*
[c]*Indiana Uni., Bloomington, IN, USA*
[*]*E-mail: luccio@bnl.gov*

Measurements of the polarization of deuterons in COSY are being carried on to prepare for similar measurements for the Electric Dipole Moment (EDM) experiment being proposed at Brookhaven. Spin tracking studies are presented, in particular the study of spin coherence time of polarization survival and possible methods to increase it.

Keywords: Deuteron polarization; spin; accelerator.

1. Strategy of the simulation — spin line width

Spin decoherence is a very important problem for the EDM experiment,[1] since it determines the time available for measurement. The accuracy of polarization measurements, during a single storage ring fill, can be expressed as

$$\sigma_s \propto 1/(PE\sqrt{N}TA),$$

with P, polarization, E, electric field in the particle rest frame, N, number of particles, T, time of measurement and A, analyzing power of polarimeter. To reach small values of σ_s, we should try, among other things, to keep the time of measurements with a high polarization as long as possible. A polarized beam is spin coherent when all the particles' spins precess in phase. Long spin coherence implies that the original polarization is retained.

The low energy deuteron storage ring COSY at Jülich, provides a very important facility for performing polarization measurements useful for the EDM. In view of polarimetry tests, a simulation study was done with the spin tracking code SPINK,[2] to specifically address the following issues: (1) what is the polarization lifetime with the current COSY lattice, (2) what are the causes of spin decoherence, (3) what options are available for lengthening the polarization lifetime? Spin decoherence was studied on a realistic beam simulated by an ensemble of particles, with finite emittance, energy spread and bunch length, in the full 6-dimension phase space, transverse and longitudinal.

Polarized deuterons of momentum and G-energy

$$pc = 0.97\text{ GeV}, \quad G\gamma = -0.161003$$

were stored in the ring at constant energy with initial vector polarization "up". G is the magnetic anomaly ($G = -0.14301$ for deuterons.) The COSY storage ring contains sextupoles for chromaticity correction. In the simulation we included magnet fringe fields that, at low energy, produce a noticeable effect on spin dynamics.

The number of spin oscillations per turn of a particle is defined as that particle's spin tune. Spin tracking of a bunch of particles produces a spin tune line. Spin coherence can be calculated by the spin tune linewidth: the narrower it is, the highest the spin coherence.

There are several ways to calculate spin coherence in simulation by spin tracking: (1) calculate the spectrum of spin oscillation frequency by Fourier analysis of spin oscillations: method used by F.Lin,[4] (2) produce a spin-flip by some device like an RF solenoid and measure the frequency at which the spin of each particle flips,[5] (3) calculate the spin tune as one of the eigenvalues of the one-turn spin matrix: method we used in this study.

2. *SPINK* formalism — one turn spin matrix

The tracking code SPINK uses the relativistic Thomas-BMT equation for spin motion in an an e.m. field, in vector and matrix form, respectively

$$\frac{d\mathbf{S}}{dt} = \mathbf{S} \times \mathbf{b}, \;\; \mathbf{S} = \mathcal{M}\mathbf{S}. \tag{1}$$

The equation describes the rotation of the spin vector $\mathbf{S}$ around an axis $\mathbf{b}$. In the absence of electric field

$$\mathbf{b} = \frac{q}{\gamma m}\left[(1 + G\gamma)\mathbf{B}_\perp + (1 + G)\mathbf{B}_\parallel\right], \tag{2}$$

q and m, charge and mass of the particle, $\gamma = E/mc^2$, Lorentz energy factor, $\mathbf{B}_\perp$ and $\mathbf{B}_\parallel$, magnetic field components perpendicular and parallel to the velocity of the particle.

$\mathcal{M}$ is a 3×3 matrix for vector polarization, with elements a function of three angles: a spin kick $\delta\mu$, latitude θ, and longitude ϕ, to define the direction of the axis **b**. $\mathcal{M}$ represents exactly one rotation. The matrix elements are a function of the coordinates of a moving particle, then the three angles are in turn a function of the phase space coordinates, calculated by simultaneous solution of the equations of motion, e.g. by TEAPOT.[3]

By multiplying all the spin matrices in one machine turn, obtain the spin One-Turn Matrix (OTM), that represents the rotation of the **S** vector. One of the eigenvalues of the OTM $\mathcal{M}_{OT}$ gives the spin tune in one turn[a]

$$\nu_s = \frac{1}{2\pi} arcos \left(\frac{Tr(\mathcal{M}_{OT}) - 1}{2} \right) \tag{3}$$

where $Tr(\mathcal{M}_{OT})$ is the trace of the matrix.

Spin tune so calculated oscillates, from turn to turn. around an average value, converging to some asymptotic value (fig. 1), which we will use in the following as the "spin tune". Equivalently, since spin and orbital motion are coupled, the spin tune could also be calculated from the matrix of a "true" spin one-turn, i.e. when the particle approximately reassumes its starting coordinates.

Spin linewidth will linearly increase with the number of turns.

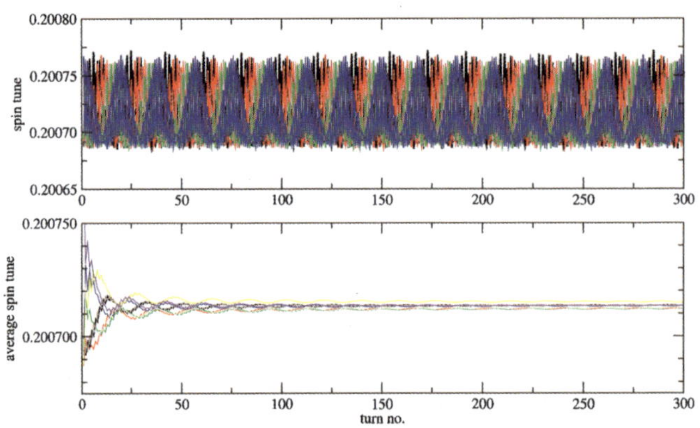

Fig. 1. Spin tune vs. turn number calculated for eight particles sitting on the contour of a phase space ellipse, corresponding to an emittance of 6.410^{-5} m rad

3. Spin decoherence of a beam of deuterons

The general causes of spin decoherence can be explained by inspecting the vector **b** in the Thomas-BMT equation (2). The spin kick depends on the

[a]Similarly, the eigenvalues of the one turn orbit matrix give the betatron and synchrotron tunes.

value of the local field and on particle energy, so the spin evolution of particles on different trajectories or of different energies is different and their spin tune will also be different.

In this study, we will define spin coherence time corresponding to the turn number where the RMS average spin detune in the beam is $\pi/2$. Let's examine effects of **emittance** on spin coherence: figure 2, left, shows the asymptotic spin tune average value for a particle on ellipses of increasing emittance. If the total beam emittance encompasses phase space ellipses up to say, an RMS value of 5.10^{-6} m rad and no more, the emittance spread, as the figure shows, is of the order of 2.10^{-6}, so the beam may remain coherent for millions of turns.

Figure 2, right, shows the spin tune line for 256 particles with Gaussian distribution in phase space with RMS emittance $= 10^{-7}$ m rad, after 30,000 turns. The spin linewidth is $\approx 1.7\ 10^{-6}$.

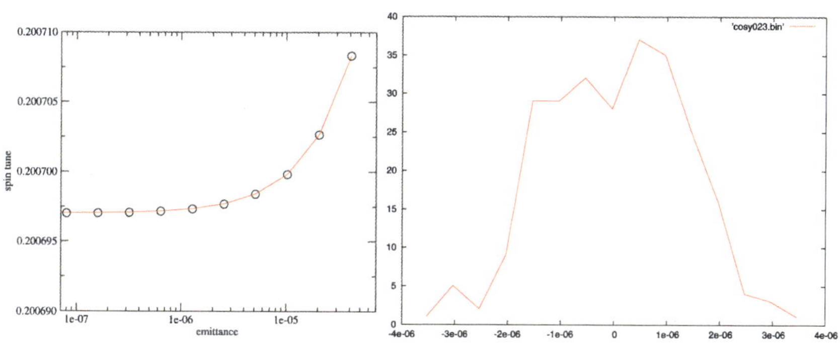

Fig. 2. Left: spin tune of a single particle on ellipses of increasing emittance. 30,000 turns. $\Delta p/p = 0, \Delta\phi = 0$. No RF (no longitudinal motion). Right: spin tune line for RMS emittance of 10^{-7} m rad, in arbitrary units.

About effects of the momentum spread of the beam on spin coherence: momentum spread affects spin coherence. Particles with different energies move on different paths and also take different times in each oscillation.

Figure 3, left, shows the spin of a coasting particle, i.e. that performs a longitudinal sheering motion in absence of any RF cavity. Figure 3, right, shows the spin tune line for a coasting bunch of 2048 particles with RMS emittance 10^{-6} and RMS of gaussian energy spread.

Let us activate the RF, which forces deuterons to perform synchrotron oscillations. Figure 4, left, shows the comparison of the evolution of spin tune vs. turn number without and with an RF. The modulation of the spin tune by the RF is evident, as caused by the modulation of the particle

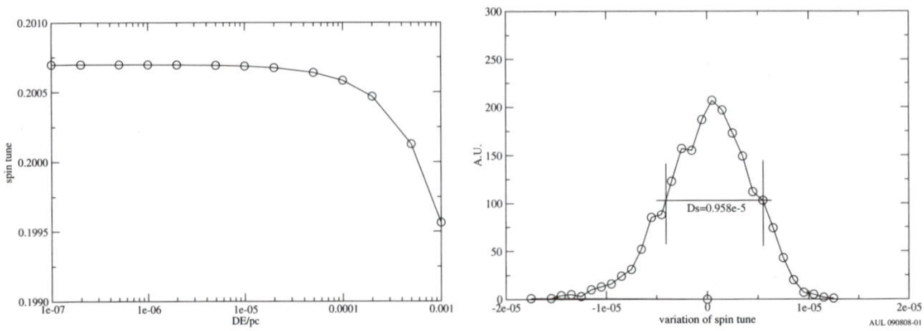

Fig. 3. Left: spin tune of one particle at $\epsilon = 10^{-7}$mrad, $\Delta E/pc = 10^{-7}$ to 10^{-3}. Coasting beam. Right: 2048 deuterons randomly extracted. Gaussian energy spread $\Delta E/pc = 10^{-4}$.

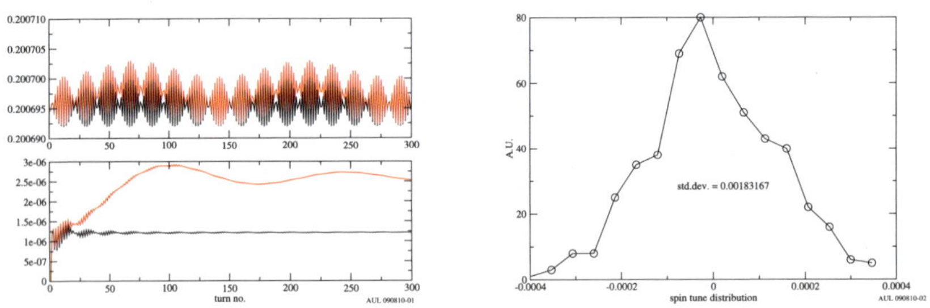

Fig. 4. Left: upper graph: evolution of the spin tune vs. turn number without RF (black) and with RF (red). Lower graph: difference of spin tune in the same conditions. Right: spin tune line of 2048 particles (RF cavity on with 30 KV). $\Delta E/pc = 10^{-4}$. RMS emittance $\epsilon = 10^{-6}$. Spin tlinewidth $\sigma = 2.4389\ 10^{-4}$.

energy. We repeated this game on a bunch of particles of emittance 10^{-6} and energy spread of 10^{-4} with the corresponding spin tune line shown in figure 4, right. The effect of the longitudinal motion on spin tune is produced by the average different amplitude of the orbit due to synchrotron-betatron coupling that, in turn, produces a dispersion in the length of trajectories. Figure 5, left, shows the distribution of trajectory lengths among 1024 random particles with no longitudinal motion, 10,000 turns, $\Delta E/pc = 0$ and with a gaussian energy distribution $\Delta E/pc = 1.10^{-4}$ and RF on. We obtained:

$$\Delta E/pc{=}0 \quad < \Delta L > = 3.739\ 10^{-7} \quad \text{std.dev. (L)} = 0.0006115$$
$$=1.10^{-4} \quad = 3.792\ 10^{-7} \quad = 0.0006158$$

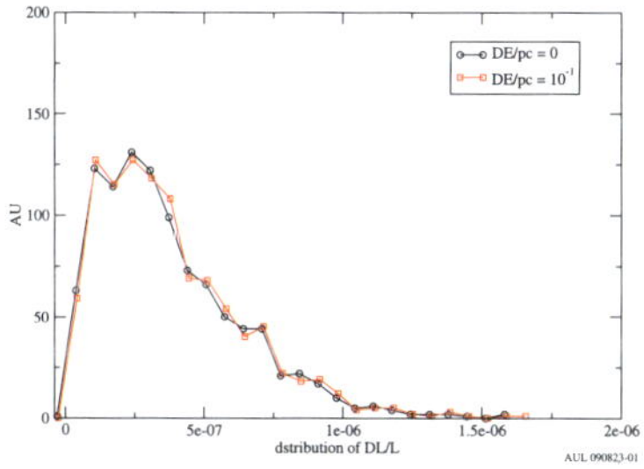

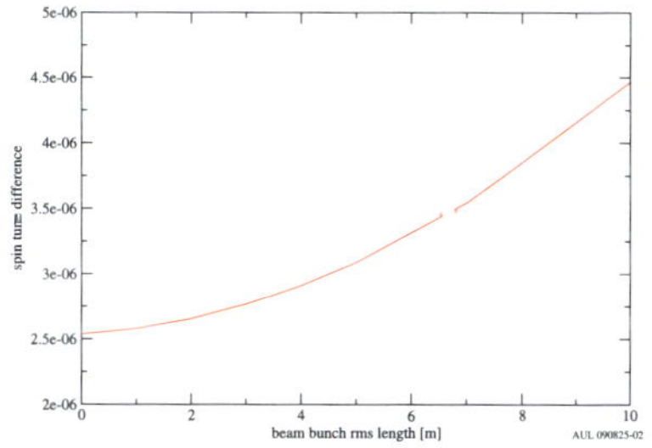

Fig. 5. Left: distribution of length variation among 1-024 particles with and without longitudinal motion. Right: spin tune difference at turn end vs. beam bunch length. The beam density has gaussian transverse and energy distribution and parabolic structure in $\Delta\phi$. Standard deviation over 128 sample random particles after 10,000 turns. Emittance $= 1.0\ 10^{-6}$, RF voltage 30 kV.

Another effect to be considered on spin decoherence is bunch length: figure 5, left, shows how the bunch length of the beam affects the distribution of length of the particle trajectory, averaged over 128 particles. The effect is rather strong.

4. Correction of spin decoherence

Data shown above show that causes of spin decoherence are: (1) finite emittance of the beam, (2) energy spread of the beam, (3) finite bunch length.

A cure for (1) is to increase the brightness of the beam. Start with a beam of low ϵ at the source and avoid mechanisms of diffusion that can dilute the emittance. At low energy, space charge forces are important and produce an increase in the emittance. Beam cooling is an effective method to reduce the emittance, but has limitations and problems. To cure (2), decrease energy spread, and for (3), make bunches short.

Since all effects cause particles to perform different trajectories in 6-D, a general correction of spin decoherence is to try and minimize the difference in orbits. It is a non-linear problem, since the BMT and the equations of motion gives rise to non-linear oscillations. The solution is like chromaticity correction in an accelerator using non-linear machine elements. Yuri Orlov[6] has proposed a solution for the EDM using sextupoles, and Fanglei Lin[4] has performed simulation by spin tracking with UAL-SPINK. Here, we apply a similar treatment using sextupoles in COSY.

The best position of the sextupoles in the lattice is where the beam transverse size is large (large values of the functions β_x and β_y and of the dispersion η_x), because the magnetic field of sextupoles increases with the square of the distance from the accelerator equilibrium orbit. COSY has 18 sextupoles used for chromaticity correction, set in good locations to serve as spin decoherence correctors. At present, we use the sextupoles where they are, without any effort to propose a "better" location for our purposes.

The location of the sextupoles in COSY is shown in figure 6. Eight of them are located in correspondence with a maximum value of the dispersion, where COSY dipoles are. Other sextupoles are in correspondence with high values of the beta function. In the simulations, we only used the sextupoles

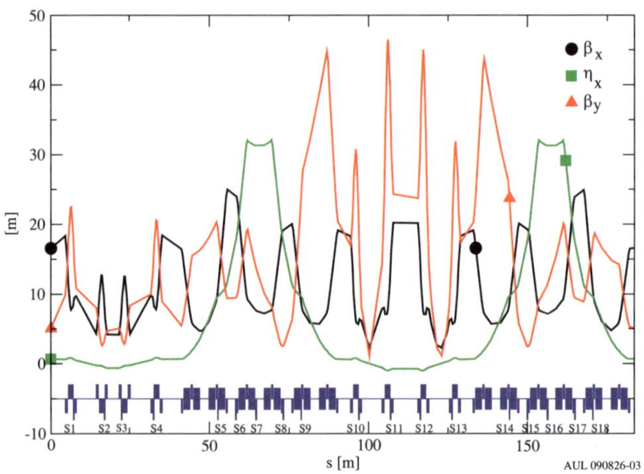

Fig. 6. Lattice of COSY and Twiss functions. The positions of the 18 sextupoles are indicated.

placed at the dispersion maxima. In a first run, with all sextupoles at equal strength, the best spin coherence was found at a value

$$K_2 = -\frac{\partial^2 B/\partial r^2}{B\rho} = 0.05 \quad [\mathrm{m}^{-3}],$$

B, magnetic field in the sextupole and $B\rho = pc/q$, "rigidity" of the beam. The result is shown in figure 7. At the minimum of the curve, for beam energy spread $\Delta E/pc = 10^{-4}$, the spin tune linewidth showed a reduction of a factor of about three with respect to no sextupoles. Assumeing spin coherence is completely lost when the spin oscillation phase ϕ_s slips by π in RMS among the particles in the beam

$$\delta\phi_s = 2\pi\delta\nu_s N_d = \pi, \quad \text{or} \quad N_d = 1/2\delta\nu_s.$$

The equation allows us to convert the coherence turn number to coherence time using the length of COSY = 183.473 m and deuteron speed $\beta = 0.459$,

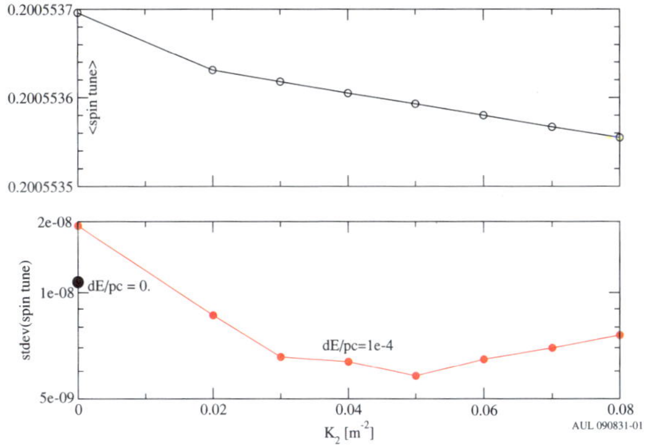

Fig. 7. Using sextupoles. Upper curve: average of spin tune as a function of sextupole strength. Lower curve: spin tune line width.

where the period of the machine is 1.33 μsec. Using the numbers of the last optimization, we can conclude that the spin coherence of a coasting beam with emittance 10^{-6} and energy spread 10^{-4} may remain coherent for

$$150 \cdot 10^6 \text{ turns,} \quad \text{or} \quad 205 \text{ sec}$$

A much more systematic and extensive work is in progress to search for optimum values of all available sextupoles in different combinations, to further increase spin coherence time.

5. Acknowledgments

We thank prof. Rudolf Mayer and his group at the Jülich Research Center for past hospitality, that gave us good hands-on knowledge of COSY and of its structure and capabilities.

References

1. Y. K. Semertzidis (spokeperson) *Search for a permanent electric dipole moment of the deuteron nucleus at the* 10^{-28} *ecm level*, AGS Proposal, BNL (2008).
2. A. U. Luccio, SPINK, A Thin Element Spin Tracking Code, in *Proc. 18th Int. Spin Physics Symposium*, AIP Conf. Proc. **1149**, 759 (AIP, New York, 2009).
3. N. D. Malitsky, R. Talman *Framework of Unified Accelerator Libraries* ICAP98 (1998).
4. F. Lin, N. D. Malitsky,A. U. Luccio, W. M. Morse, Y. K. Semertzidis, C. J. G. Onderwater and Y. F. Orlov, Study by Spin Tracking of a Storage Ring for

Deuteron Electric Dipole Moment, in *Proc. 18th International Spin Physics Symposium* AIP Conf. Proc. **1149**, 777 (AIP, New York, 2009).
5. C. J. G. Onderwater *private communication.*
6. Y. F. Orlov, Principal scheme of a deuteron edm ring with a long spin coherence time, Muon EDM Note No. 61 (2004).

SUMMARY OF THE XIII INTERNATIONAL WORKSHOP ON POLARIZED SOURCES, TARGETS AND POLARIMETRY

F. Rathmann

Institut für Kernphysik
Forschungszentrum Jülich,
52425 Jülich, Germany

The workshops on polarized sources, targets, and polarimetry are held every two years. The present meeting took place in Ferrara, Italy, and was organized by the University of Ferrara. Sessions on *Polarized Proton and Deuterium Sources*, *Polarized Electron Sources*, *Polarimetry*, *Polarized Solid Targets*, and *Polarized Internal Targets*, highlighted topics, recent developments, and progress in the field. A session decicated to *Future Facilities* provided an overview of a number of new activities in the spin-physics sector at facilities that are currently in the planning stage. Besides presenting a broad overview of polarized ion sources, electron sources, solid and gaseous targets, and their neighboring fields, the workshop also addressed the application of polarized atoms in applied sciences and medicine that is becoming increasingly important.

Keywords: Polarized beams and targets; polarized ion and electron sources; polarized solid and internal targets.

1. Introduction

The present workshop is part of a series of workshops on techniques required for experiments in nuclear and particle physics, to measure spin-dependent observables in the scattering of energetic particles.

About 80 participants registered for the workshop, which reflected all relevant areas where polarized beams and targets are presently employed, or likely to be employed in the future. The various sessions included presentations (number in brackets) on:

- Polarized Proton and Deuterium Sources (4),
- Polarized Electron Sources (10),
- Polarimetry (8),

- Polarized Solid Targets (12),
- Polarized Internal Targets (8 talks, 1 open discussion),
- Future Facilities (6).

In addition, there was one introductory talk given by E. Steffens, which presented a brief history of the workshops in this series, and one summary talk given by F. Rathmann. In total, there were about 50 presentations scheduled at the meeting.

2. Polarized proton and deuterium sources (4 talks)

Polarized hydrogen and deuterium sources are mainly used to either inject electrically-charged polarized projectiles into an accelerator, or to feed a polarized target. Progress from BNL was reported by A. Zelensky. The polarized beam intensity for the relativistic heavy ion collider (RHIC) at BNL which is produced in an optically-pumped polarized H^- ion source, is intensive enough so that the polarized proton beam intensity in the high-energy accelerator is not limited any longer by the intensity of the polarized source. The RHIC spin program benefits strongly from developments in the polarized ion source and polarized target technology.

The polarized ion source at the cooler synchrotron COSY at Jülich, Germany, delivers negatively-charged polarized protons or deuterons for investigations in hadron physics in the momentum range from 0.3 GeV/c to 3.8 GeV/c. As explained by O. Felden, the polarized ion source is based on the colliding beams principle, using an intense pulsed neutralized cesium beam for charge exchange with a pulsed highly polarized hydrogen or deuterium beam. Commissioning of a Lamb-Shift Polarimeter (LSP) is underway, H^- and D^- beams from the COSY source have already been transported to the LSP.

V. V. Fimushkin discussed the new source of polarized ions for the JINR accelerator complex, which will make it possible to increase the polarized deuteron beam intensity up to the level of 10^{10} d/pulse. The universal high-intensity source of polarized deuterons (or protons) uses a charge-exchange plasma ionizer. The operation of the ionizer with a storage cell at room temperature is planned for the fall of 2010.

The effect of nuclear spin dichroism, predicted by theoretical studies as the appearance of tensor polarization in initially unpolarized beams behind unpolarized or spinless targets, has been studied at the Cologne tandem accelerator using 9.5 to 18.7 MeV unpolarized deuteron beams, impinging on graphite targets of areal densities ranging from 36 to 188 mg/cm^2.

As reported by H. Seyfarth, distinct deviations from the predicted weak effects were observed, with a maximum value of $p_{zz} = -(0.28 \pm 0.03)$ measured behind a 129 mg/cm^2 carbon target at 14.8 MeV initial beam energy. One implication of this finding is that it will allow one to produce tensor-polarized deuteron beams with p_{zz} about -0.30 or $+0.25$ from an initially unpolarized deuteron beam using a graphite target of appropriate thickness.

3. Polarized electron sources (10 talks)

Two innovations, the energy recovery LINAC and the CW operation of superconducting structures at gradients of up to 20 MV/m are proposed to be combined in a project called MESA (Mainz Energy recovering Superconducting Accelerator), that was presented by K. Aulenbacher. MESA is considered an extension to the experimental facilities at the institute for nuclear physics at Mainz that offers unique conditions for several experiments in particle and hadron physics, and also in applied science. The parity violating experiment will require a 20 cm long hydrogen target, yielding a luminosity of almost 8×10^{38} cm^{-2}s^{-1}. Future work will concentrate on detailed design studies to be completed within the next two years. The operation of MESA could start in 2015.

Y. Poltoratska reported on the status of the Darmstadt polarized electron injector. A source of polarized electrons was developed for the superconducting Darmstadt electron linear accelerator (S-DALINAC) in Darmstadt. It has been set up, characterized, and operated at a test stand. Installation at the S-DALINAC is scheduled to start in January 2010. Experiments with polarized electrons and photons at the S-DALINAC may commence as early as the middle of 2010.

The Mott polarimeter at MAMI in Mainz, Germany, presented by V. Tioukine, uses two double-focusing magnet spectrometers to collect elastically back-scattered electrons from gold targets. The polarimeter provides high efficiency for almost all beam intensities used at MAMI, aiming at an absolute accuracy well below 0.02 in the near future.

L. Rinolfi presented a progress report on the electron and positron polarized sources for CLIC (Compact LInear Collider), where in particular the generation of polarized positron presents an enormous challenge for which different schemes are studied. All proposed schemes for polarized positrons need substantial R&D to fulfill the requested CLIC performance.

Although CEBAF delivers only polarized electrons to its users, recently, an interest was expressed to also provide polarized positrons on the CEBAF footprint, as reported by J. Grames. A physics workshop in 2009 identified a number of key issues where (polarized) positrons would be useful: studies of generalized parton distributions, investigations of 2 gluon exchange in elastic scattering, studies of the Coulomb distortion in the inelastic regime, searches for a light dark matter gauge U-boson, measurements of the C_{3q} neutral weak coupling, and studies in positron annihilation spectroscopy. The requirements are e^+ beam currents in excess of 100 nA in cw mode, and as large a e^+ beam polarization as possible. The production schemes presently considered involve polarized positrons produced either from polarized bremsstrahlung, or by polarized pair production.

E. Tsentalovich presented the status of the high intensity polarized electron gun for the eRHIC project, developed by MIT-Bates in collaboration with BNL. The gun implements a large area photocathode, a ring-shaped beam, and active cathode cooling. In order to achieve in eRHIC a luminosity of 10^{33} cm^{-2}s^{-1}, an average electron current of at least 50 mA is required. The highest average currents produced in existing polarized electron guns presently reach about one mA.

I. Ben-Zvi reported on the planning for the electron-ion collider eRHIC at BNL, where as a first stage MeRHIC (medium energy eRHIC) is envisioned. The polarized electron beams for these facilities will be provided by either a DC or RF gun, and then accelerated by a multi-pass superconducting energy recovery linac to collide with polarized protons provided by RHIC. MeRHIC requires a polarized electron beam current of 50 mA, while eRHIC may require as much as 260 mA. In order to test the feasibility of a high-current polarized electron source, an R&D program together with MIT/Bates and JLab is carried out, focusing on:

- demonstration of the feasibility of a polarized electron cathode in a superconducting RF gun and
- development of a 50 mA polarized electron gun based on a funnel scheme of multiple low-current photocathodes in a "Gatling gun" scheme.

The upgrade plans for the 50 keV GaAs source of polarized electrons operated at electron stretcher accelerator (ELSA) in Bonn, Germany were discussed by D. Heiliger. For about 10 years, an inverted source of polarized electrons has been operated at ELSA, providing a pulsed beam with a current of 100 mA and a polarization of about 0.80, emitted in space-charge limitation. Measurements of the photo-emission current and numerical simulations of the space-charge dominated beam transport indicate that an intensity upgrade to 200 mA is feasible.

The report given by E. Riehn described studies of the effects of intense laser irradiation on the lifetime of superlattice photocathodes with and without Distributed Bragg Reflectors (DBR). Cathodes of both types were exposed to different laser intensities in the range of 30 mW to 800 mW at a wavelength of 808 nm. With a reflectivity close to unity, the DBR prevents light from entering the substrate and reduces effects ascribed to cathode heating. The advantages are two-fold:

(1) the DBR cathode allows for a factor of ≈ 3 more laserpower (or beam current) at a given lifetime; and
(2) for a given laser power, the DBR lifetime at 300 mW is larger by a factor of ≈ 7.

The development of a polarized electron source based on a Superconducting RF (SRF) gun at Forschungszentrum Dresden was described by R. Xiang. The SRF gun is able to produce a 1 mA cw beam with 9.5 MeV energy and an emittance of 1 mm mrad. Based on the successful operation during the last two years, an SRF gun equipped with an GaAs-type cathode is considered to be a promising alternative for current polarized guns.

4. Polarimetry (8 talks)

In his presentation on the proton beam polarimetry at RHIC, Y. Makdisi explained that polarimeters in each of the Blue and Yellow rings utilize the analyzing power in p-carbon elastic scattering in the Coulomb Nuclear Interference (CNI) region to measure the absolute beam polarization. The carbon polarimeters are calibrated by the polarized hydrogen jet target that measures the absolute beam polarization in pp elastic scattering in the CNI region. For these measurements, which up to now have been carried out at 24, 31.2, 100, and 250 GeV, R&D is underway to test an improved set of silicon detectors that will provide better energy resolution, rate capabilities, and allow access to larger analyzing powers.

For the experiments STAR and PHENIX at RHIC, local polarimetry capabilities have been developed independently, as discussed by M. Togawa. PHENIX employs asymmetries of very forward neutrons, while STAR uses asymmetries from forward charged particles. These methods were used to monitor the spin direction for the 62 and 200 GeV runs. In 2009, the first 500 GeV polarized pp run was carried out, and large analyzing powers A_N for leading neutron production were found, with the observation that $A_N(62\,\mathrm{GeV}) < A_N(200\,\mathrm{GeV}) < A_N(500\,\mathrm{GeV})$.

Longitudinal polarization of the lepton beams was a key ingredient in the success of the world's unique $e^{\pm}$ p ring collider HERA, as reported in the presentation of B. Sobloher. The beam polarization in HERA was produced via radiative polarization, which has been first described theoretically by Sokolov and Ternov. The beam polarization was measured routinely with two polarimeters, using polarization-dependent Compton scattering. The Transverse POLarimeter (TPOL) detected the tiny up-down asymmetries associated with the vertical polarization, while the Longitudinal POLarimeter LPOL utilized the energy asymmetry caused by the longitudinal polarization. The preliminary estimation of the systematical uncertainties for TPOL amounts to about 2.9 % and for LPOL to 2 %. The two polarimeters show a varying behavior over time which is not yet understood. A third option to measure the beam polarization using a high finesse Fabry-Perot cavity has been established at HERA, successfully operating with increasing data-taking frequency towards the end of the HERA running period.

As reported by S. Nanda in his presentation on the electron beam polarimetry at Jefferson Lab (JLab) hall A, Møller and Compton polarimeters have been in operation since 1999. Møller polarimetry achieves 2–3 % uncertainty up to 6 GeV beam energy, while the Compton polarimeter achieves 1–2 % uncertainty in the range from 3–6 GeV. Both polarimeters are presently undergoing performance upgrades for operation at 6 GeV to improve the accuracy to 1 %, and to extend the coverage towards lower energies. The upgrade of JLab to 12 GeV beam energy also includes upgrades of both polarimeters, for which the design has already been completed; construction will be initiated in the near future.

Electron beam polarimetry at JLab-Hall C, discussed by D. Gaskell, presently uses a single device for measuring the electron beam polarization, namely a Møller polarimeter. Although the systematic precision is better than 1 %, the smallest quoted relative uncertainty for a particular experiment is $\Delta P/P = 1.3$ %. For the upcoming experiments, the beam polarization must be determined to $\Delta P/P = 1$ %. The hall C strategy for achieving 1 % polarimetry consists of the following items:

(1) use of the existing Hall C Møller polarimeter to measure absolute beam polarizations to better than 1 % at low beam currents.
(2) Building of a new Compton polarimeter to provide continuous, non-destructive measurements of beam polarization.
(3) The Compton polarimeter will initially provide relative measurements of beam polarization, but will eventually yield measurements of higher precision, similar to the ones obtained with the Møller polarimeter.

In his talk on polarization measurement at the International Linear Collider (ILC) with a Compton polarimeter, C. Bartels presented an overview of the conceptual design of the polarimeters foreseen at the ILC, including the analysing power calibration and the data-driven polarization measurement. At ILC, it is planned to collide electrons and positrons at center-of-mass energies in the range of $\sqrt{s} = 200 - 500$ GeV. Polarimeters located up- and downstream of the main e^+e^- interaction point shall reach relative accuracies of $\Delta P/P = 0.25$ %.

Since future linear colliders, such as the ILC, plan to collide polarized beams and the planned physics reach requires knowledge of the state of polarization as precisely as possible, the time evolution of ground motion-dependent depolarization at linear colliders was discussed by A. Hartin. Polarized beams usually undergo depolarisation due to various mechanisms. Spin tracking in the Beam Delivery System (BDS) was achieved using the BMAD subroutine library, and the CAIN program was used to do spin tracking through the beam-beam collision. Assuming initially aligned beam-line elements in the BDS, a ground motion model was applied to obtain realistic random misalignments over various time scales. Depolarization at the level of 0.1 % occurs within a day of ground motion at a noisy site. Depolarisation at the IP also exceeds 0.1 % for the nominal parameter sets for both the ILC and the Compact LInear Collider (CLIC). The coverage of these studies needs to be extended to include further parts of the machine in order to obtain a full understanding of the spin transport.

R. Barday discussed electron beam polarimetry at low energies and its applications, describing experiments to determine the degree of polarization at the source of polarized electrons for the superconducting Darmstadt electron linear accelerator S-DALINAC. While low energy Mott scattering polarimetry ($E_k \sim 100$ keV) is a widely established technique to measure the polarization of an electron beam, the feasibility of Mott scattering at energies up to 20 MeV is discussed. For further studies of the electron spin dynamics in the scattering process, a correlation between the linear polarization of bremsstrahlung radiation and the electron beam polarization has been measured for the first time using a planar High Purity Germanium (HPGe) Compton polarimeter at the 100 keV source of polarized electrons at TU Darmstadt, Germany.

5. Polarized solid targets (12 talks)

In his presentation on recent progress and future prospects in solid polarized targets, C. D. Keith emphasized that polarized targets, both solid and gaseous, are in ever-increasing demand for nuclear scattering experiments. In addition, the technology for producing these targets is now being applied in other fields, such as materials science and medical research.

X. Wei, in his presentation on HDice (the frozen-spin solid HD target for CLAS at Jefferson Lab), explained that the HD target has many proven advantages for nuclear spin physics experiments with a photon beam:

- low holding field and relatively high temperature.
- Low background, and small dilution factor, providing a short running time.
- The portable production facility can be separated from the experimental site.

The HD target facility has been relocated from BNL to JLab for use with the CLAS detector, where a target lab is currently under construction. A new in-beam cryostat is being designed. The first γ+HD run is scheduled to start in fall 2010. The major unknown for using the target with an electron beam is the radiation damage. A test of the target is scheduled for spring 2011. If this test is passed, the e+HD run will start in the winter of 2011.

HD gas distillation and analysis for HD frozen spin targets was discussed by A. D'Angelo. The production of HD targets relies on the longitudinal relaxation time in solid HD samples, which depends strongly on the concentration of ortho-hydrogen and para-deuterium in pure HD. At low temperatures these contaminants decay into H_2 and D_2 molecular ground states and the reduction of their concentration causes a dramatic increase in the longitudinal relaxation time of H and D in the HD solid. This is obtained by aging the target sample and keeping it at about 10 mK temperature while a magnetic field of 15–17 T is applied. An accurate technique to analyze the HD gas before and after the polarization procedure, based on gas chromatography and Raman scattering, was set up to optimize the aging time.

In his talk on the study of Dynamic Nuclear Polarization (DNP) of UV-irradiated crystals aimed for polarization of solid HD, T. Kumada presented X-band electron spin resonance studies of H, CH_3, C_2H_5, and C_2D_5 radicals trapped in solid normal-H_2, para-H_2, and HD to establish their suitability for DNP. The spin-lattice relaxation time T_{1e} of H-atom radicals of $\approx$ 10 min in highly purified solid para-H_2 and HD is much larger than that required for DNP (milliseconds). T_{1e} of the H-atom radicals varies with the concentration and temperature of the ortho-H_2 molecules, and in a similar way as the spin-lattice relaxation time T_{1n} of protons. This suggests that it is very difficult to provide both a short T_{1e} and a long T_{1n}, required for DNP.

T. Nakajima presented a theoretical study on how one can polarize nuclear-spins using a short laser pulse where the nuclear-spin polarization is realized among a hyperfine manifold of an atomic bound state. The change of the degree of nuclear-spin polarization was investigated upon photoionization using short pump and probe pulses. The proof-of-principle experiment using Yb atoms is currently underway.

In his presentation on polarization and relaxation time measurements with pulsed NMR, D. Kammer discussed how relaxation time and degree of polarization of solid state targets are conventionally determined via cw NMR. It is, however, also possible to measure these parameters using pulsed NMR. The major advantages of this technique consist in a shorter measurement time and the possibility of acquiring the spectrum within a single measurement, without the necessity to sweep through the whole spectrum.

In his talk on radiation damage and recovery in polarized $^{14}NH_3$ ammonia targets at Jefferson Lab, J. D. Maxwell, presented investigations of the polarization performance and radiation recovery of ammonia targets during spring 2009 in the experiments taking place in JLab hall B (SANE, Spin Asymmetries on the Nucleon Experiment) and hall C (eg1-dvcs, Deeply Virtual Compton Scattering). The $^{14}NH_3$ used in the SANE and eg1-dvcs experiments behaved much like $^{15}NH_3$ used in previous experiments at SLAC and JLab. In-beam peak polarizations exceeded 0.9 for both experiments and material exhaustion was observed above doses of 25×10^{15} e^-/cm^2.

A polarized proton solid target has been constructed for use in radioactive nuclear beam experiments at the Center for Nuclear Study at University of Tokyo, Japan, as described by T. Uesaka. The proton polarization is based on the electron polarization in photo-excited triplet states of aromatic molecules. The target system works in a low magnetic field of 0.1 T and at high temperature of 100 K and has been applied to measurements of elastic scatterings between a proton and neutron-rich helium isotopes, conducted at the radioactive ion beam separator at RIKEN. The maximum proton polarization of 0.2 is limited by a lack of photo-excitation power, and T. Uesaka stated that using a more powerful light source produces a higher proton polarization.

Studies on the pulse structure dependence of the produced proton spin polarization rate of the aforementioned solid polarized proton target were presented by T. Kawahara. The method employs a continuous wave Ar-ion laser which is pulsed by an optical chopper. The obtained proton polarization rate was measured by changing the duty factor from 5 % to 50 % and the repetition frequency from 0.75 kHz to 10.6 kHz. At a duty factor of 50 % and a repetition frequency of 10.6 kHz, the polarization rate was improved by a factor of five compared to previous works.

The development of a Q-meter module for the polarization measurement with cw NMR was presented by S. Schrauf. The work descibes a redesign of the widely used Liverpool Q-meter NMR module by the Bochum group, with the goal of providing a modular design, in particular in the HF and NF compartments, and to bring several electronic components up to modern standards.

The proton NMR system of the COMPASS $^{14}NH_3$ target was presented by J. Koivuniemi. COMPASS uses a polarized proton target of irradiated granular ammonia, polarized with the dynamic nuclear polarization method using 4 mm microwaves in a 2.5 T field. The nuclear polarization up to 0.90–0.93 is determined with CW NMR. The presentation focused on properties of the observed ammonia proton signals, which are described. Results of spin thermodynamics studies in high fields were presented as well.

The DNP process requires paramagnetic radicals, which can be introduced into the solid target materials by chemical or radiation methods. In his presentation, L. Wang, described chemical doping with TEMPO and trityl radicals in fully deuterated polystyrene samples. The deuteron polarizations and the behavior of paramagnetic centers have been investigated; 0.073 deuteron polarization with TEMPO has been obtained at 2.5 T and 1 K and a deuteron polarization of 0.123 with a trityl radical.

6. Polarized internal targets (8 talks, 1 open discussion)

C. Barschel presented an overview of the target section for spin-filtering studies at COSY and CERN/AD, an experiment pursued by the PAX collaboration (Polarized Antiproton eXperiment) that aims to polarize a stored antiproton beam by spin-filtering. The setup requires a cell as a Polarized Internal Target (PIT), which is fed by the Atomic Beam Source (ABS) previously used at the HERMES experiment at HERA/DESY. The target cell is surrounded by silicon detectors. The working principle of the Breit-Rabi Polarimeter (BRP), including the calibration procedure, and first results of the analysis of the recorded BRP signals were presented.

First experiments with the polarized internal gas target at ANKE/COSY were discussed by M. Mikirtychyants. The PIT is utilized at the ANKE spectrometer at COSY, where after commissioning of the silicon tracking telescopes, a storage cell made from aluminum coated with teflon was used. The Lamb-shift polarimeter was mounted below the target chamber to allow online tuning of the transition units and monitoring of the ABS jet polarization during the experiments. The results of a first double-polarized experiment, performed in January 2007, were presented.

Extra physics with an ABS and a Lamb-shift polarimeter was presented by R. Engels. The polarized internal gas target of the ANKE experiment is used only for a few months per year for hadron physics experiments at the cooler synchrotron COSY. In the meantime, the setup consting of ABS and Lamb-shift polarimeter can be used for other experiments, such as nuclear fusion, atomic and molecular physics, or even in neutrino physics experiments.

L. Barion discussed systematic studies for the development of high-intensity ABSs in his presentation. In particular the effect of the dissociator cooling temperature was studied in order to better understand why the RHIC atomic beam source provides such a high intensity. Studies on trumpet-shaped nozzles using the Ferrara test bench, compared to the standard sonic nozzle were presented as well.

There is a prevailing demand for the development of more intense ABSs, in particular for the feeding of internal targets in future generation experiments. In a round table discussion, organized by A. Nass, this issue was addressed from two different perspectives: (i) understanding of exisiting sources, and (ii) the development of new ideas. The known factors presently limiting the intensity of existing sources are beam attenuation due to collisions with the residual gas, and intrabeam scattering. An additional influence might be attributed to the characteristics of the beam formation system. The most intense source presently operating is the ABS used for the $\vec{\mathrm{H}}$-jet polarimeter at RHIC. Despite its performance, this source is presently not well characterized, therefore dedicated measurements to investigate the abovementioned effects in this particular source should be performed. At Ferrara, a parallel program is being developed to study these effects using a test bench setup. The question of whether superconducting sextupole magnets should be used instead of permanent magnets is still an unanswered one. The development of new techniques should be pursued: direct simulation Monte Carlo codes could provide the capability to also include the magnetic forces on the atoms, and thus lead to a better understanding of the underlying limitations in the present sources. Storage cells trapping the polarized atoms more efficiently should be explored as well.

S. Karpuk discussed the status of spin-polarized ^{3}He, from basic research to medical application. Techniques and practices for gas production and delivery include a central polarized ^{3}He production facility, capable of providing $\approx 60 - 70$ bar liters/day with a polarization of $P = 0.6$. The main technical problems of storage and transport ($T_1 > 100$ h), polarimetry, gas administration, and the recovery of ^{3}He are solved. In a clinical study during 2002–2004, 116 patients and 37 volunteers were treated. The technology is applied in morphological magnetic resonance imaging (MRI) studies and dynamic imaging of lung functions. The clinical infrastructure includes a standard (1.5 T) MR scanner and a low-field (0.1–0.5 T) scanner, transmitter/receiver coils, a gas administration unit, and suitable software for the implementation of ^{3}He imaging sequences.

Major advances in Spin-Exchange Optical Pumping (SEOP) of polarized ^{3}He targets include the introduction of line-narrowed lasers, hybrid-alkali alloys, and convection driven recirculation inside the ^{3}He cell, were discussed by P. Dolph. SEOP uses circularly-polarized laser light to polarize an alkali metal; the alkali metal in turn transfers its polarization to noble-gas nuclei such as ^{3}He or ^{129}Xe. Until recently, SEOP usually employed rubidium (Rb) vapor and broadband lasers (2.0 nm FWHM) and ^{3}He polarizations of ≈ 0.4 were achieved in large target cells of 1–3 liter volume. Recent advances including the introduction of hybrid alkali mixtures and spectrally narrow lasers (0.2 nm FWHM) have produced polarizations in excess of 0.7.

The goal of the study of polarized metastable ^{3}He beam production, presented by Yu. A. Plis, is to produce a source of polarized $^3He^{++}$ ions on the basis of the polarized deuteron source for the JINR accelerator complex. The RF dissociator is fed with helium-3 gas to produce ^{3}He atoms in the metastable 2^3S_1 state. Stern-Gerlach separation in a sextupole magnet system and RF transitions in a weak magnetic field are used to produce nuclear polarization in the metastable atoms. It seems feasible to provide a polarized beam with rather high polarization and a $^3He^{++}$ intensity of up to 2×10^{11} ions/pulse of 8 μs duration. Depolarizing effects in the polarized ion source are expected to be small.

Polarized ^{3}He from metastability exchange optical pumping, used for various double polarized experiments for real and virtual photons at the MAinz MIcrotron (MAMI), was discussed by J. Krimmer. Usually, more than 0.7 initial target polarization has been obtained at the experimental area. The performance of the target for electron beam experiments, and details of the newly developed target for photon beams inside the Crystal Ball (CB) detector were presented.

7. Future facilities (6 talks)

Polarized antiprotons provide access to a wealth of single- and double-spin observables, thereby opening a window to physics uniquely accessible with the HESR at FAIR. The most promising approach to provide a beam of polarized antiprotons, adopted by the PAX collaboration, is based on spin-filtering using an internal polarized hydrogen gas target – a method that has been shown to work with stored protons. The necessary studies will be carried out at COSY/Jülich and AD/CERN, as discussed by F. Rathmann. A program at COSY is underway to test and commission the equipment required for the spin-filtering experiments at the AD, i.e. the polarized internal target and the new low-β section, efficient polarimeters to determine target and beam polarizations, and a Siberian snake to maintain the longitudinal beam polarization. The experimental setup for the spin-filtering studies at COSY and AD was presented in detail by A. Nass.

A recent experiment carried out by the ANKE and PAX collaborations at COSY revealed that ep spin-flip cross sections are too small to cause a detectable depolarization of a stored proton beam. This measurement rules out a proposal to use polarized positrons to polarize an antiproton beam by $e^+\bar{p}$ spin-flip interactions, as presented by D. Oellers.

Ideas for polarized electrons and nucleons at FAIR were presented by D. Eversheim, addressing mainly accelerator related aspects and consequences for the involved sources. In order to better understand the nucleon spin structure, an electron-nucleon collider working group presently discusses the implementation of a 3.3 GeV accelerator for polarized electrons inside the HESR tunnel. HESR and all upstream accelerators have to provide transport of polarized protons and deuterons from a high intensity polarized source.

In his presentation on the Electric Dipole Moment (EDM) measurement, G. Onderwater discussed that EDMs are very sensitive probes for new physics, and that storage rings allow one to search directly for EDMs in charged systems. Although extremely successful in many aspects, the Standard Model of particle physics is not capable to explain the apparent matter-antimatter asymmetry of our universe. It provides too little CP-violation, and thus fails to explain the basis for our existence. The searches for permanent electric dipole moments of protons and deuterons, which violate both time reversal and parity invariance (and are thus CP-violating), promise at present the highest accuracy, constituting long-term projects of enormous physics potential.

One of the obstacles of the aforementioned EDM searches using charged particles in storage rings is that one has to understand how to manipulate and increase the spin coherence time of polarization survival, as discussed by A. U. Luccio. The spin coherence time determines the available time per fill for the EDM measurement. Recent results of tracking studies applied to an existing machine such as COSY were presented.

8. Conclusions

The workshop provided a broad overview of the ever-increasing set of spin tools used nowadays, with many inspiring presentations and insightful discussions. Chairman Paolo Lenisa and his colleagues did a fantastic job putting together this workshop on polarized sources, targets and polarimetry.

Acknowledgments

The author would like to thank Paolo Lenisa for a careful reading of the manuscript.

ACKNOWLEDGEMENTS

The Organizers would like to acknowledge

- MIUR, VI-QCD and INFN for the finacial support,
- L. Barion, S. Bertelli, G. Guidoboni, L.L. Pappalardo, M. Statera and the secretaries L. De Marco and P. Fabbri for their help in logistics and organization,
- Comune di Ferrara, Provincia di Ferrara, Camera di Commercio di Ferrara, Consorzio Ferrara Ricerche and Soprintendenza per i Beni Archeologici dell'Emilia-Romagna,
- the companies Varian, Rial Vacuum, Adixen, SpringerVerlag-Italia and World Scientific.

Ciullo Giuseppe — Università degli Studi di Ferrara and INFN
Contalbrigo Marco — INFN sezione di Ferrara
Lenisa Paolo — Università degli Studi di Ferrara and INFN

2nd february, 2010

AUTHOR INDEX